A Traveler's Guide to the

Geology of the Colorado Plateau

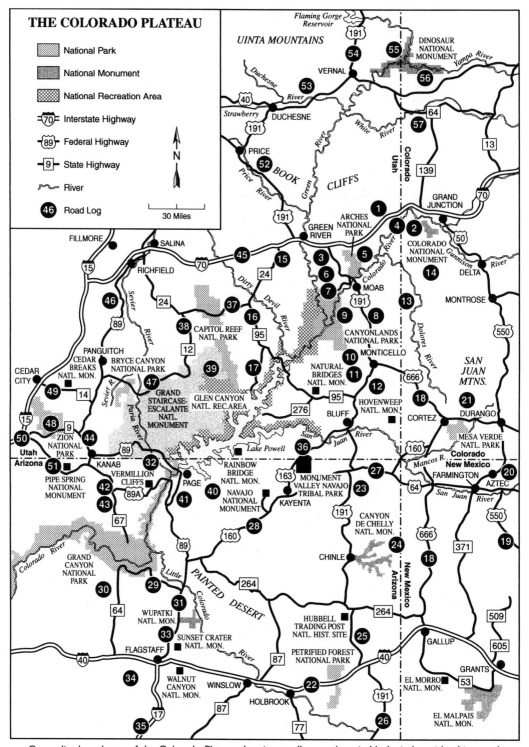

THE COLORADO PLATEAU

National Park

National Monument

National Recreation Area

70 Interstate Highway

89 Federal Highway

9 State Highway

River

46 Road Log

N

30 Miles

UINTA MOUNTAINS

Flaming Gorge Reservoir

191

54

55

DINOSAUR NATIONAL MONUMENT

56

VERNAL

Yampa River

53

Duchesne

River

40

White River

64

57

13

DUCHESNE

191

Strawberry

Green River

Colorado Utah

139

GRAND JUNCTION

70

PRICE

52

BOOK

CLIFFS

Price River

191

GREEN RIVER

1

ARCHES NATIONAL PARK

4

2

50

COLORADO NATIONAL MONUMENT

Gunnison River

FILLMORE

SALINA

45

15

3

5

14

DELTA

River

15

70

RICHFIELD

24

6

Colorado River

MOAB

MONTROSE

46

24

37

7

191

550

9

8

13

38

CAPITOL REEF NATL. PARK

16

95

CANYONLANDS NATIONAL PARK

Dolores River

SAN JUAN MTNS.

89

Dirty Devil River

17

10

MONTICELLO

CEDAR CITY

PANGUITCH CEDAR BREAKS NATL. MON.

BRYCE CANYON NATIONAL PARK

39

NATURAL BRIDGES NATL. MON.

11

666

18

21

DURANGO

49

14

47

GRAND STAIRCASE-ESCALANTE NATL. MONUMENT

GLEN CANYON NATL. REC. AREA

276

12

95

HOVENWEEP NATL. MON.

CORTEZ

15

48

9

BLUFF

160

MESA VERDE NATL. PARK

50

ZION NATIONAL PARK

44

89

Lake Powell

36

San Juan River

Colorado New Mexico

Utah Arizona

51

KANAB

32

RAINBOW BRIDGE NATL. MON.

163

MONUMENT VALLEY NAVAJO TRIBAL PARK

27

64

666

FARMINGTON

20

AZTEC

PIPE SPRING NATIONAL MONUMENT

VERMILLION CLIFFS

PAGE

40

NAVAJO NATIONAL MONUMENT

23

San Juan River

550

42

89A

41

KAYENTA

191

CANYON DE CHELLY NATL. MON.

43

67

160

28

19

89

CHINLE

24

18

371

Colorado River

GRAND CANYON NATIONAL PARK

Little Colorado River

PAINTED DESERT

264

New Mexico Arizona

30

29

264

509

31

64

WUPATKI NATL. MON.

HUBBELL TRADING POST NATL. HIST. SITE

25

GALLUP

605

33

SUNSET CRATER NATL. MON.

River

FLAGSTAFF

87

PETRIFIED FOREST NATIONAL PARK

40

GRANTS

34

WALNUT CANYON NATL. MON.

WINSLOW

22

EL MORRO NATL. MON.

53

40

17

87

HOLBROOK

191

35

26

EL MALPAIS NATL. MON.

77

Generalized road map of the Colorado Plateau showing roadlog numbers in black circles with white numbers.

A Traveler's Guide to the
Geology of the Colorado Plateau

Donald L. Baars

The University of Utah Press

Salt Lake City

© 2002 by
The University
of Utah Press

Library of Congress Cataloging-in-Publication Data

Baars, Donald L.
   A traveler's guide to the geology of the Colorado Plateau /
Donald L. Baars.
      p.   cm.
Includes bibliographical references.
   ISBN 978-0-87480-715-8
   1. Geology—Colorado Plateau—Guidebooks.
2. Colorado Plateau—Guidebooks. I. Title.
   QE79.B33 2002
   557.91'3—dc21
                                   2002001065

Contents

V. The Plateau Interior — 181

VI. The High Plateaus — 223

VII. The Uinta Basin and Dinosaur National Monument — 249

Introduction

So what is the Colorado Plateau, anyway? We have all been told that the magnificent splendors of Grand Canyon, Zion Canyon, Bryce Canyon, Capitol Reef, Canyonlands, Arches, and Mesa Verde National Parks lie within the Colorado Plateau, but where does the Colorado Plateau begin and end?

To the geographer, the Colorado Plateau Province is a broad expanse of generally flat country, dotted here and there by high mesas, pinnacles, and smaller plateaus and incised in places by deep canyons. The region is punctuated with random and isolated rounded mountain ranges, known as laccolithic ranges, formed by molten rock that squeezed into the stratified rocks. The province is generally high, averaging something like 5,000 feet in elevation with low average precipitation. Thus, it is a high, semiarid desert. The Colorado Plateau constitutes a large part of the drainage basin for the Colorado River and its many tributaries. It is not located in Colorado alone but lies in Utah, Arizona, and New Mexico as well. The surrounding provinces constitute various complex mountain chains.

To many it is a bleak and threatening region: summers are very hot, winters are very cold, and water is almost nowhere. For those folks it is a place to be avoided or, at best, to get through as quickly as possible. For them the interstate highways are a blessing. Still others are fascinated by the stark beauty of it all, especially those brave folks who wander off the beaten paths. The highways that interlace the larger towns and cities were built on the flattest and most drab-colored strata, avoiding topographic relief such as mesas and canyons for economic reasons. When one ventures onto roads that go "nowhere," the excitement begins. Red rock badlands are everywhere. Pinnacles and buttes of spectacular proportions dominate the landscape. And canyons, large and small, are ubiquitous. Beauty reigns. Unbelievable shapes and colors abound. As an ancient Navajo poem concludes: "Beauty all around us. With it I wander."

There are as many different boundaries of the Colorado Plateau Province as there are experts in the fields of geomorphology (the study of the origin of landforms), geology (the study of rocks), and geography (the study of the surface of the earth). Generally speaking, the Colorado Plateau is a vast expanse of relatively flat-lying strata sandwiched between regions of complexly organized mountainous terrains. But the boundaries of these flat-lying rocks are fuzzy at best. It is easy to say that mountainous regions surround the Colorado Plateau, and thus define it, but where does a mountainous region end and a generally flat plateau begin? The answer is that the transitions are gradual and the boundaries are broad zones that lack distinctive lines of demarcation; thus, the boundaries are subject to varying interpretations by different authorities. But the real question is, Why? What has caused the Colorado Plateau to stand alone in the midst of all this mountain-building activity? The answer lies in an understanding of the geologic features and processes that have formed the earth as we see it today.

The region known as the Colorado Plateau is a geologic entity unto itself, a relatively stable block of the earth's crust set apart by major, ancient fracture zones. Some would blame this anomalous block on some quirk of plate tectonics, the idea that continental "plates" migrate around on the crust of the earth, first gurgling up at the oceanic ridges only to slip away into the depths of the earth or crash into one another. Although plate tectonics is a popular concept, it does not necessarily apply to the geology of continental interiors.

An alternative to the understanding of continental geological processes proposes that major fracture zones in the very ancient Precambrian basement (rocks that originated more than 550 million years ago) control the location and orientation of the various major geologic features. These ancient fault zones have apparently controlled geologic features since about 1.7 to 1.6 *billion* years ago and have been reactivated repeatedly throughout geologic time. It should be noted here that a *fault* is a fracture in the earth's crust along which movement has taken place. There are several varieties of faults, but in this case we are dealing with faults along which movement has been mainly lateral (strike-slip), much like what is happening in the San Andreas fault zone in southern California today. These faults usually occur in complex swarms and are known as *wrench faults*.

The pattern of the major basement wrench fault zones on the Colorado Plateau is one of northwest-southeast–directed faulting and northeast-southwest–oriented fault zones, the two trends intersecting at regular intervals and usually offsetting each other at the intersections, forming "trellis" patterns. In addition, there are the occasional north-south or east-west faults that punctuate the overall pattern (Figure 2).

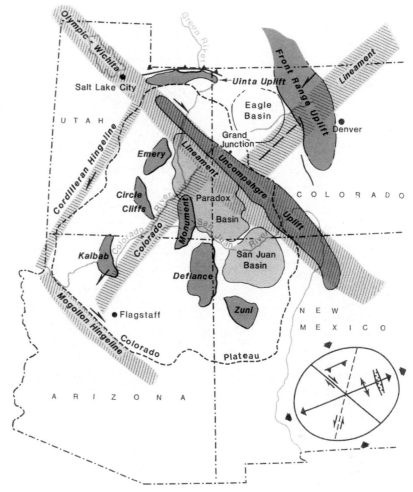

The geologically simplified Colorado Plateau Province is bounded by
basement fault zones as just described. The southern boundary of the
province lies adjacent to a fracture zone known as the Texas Lineament, or
the Mogollon Hingeline (Figure 2). The fracture zone apparently extends
northwestward through Nevada, forming the eastern margin of the Sierra
Nevada. Topographically, the fault zone bounds the margin of relatively
flat-lying sedimentary rocks, called the Mogollon Rim. Along this topographic break are several volcanic features, such as the White Mountains
and San Francisco Volcanic Field, and numerous volcanic cinder cones and
lava flows of Tertiary to Recent origin. South of the Mogollon Rim are numerous fault-bounded mountain ranges that consist of Precambrian crys-
talline basement rocks and much-thickened sections of Paleozoic and
Mesozoic rocks (see time scale in Figure 5).

The western border of the Colorado Plateau lies generally along the

Figure 3. Faults associated
with the Wasatch Line in
central Utah. Names of
the various normal faults,
shown with plain lines,
and thrust faults, shown
with barbs that point
away from the direction
of movement, are desig-
nated by letter codes.
The fault zone forms the
western boundary of
the Colorado Plateau
Province.

A, Absaroka;
AE, Ancient Ephraim;
B, Bannock;
BC, Broad Canyon;
BM, Blue Mountain;
C, Crest; Ca, Cache;
CM, Cedar Mountain;
CN, Charleston-Nebo;
Cr, Crawford;
CV, Cache Valley;
D, Darby; F, Frisco;
G, Glendale;
GP, Glass Peak;
GW, Grand Wash;
H, Hurricane;
HL, Hogsback-Labarge;
J, Jackson; K, Keystone;
LVSZ, Las Vegas shear zone;
MM, Muddy Mountains;
MR, Mineral Range;
N, Needles;
NF, North Flank;
P, Paunsagunt;
Pv, Pavant; S, Sevier;
SF, South Flank; T, Teton;
To, Toroweap;
UB, Uinta boundary;
W, Wasatch; Wi, Willard;
WP, Woodruff-Paris;
WW, Wah Wah;
Wg, Washington.

From Stokes 1986.

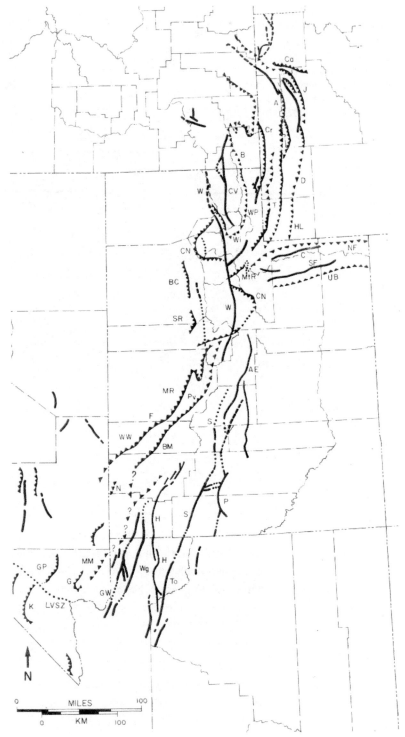

Wasatch Line, or the Cordilleran Hingeline (see Figure 3). This significant trend may have been the western edge of the North American continent, formed by rifting and sagging of a former landmass to the west in Late Precambrian time. Along the broad fault zone, 50 to 100 miles wide, are drastic changes in thickness of Paleozoic and early Mesozoic strata. Layered rock units to the east may range from a few hundred to a couple of thousand feet thick, as compared with several thousand feet to the west in the Great Basin, or Basin and Range Province. West of the Wasatch Line are numerous mountain ranges and intervening basins, whereas east of the line are relatively flat-lying sedimentary and volcanic rocks. Surface faults of relatively recent age, Tertiary, occur along the Wasatch Line, bounding the various units of the High Plateaus Subprovince (see Figure 3). These Tertiary-age faults bound several high plateaus, such as the Paunsagunt, Pavant, Sevier, Toroweap, and Wasatch. Many are capped by volcanic rocks of Tertiary age. Thick fault-bounded deposits of salt and gypsum mark the transition zone.

The Wasatch Line is interrupted in north-central Utah by the Uinta Mountains, a maverick east-west-trending mountain range along the Utah-Wyoming border. These high, rugged mountains are composed of a core of Late Precambrian quartzitic rocks along the crest of a monstrous upfold (anticline) bounded on either flank by thrust faults. Because of topographic considerations, the Uinta Mountains constitute a well-defined northern boundary of the Colorado Plateau Province.

The eastern edge is the most arbitrary and most manipulated boundary of the Colorado Plateau Province. If we consider the Uinta Basin, south of the Uinta Mountains, to be part of the Colorado Plateau, then the northwest-trending basement highland, the Uncompahgre Uplift, would best define the eastern boundary in northwestern Colorado. But then what? If we consider the San Juan Mountains of southwestern Colorado to belong to the Southern Rocky Mountains (see Figure 4), as do many authors, then an arbitrary line must be drawn from the southern Uncompahgre Uplift to the northern edge of the San Juan Mountains. Certainly, the San Juan Mountains are truly mountains, sometimes called the American Alps, but their geology is much more closely related to Colorado Plateau geology than to that of the Southern Rockies. I have previously included the San Juans as part of the Colorado Plateau, but in this case we will follow conformity and exclude this beautiful and significant mountain range. From there, the southeastern boundary most likely should follow a prominent fault zone, apparent both in the Precambrian basement and at the surface: the Rio Grande Rift and related components.

The Rio Grande Rift is a very noticeable faulted valley system that extends from south of Albuquerque, New Mexico, northward into west-

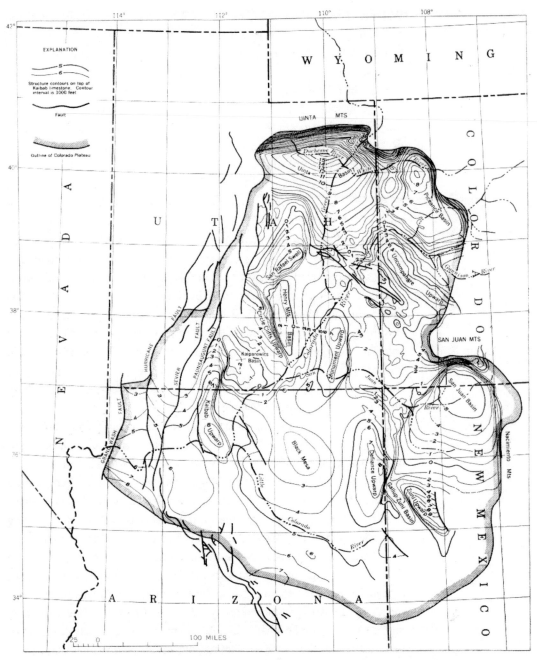

Figure 4. Structural geology of the Colorado Plateau Province. Contour lines connect points of equal elevation in thousands of feet drawn on top of the rocks of Permian age. The entire Plateau was tilted toward the Uinta Mountains to the north in about mid-Tertiary time. Major structural features are identified by name. From Hunt 1969.

central Colorado. The broad valleys resulted from north-south faulting that extended, or pulled apart, a zone in the earth's crust. The rift has been documented as active for the past 17 million years or so, but there is substantial evidence that it was active even during Paleozoic time and probably originated in Precambrian time. The margin of the Colorado Plateau is usually placed along the Nacimiento Mountains (see Figure 4), an active segment of the rift system in Paleozoic time.

From about Socorro, New Mexico, toward the southwest, one must use considerable imagination to connect the Rio Grand Rift to the White Mountains volcanic field, and thus the Mogollon Hingeline, completing the circuitous boundaries of the Colorado Plateau. This imaginary line must be drawn such that the Zuni Uplift of northwestern New Mexico is separated from the fault-bounded basin ranges of west-central New Mexico. This southeastern margin of the plateau is somewhat evasive, lying along a trend of volcanic features and mineralized zones called the Jemez (pronounced *hey! mess*) Lineament.

Time Enough

One of the most fundamental problems in geology is comprehending the enormous period of time involved in the formation of the earth as we see it today. After the earth was formed some *4.5 billion* years ago, a solid crust of metamorphic rock (highly altered rock) enclosed its molten core and was buried by thousands of feet of sedimentary rock and some volcanic rock. Dating individual events along the way is a difficult task at best. For the most part, geologists can determine dates only in a relative sense. In other words, if rock layers A, B, and C occur together, the bottom layer (A) is the oldest because it had to be present before the next layer (B) could be deposited on it, and the top layer (C) would have been deposited last in the sequence. So in relative dating of rocks, unit B is younger than unit A and older than unit C.

In dating geologic events in a relative sense, the old-time geologists devised a scheme in which the layered rocks were arranged in a relative sequence, as accurately as possible, and then subdivided into packages of closely related rock layers. The oldest package of rocks was given a name of a location where the particular rocks could be studied: in this case, the Cambrian System was named for Cambria, an ancient name for Wales in the British Isles. Rocks older than Cambrian were called simply pre-Cambrian, now spelled *Precambrian*. Because rocks of Cambrian age contain the oldest fossil animals with hard "shells," this geologic period was placed at the base of the Paleozoic Era (meaning "time of early life"). Rocks that contain similar fossils are said to have been deposited during the Cambrian Period, wherever they may be found throughout the world.

Younger geologic periods were established in much the same manner, the names being derived mostly in Europe, as that is where the discipline of geology began. Rocks formed during the Paleozoic Era include the Cambrian, Ordovician, Silurian, Devonian, Carboniferous (Mississippian and Pennsylvanian in North America), and Permian Periods, listed from oldest to youngest. Strata directly above and younger than rocks of Paleozoic age were placed in the Mesozoic Era ("time of middle life"). The Mesozoic Era includes, from older to younger, the Triassic, Jurassic, and Cretaceous Periods. The youngest strata, deposited above the Mesozoic rocks, are in the Cenozoic Era ("time of late life"), which includes the Tertiary and Quaternary Periods, topped off by deposits of Recent age (see Figure 5).

The geologic time scale could be used only in a relative sense, dated by the fossil content of the rocks, until much later, when radioactive dating was developed. With the discovery that radioactive elements (uranium, potassium, rubidium, and the like) decay with time to form other elements, absolute dating became possible: dating rocks containing radioactive elements in terms of years before the present. Each of the radioactive elements has a specific decay rate in years, and by carefully comparing the amount of "parent element" in a specimen with the amount of "daughter element," and knowing the decay rate, one can determine the time of crystallization of the radioactive element. The term *absolute dating* is somewhat misleading in that dates thus obtained have a considerable margin of error, but these dates are far better than the guesswork of relative dating. Through decades of study it has been possible to determine the ages of the geologic periods in years, although the numbers vary somewhat from author to author.

It would be wise to flag Figure 5 for future reference, as these geologic time terms are used throughout the book, and it can be difficult to remember such unfamiliar names.

Most of the spectacular scenery on the Colorado Plateau has resulted from severe erosion and the consequent exposure of sedimentary rocks. These layered rocks were derived from particles of sediment of varying size and composition, deposited in great, geographically extensive sheets under varying conditions, and later cemented into rock. Layers of mud, compacted and perhaps cemented by calcite or silica, are known as shale or mudstone. Deposits of sand are usually cemented by calcite, silica, or iron minerals to form sandstone. Coarser material such as gravel and cobbles may become conglomerate when cemented. Limestone is somewhat different, as the relatively hard rock consists of fragments of any size of fossil debris of animals and calcareous plants, such as algae, that were cemented by calcite shortly after deposition in a warm, shallow sea. Less commonly, limestone strata formed in freshwater lakes. Recognizable fossils are commonly found in limestone, but much of the rock consists of sand- or mud-size particles derived in one way or another from organic material, forming various textures in the rock. From a distance, the variations in rock type can be distinguished rather well, as the harder, more resistant sandstone and limestone layers form steep cliffs, and the softer shale layers form slopes.

GEOLOGIC TIME SCALE

ERA	PERIOD	MILLIONS OF YEARS AGO	FOUR CORNERS FORMATIONS
CENOZOIC	Quaternary	0-1.6	soil, sand, gravel, terrace gravels
	Tertiary	1.6-65	Diatremes
			Igneous intrusives
			Chuska Formation
			Animas Formation/
			San Jose Formation
			Nacimiento Formation
MESOZOIC	Cretaceous	65-141	McDermott Formation
			Kirtland Shale
			Fruitland Formation
			Pictured Cliffs Sandstone
			Lewis Shale
			Mesaverde Group
			Cliff House Sandstone
			Menefee Formation
			Point Lookout Sandstone
			Mancos Shale
			Dakota Sandstone
			Cedar Mtn./Burro Canyon Formation
	Jurassic	141-208	Morrison Formation
			Summerville/Wanakah
			Entrada Sandstone
			Carmel Formation
			Navajo Sandstone
			Kayenta Formation
			Moenave/Wingate Ss.
	Triassic	208-245	Chinle Formation
			Moenkopi Formation
PALEOZOIC	Permian	245-290	Cutler Group:
			White Rim/ DeChelly Ss.
			Organ Rock Shale
			Cedar Mesa Sandstone
	Pennsylvanian	290-322.8	Elephant Canyon/Halgaito Fm.
			Hermosa Group:
			Honaker Trail Fm.
			Paradox Formation
		(Rocks below not exposed)	
			Pinkerton Trail Fm.
			Molas Formation
	Mississippian	322.8-362.5	Leadville/Redwall Fm.
	Devonian	362.5-408	Ouray Limestone
			Elbert Formation
	Silurian	408-438	Rocks misssing
	Ordovician	438-510	Rocks missing
	Cambrian	510-570	Ignacio-Lynch Fms.
PRECAMBRIAN	Upper	570-2,500	Quartzite/granite/metamorphic
	Lower	2,500-4,500?	Metamorphic/granite

Figure 5. Geologic time scale (ages in millions of years).

Variations in the bedding of sedimentary rocks may be used to determine the nature of the depositing agent, whether wind or water, and the kind of environment in which the sediments were deposited, perhaps streams, oceanic beaches, or windblown sand deserts. Of course, fossils that the rock may contain are most useful in this regard. The nature of cross bedding may be key to the environment of deposition, as sediments deposited by streams have much different bedding features than those deposited by wind, and so forth. Detailed markings on bedding surfaces, such as mud cracks, ripple marks, and raindrop impressions, offer definite clues as to the origin of sedimentary rocks.

Most folks not trained in geologic processes consider a rock "formation" to be some strangely shaped erosional feature that looks like an owl, an elephant, or a goblin. Geologists, however, distinguish a layer of a single sedimentary rock type, or a series of closely related rock types, as a formation. Geologic formations are distinctive in appearance from other rock layers above or below and usually are geographically widespread in distribution. Erosion may remove small sections of the formation in a canyon, but a formation or sequence of formations on either side of the canyon was at one time connected in a single sheet. Geologic formations are given specific and unique names, the name being derived from a location at which the formation, as formally described, can be studied (the *type section*).

The study of the layered rocks is known as *stratigraphy*, and those folks who study the layered rocks are called *stratigraphers*. The fundamental rock unit in stratigraphy is the formation, which may be subdivided into *members* or lumped with other, similar formations into *groups*. For example, the Carmel, Entrada, Curtis, and Summerville Formations each have two or more members and are lumped into the San Rafael Group. It sounds complicated, but the system, when followed rigidly, usually works rather well for stratigraphers. A formation is assigned to a geologic age, according to the geologic time scale, making it relatively easy to understand the stratigraphic relationships in a time framework. Thus, the Middle Member of the Entrada Sandstone of the San Rafael Group is Middle Jurassic in age. Note that formally accepted names are capitalized in print. The formal name of a single rock unit may vary from one locality to the next because when it was first studied in the old days, the geologist didn't necessarily know that the Hermit Shale of the Grand Canyon was the same layer named by another geologist the Organ Rock Shale in Canyonlands. When such duplication of names is later realized, the first applied name should be used throughout the geographic extent of the formation. Geologists, like most folks, tend to be very territorial, however, seldom accepting the name used by someone else in a different area. Consequently, we have to devise correlation charts to understand that the Hermit Shale is the same bed as the Organ Rock Shale, only in widely separated locations. Such monkey business is the reason stratigraphers, like this author, are such a strange breed; it could all drive anyone crazy.

The layered rocks of the Colorado Plateau are divided into numerous formations and two or three times as many stratigraphic names. In this book formation names are often indicated by local names, with other, related names in parentheses: for example, Organ Rock Shale (Hermit Shale of Grand Canyon and Abo Formation of New Mexico). A correlation chart or two, appropriately included, should clarify the relationships. But without formal formation and age assignments, a description of the geology of the Colorado Plateau would be impossible, so bear with the system.

Colorado Plateau Structures

Although we have defined the Colorado Plateau as a region of relatively flat-lying strata, the province contains numerous geologic structures of major proportions. The key word in this apparent contradiction is *relatively*, as the structure here is simple in comparison with surrounding provinces.

Structural geology is the study of the various kinds of folds and faults that may occur in crustal rocks of the earth. We already know that faults are fractures in the earth's crust along which there has been movement, but there are several possible variations. Folds, which cause strata to wrinkle, likewise occur in several varieties.

Faults A *normal fault* is a fault along which one side has moved down a steeply inclined fracture surface relative to the opposite side. Normal faults occur where the crust of the earth has undergone extension, or pulling apart, thus lengthening its surface (see Figure 6).

A *reverse fault* is one in which one side has been pushed upward along a steeply inclined surface relative to the other side. These faults result from compression in the earth's crust, causing the crust to be shortened. Yet another variety of faulting is the *thrust fault*, in which one side is pushed up and over the other side along a low-angle surface. This kind of fault is due to extreme compression in the crustal rocks, resulting in drastic shortening of the crust. Thrust faults are almost nonexistent on the Colorado Plateau, although they are rather common in the more complex structures of adjacent provinces.

Wrench faults, or strike-slip faults, have lateral movement of the opposite sides rather than vertical movement. They are very complex in detail and usually occur in swarms rather than along a single line of faulting, with individual faults often intersecting one another in meandering configurations, creating anastomosing patterns. The relative sense of lateral movement must be determined by considering the relative motion from opposite

Figure 6. Types of faults.
A. Normal fault.
B. Reverse fault.
C. Thrust fault.
E. Strike-slip, or
wrench, fault.

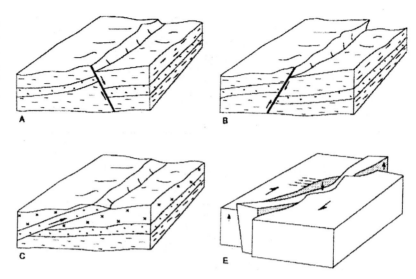

sides of the fault pattern rather than movement along any single fault. The overall sense of movement is termed either *right-lateral* or *left-lateral*. When one stands facing the fault zone, from either side, if the opposite side has moved relatively to the right, the fault zone is said to be right-lateral. If the opposite side has moved to the left, it is said to be a left-lateral fault zone. The well-known San Andreas fault in southern California is a right-lateral wrench fault zone.

Folds   *Anticlines* (upfolds) and *synclines* (downfolds) are a wrinkling of the strata usually resulting from compression, or shortening, of the earth's crust but on a much gentler scale than faults. Anticlines and synclines almost always occur together, as there must be a downfold between every upfold (see Figure 7). If the anticlines are of large proportions, they may be called uplifts, upwarps, swells, or highlands. When synclines are very large they are called basins, downwarps, or perhaps sags. All these terms may be found on geologic structure maps. Folds of various kinds and sizes are very well exposed and expressed in the rather naked landscapes of the Colorado Plateau.

Special kinds of folds, called *monoclines*, are typical of Colorado Plateau geology. They are formed where flat-lying strata are draped across deep-seated faults in what may be thought of as half-folds. They look something like a rug draped over a stair step, with a relatively flat upper surface of strata, a strong downfolding above a deep-seated fault, and a relatively flat surface on the lower side (see Figure 7). These very long, very prominent folds on the Colorado Plateau vary from the strict definition of a monocline in that neither the topside nor the downside is really flat. Although most are in actuality asymmetrical anticlines, with a long, gentle slope forming one

Figure 7. Types of folds: Monocline, anticline, and syncline.

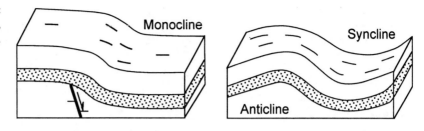

flank of the structure and a very abrupt, steep fold on the other flank, the scale is so huge that they appear to be real monoclines. These structures vary in size but generally are a few tens of miles in cross section and several tens of miles in length and occur in northwest or northerly arrangements.

Salt-Intruded Anticlines

Unusual kinds of geologic structures occur in regions underlain by thick deposits of salt (halite; NaCl). Such deposits occur in nature in basins where marine circulation is restricted; seawater may enter the basin and evaporate rather quickly in an arid climate. The saline waters may dry up completely to precipitate layers of the contained minerals, such as carbonates, gypsum, salt, and even perhaps potash salts. Under some rather rare circumstances, marine water may enter a narrowly constricted basin, only to begin intense evaporation and the concentration of brine that sinks to the bottom and cannot escape the narrow and shallow entryway of the basin. Salt hoppers, or "seed crystals," form at the surface and, as they sink, grow into larger crystals of salt, settling through the brine layers to form layers of bedded salt on the sea floor. Fresh seawater may be entering the restricted basin at all times but cannot escape the basin, thus producing salt deposits in relatively deep water rather than forming dry lake beds. Because many different kinds of minerals occur in seawater, the salt beds are almost always found to be associated with gypsum ($CaSO_4 \cdot H_2O$), carbonate rocks such as limestone and dolomite, and perhaps potash salts (various forms of KCl). Geologists group such mineral associations into the generalized term *evaporites*.

Deep-water salt and gypsum deposits are forming today in the Persian Gulf and formed during Middle Pennsylvanian time (some 300 million years ago) in what is now the eastern Colorado Plateau Province. The evaporite basin is called the Paradox basin and is centered in a general way around Moab, Utah. The outline of the region underlain by salt and other evaporites is shown in Figure 8.

A peculiar property of salt is that it flows like toothpaste or wet putty when squeezed under high confining pressure, such as a thick overburden of younger sedimentary rock. Salt flowage is directed away from areas of high pressure toward areas of low pressure, and if a local point of low over-

Figure 8. The eastern Colorado Plateau and the relationship of the Paradox basin to zones of possible basement fracture systems, here called lineaments, indicated by broad bands. The limits of salt deposition in the Paradox basin are shown by a dotted line, and the salt-intruded structures are shown by hatchured patterns in the northeast part of the map. The heavy dashed line is the approximate limit of coarse-grained sedimentary rocks that were derived from the adjacent Uncompahgre Uplift and restricted to their proximity to the uplift by the growing salt anticlines. The Cataract lineament is also known as the Colorado lineament.

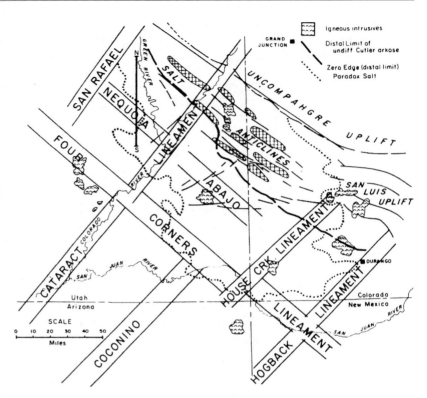

burden pressure exists, the salt flows upward, even reaching the surface in some cases. Such a location may be seen today in the Zagros Mountains of Iran. When salt flows upward toward the surface, it usually does so in globular masses to form semicircular salt domes, known in technical jargon as *diapirs*. Although salt domes are not always noticeable at the surface of the earth, they are common features along the American Gulf Coast and throughout many other parts of the world.

Structures formed by the upward penetration of salt and other evaporites in the Paradox basin of the eastern Colorado Plateau are linear in shape, for the most part, and trend almost invariably in northwest-southeast orientations. They range in length up to about 80 miles. The Paradox basin sagged into existence in Pennsylvanian time along northwest- and northeast-trending basement faults, the northwest grain being predominant in forming most of the surface structures. Several of the northwesterly faults underlie the salt basin (Figure 8). Salt deposition began first in the deep-seated half-grabens formed along the basement faults and was thickest in these structural depressions. As the salt flowed southwestward away from the very thick deposits of sand, gravel, and boulders washing down from the Uncompahgre Uplift on the east, the mushy salt encountered the deep-seated faults and was diverted upward along the fault planes. Elongate salt-flowage anticlines (upfolds) resulted. As the mobile salt neared the surface, fresh groundwater attacked the very soluble evaporites and dissolved the salt, leaving a residual cover of gypsum, dolomite, and black shale behind to cap the salt structures. Massive collapse of the overlying sedimentary rocks ensued to fill the voids left by the dissolved salt, producing the topless anticlinal structures, such as Moab Valley, we see clustered in the eastern Paradox basin today.

In the early days of drilling for oil, it was believed that all oil occurred where it was trapped at the crests of anticlines. Numerous wells were drilled on top of the salt-intruded anticlines because the nature of these structures was not understood. Some 500 to 1,000 feet of the gypsiferous caprock would be drilled, only to enter salt, salt, and more salt. Some wells drilled a few thousand feet of salt before they were abandoned. In the late 1940s Lee Stokes, professor of geology at the University of Utah, made a detailed map of the Gypsum Valley structure, documenting its salt-intruded nature and the fact that salt had flowed from Middle Pennsylvanian time through Jurassic time, some 150 million years of earth history. To finalize the scale of these structures at depth, in the late 1950s Shell Oil Company drilled a hole on the southern flank of the Paradox salt-intruded anticline and found *no* salt along the flank; it had all flowed into the salt structure. Shortly thereafter Conoco drilled an experimental hole down the throat of the same structure a few miles to the northeast of the Shell well. Including the gypsiferous caprock, the well drilled nearly 15,000 feet of salt before encountering older rocks of Mississippian age beneath the salt. The vertical movement along the flanking fault between the two wells was almost 6,000 feet. These are truly salt-flowage structures on a grand scale.

Basement Fault Systems

A common practice for today's structural geologists is to study the folding and faulting that occurred during the latest mountain-building episode, or *orogeny*. Such a study provides information about the structures we see at the surface today but may disregard the structural features of earlier events. To understand *why* the surface structures occur as they do, one must first determine the nature of older structures to see what the latest structural episode had to deal with when it began. A few structural geologists ignore previous geologic structures as if they did not exist and, therefore, miss completely the origins and orientation controls of the last event. Many surface structures seen on the Colorado Plateau are reactivated and enhanced basement structures that originated in Precambrian times. These older structures are mainly wrench fault zones that originated sometime around 1.6 to 1.7 *billion* years ago, with reactivated movements occurring throughout the remainder of geologic time.

Figure 9. The various up-
lifts and related deposi-
tional basins of the Ancestral Rocky Mountains as
they appeared in Middle
Pennsylvanian time. The
mountainous segments
formed along reactivated
basement faults of Pre-
cambrian age and shed
vast accumulations of
coarse-grained sediments
to the adjacent basins by
normal erosional
processes, as shown by
the map patterns. The San
Luis Uplift was a source
of sandy sediments during
earliest Pennsylvanian
time, somewhat prior to
the main Uncompahgre
Uplift. Both the Paradox
and Eagle basins were the
sites of evaporite
sedimentation.

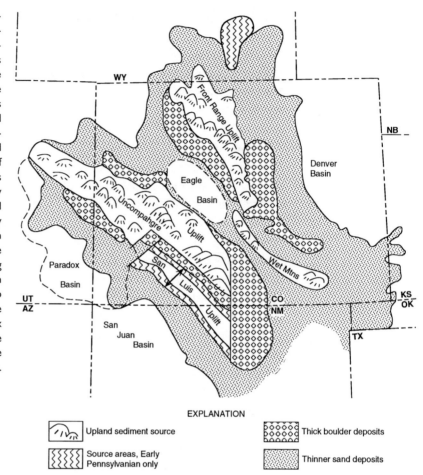

EXPLANATION

Upland sediment source

Source areas, Early
Pennsylvanian only

Thick boulder deposits

Thinner sand deposits

One of these northwest-trending fault zones passes through the eastern
Paradox basin, being offset locally by northeast-trending faults that occur
approximately beneath the course of the Colorado River. The northwesterly
faults show right-lateral displacement (rocks on either side of the fault zone
are offset to the right), and the northeasterly set of faults show left-lateral
displacement. The earliest movement, dated in the San Juan Mountains of
southwestern Colorado, occurred at about 1.7 to 1.6 billion years ago.
Movement along the fault zones continued episodically throughout geologic time. The orientation and sense of displacement along the faults is
predictable through the study of rock mechanics, indicating that the globe
was under north-south compressional forces from Late Precambrian
through Paleozoic times.

Recurrent movement along the Precambrian fault zones was on a minor,
episodic scale during early Paleozoic time, but a major resurgence of activity occurred in Middle Pennsylvanian time, some 300 million years ago,

when large faulted mountain ranges were uplifted between adjacent basins along the northwesterly fault zones. This mountain-building event, known as the Ancestral Rocky Mountain Orogeny, completely changed the earth's crust in western North America during the Pennsylvanian and Permian geologic periods (see Figure 9). Tectonic theorists believe the fault movements resulted from the collision of a volcanic island arc with the North American Plate somewhere in the vicinity of the Ouachita Mountains in southern Oklahoma. Whether or not that was the stimulus for the mountain-building event, renewed movement along the fault zones occurred along preexisting faults that were emplaced during Precambrian time.

Colorado Plateau Structural Style

The surface geology of the Colorado Plateau is typified by great monoclinal folds and their associated huge uplifts. These are striking features indeed when viewed from the air and constitute major topographic features even to the casual eye. Although the giant folds are generally oriented north-south, the underlying faults over which the sedimentary rocks are draped fit the rock-mechanics model described above as normal faults in the stress field of the province, that is, faults formed by west-east extension of the crust. The steep flanks of the monoclines usually face east, although a few face west.

Some of the more prominent of these spectacular folds are the East Kaibab monocline, the Waterpocket Fold (Capitol Reef), the San Rafael Reef, the Circle Cliffs Uplift, and the Comb Ridge monocline (Figure 10). All have their steep flanks facing east, all have formed over nearly vertical normal faults in the basement, and the last three have sharp, jagged ridges of partially eroded, strongly upturned Navajo Sandstone that guard the large anticlinal folds to their west. Because of these ragged fence-like ridges along the monoclines, trail and road crossings are rare and widely spaced. Pioneers traveling the region in the mid-1800s referred to their wagons as "prairie schooners." Coming upon these impassable rock barriers, they called the jagged ridges "reefs." Thus, the names Capitol Reef and San Rafael Reef refer to the rocky barriers to travel, just as coral reefs are navigational barriers in the oceans.

Between these large uplifted structures are major structural basins that may not be as noticeable to the casual traveler. For example, the Kaiparowits Basin lies between the East Kaibab monocline on the west and the Circle Cliffs Uplift to the east; the Henry Basin lies between the Circle Cliffs and Capitol Reef to the west and the Monument Upwarp to the east; the Black Mesa Basin is between the Echo Cliffs monocline and the Defiance Uplift; and the San Juan Basin is found between the Defiance Uplift and the Nacimiento Uplift in northwestern New Mexico (Figure 10). Although all these basins and uplifts were present as low relief features in Paleozoic time,

Figure 10. Major geological structures on the eastern Colorado Plateau. Heavy lines with barbs are on the down-thrown side of the large monoclines that form the margins of the major uplifts. Solid black areas are intrusive igneous (laccolithic) mountain ranges, and stippled patterns near Moab, Utah, are salt-intruded anticlines.

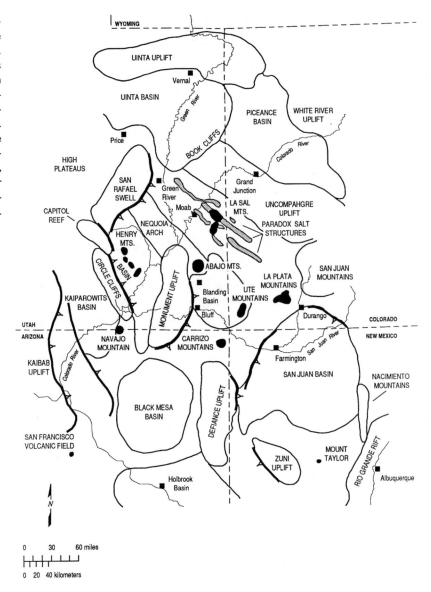

they became major depositional basins in Mesozoic time. In contrast, the Paradox basin of the eastern Colorado Plateau, lying between the San Rafael Swell and Monument Upwarp on the west and southwest and the Uncompahgre Uplift on the east, was strictly a Paleozoic basin of deposition associated with the Ancestral Rocky Mountains Orogeny. (Because the Paradox basin is entirely below the earth's surface, *basin* is not capitalized in its name, unlike the case for the other basins just described.)

Paradox Basin   Perhaps the least obvious basin of the Colorado Plateau, yet one of the most important, is the Paradox basin. It is seen only in the subsurface with geophysical studies and deep drilling for petroleum but is unique in the region as a basin of thick salt and related gypsum, dolomite, and black shale layers. The elongate depression originated in Middle Pennsylvanian time in association with the Ancestral Rocky Mountain Orogeny and associated large-scale movement along the several basement faults (see Figure 8). Although the basin is not visible at the surface, the presence of thick salt beds in the subsurface was responsible for numerous collapse features, such as seen in Canyonlands National Park, and salt-flowage structures, such as Moab Valley and Salt Valley (Arches National Park) of the eastern Colorado Plateau. Thus, the presence of deep-seated salt has played a major role in the formation of geologic features seen at the surface.

As the Uncompahgre segment of the Ancestral Rocky Mountains began to rise in Middle Pennsylvanian time, the adjacent Paradox basin sagged into existence immediately to the west of the mountain chain. Seawater poured into the huge depression and was trapped there between surrounding higher grounds. A narrow inlet through the present-day San Juan Basin fed salt water into the basin, where evaporation proceeded to deplete the supply of seawater at an enormous rate in the hot, arid climate of the time. As the high rate of evaporation took its toll on the restricted marine basin, salt brines formed and sank to the sea floor and could not escape through the narrow, shallow inlet to the southeast. Sea level fluctuated greatly during Pennsylvanian and Permian times, apparently owing to cyclic glacial episodes in the polar regions. At low sea-level stands, evaporation of the seawater dominated the scene, and salt was deposited in the basin. During high stands of sea level, fresh seawater flooded the basin through the inlet, and black organic shale beds were deposited between the salt beds. These cyclic deposits of salt and black shale (with associated gypsum and dolomite) were repeated at least 29 times until some 3,000 to 4,000 feet of salt accumulated in the heart of the basin. A nearly identical analogue is the present-day Persian Gulf in the Middle East.

Meanwhile, voluminous amounts of sand, gravel, and boulders were being washed down from the Uncompahgre Uplift by streams draining the mountain range. The coarse sediments intermingled with salt deposits along the eastern flank of the Paradox basin. Eventually, the sand and gravel (the Cutler Formation of Late Pennsylvanian and Permian age) buried the salt along the eastern basin margin, providing a very heavy overburden on the salt deposits. As the salt began to flow away from this overburden toward the west, it encountered major faulted ridges in the basement and began to rise toward areas of lesser pressure. The rising, plastic salt forced

Figure 11. Idealized cross section through the Moab Valley salt-intruded anticline. Salt flowed upward into the structure along deep-seated basement faults from Middle Pennsylvanian through Jurassic time until the supply of available salt was exhausted. The straight vertical lines represent key bore holes drilled in the search for petroleum.

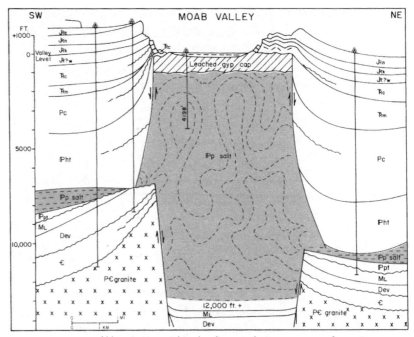

Abbreviations within the diagram designate ages or formation names: PCgranite = Precambrian granite, C = Cambrian, Dev = Devonian, Ml = Mississippian, IPpt = Pennsylvanian Pinkerton Trail Formation, IP salt = Paradox Formation salt (shown in gray), IPht = Honaker Trail Formation, Pc = Permian-Pennsylvanian Cutler Formation, TRm = Triassic Moenkopi Formation, TRc = Chinle Formation, JRw = Jurassic Wingate Sandstone, JRk = Kayenta Formation, JRn = Navajo Sandstone, and JRe = Entrada Sandstone. From Baars and Doelling 1987.

overlying beds of sedimentary rock to bulge upward and eventually pierced the thick sedimentary cover, thus initially forming the salt-piercement structures (diapirs), such as in Moab, Salt, Paradox, Gypsum, Castle, Onion Creek, Lisbon, and Sinbad Valleys. Salt flowage continued from Late Pennsylvanian through Jurassic times until salt deposits along the flanks of the structures were depleted. The salt anticlines were later buried by thick marine rocks of Cretaceous and perhaps Tertiary age. Finally, fresh near-surface groundwater dissolved the salt from the tops of the salt-intruded structures, and the Cretaceous beds collapsed into the resulting voids, thus creating the collapse valleys we see at the surface today. As seen in deep drill holes, the entire salt section is missing along the crests of the deep-seated fault blocks but reaches thicknesses of nearly 15,000 feet within the cores of the salt structures. Moab Valley is a classic example (Figure 11).

Along the shallow-water shelves of the Paradox basin to the south and west, warm seawater of nearly normal salinity bathed the sea floor, at times almost synchronously with salt deposition in the deep basin. Marine organisms flourished along these shallow banks to become limestone deposits as the sediments hardened. A special breed of green algae ("seaweed") that secretes $CaCO_3$ (lime deposits) loved the conditions along the margins of the salt basin. The algae were so happy in the warm, shallow banks that they built great mounds of their fossil debris along the shelf regions of the basin margins. Individual algal fossils look something like cornflakes in size and shape, and a rock composed of these fossils looks like a solidified box of cornflakes without the wrappings. Spaces between the flakes often remain as pores in the rock, providing ample room and communication to contain and produce great volumes of oil and natural gas. Major oilfields in the Four Corners region, such as Aneth Oilfield in southeastern Utah, are some of the country's largest producers of petroleum.

Laccoliths   Laccolithic mountain ranges, which may be seen scattered about the province, are another characteristic feature of Colorado Plateau geology. A laccolith is an intrusive igneous body, in other words, molten material within the earth that cooled and hardened without reaching the surface. Because of the slow cooling rate beneath the earth's surface, the resulting rock is usually coarse-grained in texture. Granite is a common example of an intrusive igneous rock. Lava flows, in which molten material flows onto the earth's surface, produce extrusive igneous rocks. As the rate of cooling is much faster at the surface, fine-grained rocks are formed, such as basalt.

Laccoliths are special kinds of intrusive igneous bodies. The molten rock rises through the earth's crust in a more or less circular pipe, only to spread out between bedded sedimentary rocks to form dome-topped sills. A classic laccolith looks something like a mushroom. As the domed top of the intrusion is later eroded, the sedimentary rock layers are usually stripped from the top and form sloping layers that dip away from the igneous core. On the Colorado Plateau these features of laccolithic mountain ranges are best exposed in the Henry Mountains of Utah. The mushroom-shaped igneous bodies are not always perfectly formed and may overlap in complex intrusive relationships. There are also many variations on the shapes of these igneous intrusions, all of which have unpronounceable and indigestible names.

The many isolated mountain ranges on the Colorado Plateau with intrusive igneous rock cores are collectively considered laccolithic ranges, although few igneous rock bodies in these outstanding mountains fit the original definition. Mountain ranges that fall into this category are the Henry Mountains south of Hanksville, Utah; the Sleeping Ute Mountains west of Cortez, Colorado; the La Sal Mountains near Moab, Utah; the Abajo Mountains near Monticello, Utah; the La Plata Mountains west of Durango,

Colorado; and the Carrizo Mountains in the Four Corners region (see Figure 10). Some of these ranges appear to occur along basement faults or at intersections of basement fault zones. The igneous rocks in most of these ranges date at about middle Tertiary time, but some are Cretaceous in age. Several of the laccolithic ranges are of sufficient elevation to have been glaciated during the ice ages of the past couple of million years. Navajo Mountain on Lake Powell is considered a laccolithic mountain, although only a tiny exposure of igneous rock has been located, with no apparent characteristic shape. These prominent mountain ranges serve as ideal navigational markers for pilots of small aircraft on the Colorado Plateau.

Volcanoes   Large volcanoes erupted along the southern margin of the Colorado Plateau in Late Tertiary to Pleistocene time, forming high, conical, and isolated mountains. One such volcano, Mount Taylor, near Grants, New Mexico, is cherished by the Navajo people as the sacred mountain of the south. Another, Mount Humphreys of the San Francisco Peaks near Flagstaff, Arizona, serves as the Navajos' sacred mountain of the west. A collapsed volcano, the Jemez Caldera near Los Alamos, New Mexico, has been studied as a source of hydrothermal energy. The White Mountains, at the southeastern end of the Mogollon Rim in east-central Arizona, are a complex of large volcanoes. All lie along or near the structural boundaries of the Colorado Plateau, where deep-seated fault zones provide ready access to the surface for fiery extrusions. They are mostly of Late Tertiary age, with some succeeding activity extending into the Pleistocene Epoch of the past couple of million years.

Swarms of well-preserved cinder cones and basaltic lava flows occur in association with the large volcanoes, generally being found to the north of the southern margin of the Colorado Plateau. The lack of erosion of the classic cinder cones, such as the well-developed cone at Sunset Crater National Monument near Flagstaff, Arizona, suggest that they formed just in the past few hundred or few thousand years. Valley-fill lava flows appear to have cooled just in the past few days, although they were present when this author first saw them in the early 1950s. An excellent example of very young lava flows is in the Malpais near Grants, New Mexico. Another is in Wupatki National Monument near Flagstaff, where lava flows postdate the construction of Indian ruins. The lava dams of the western Grand Canyon formed when lava flows cascaded down the canyon walls about a million years ago.

A special kind of volcanic feature, found scattered along the southern Colorado Plateau, generally south of the San Juan River, is the *diatreme*. Such features, usually in the form of volcanic vents, or pipes, formed when highly explosive, gaseous eruptions occurred in about Middle Tertiary time, approximately coincidental with the intrusion of the laccolithic mountain ranges. The results primarily take the form of circular rock peaks, such as Ship Rock in the Four Corners region, or rounded mounds of strange and uncommon rock occurrences: random boulders, some the size of trucks or houses, in a mixture of dark green; glassy rocks that have come up from the earth's mantle; granite and metamorphic rocks carried up from the basement; and limestone and sandstone blocks carried upward from the older sedimentary rock layers. These in turn are interspersed with irregular blocks of previously overlying sedimentary formations, now eroded away, that collapsed back into the explosive volcanic vent. The gaseous eruptions originated deep in the earth within the mantle and ripped upward through all existing rock units to spew debris across tens of square miles, only to collapse back in when the violence subsided.

The diatremes are usually found along and near the monoclinal folds, having used basement fault zones as easy pathways to the earth's surface. Examples of the basement-controlled vents are Ship Rock, the Mule Ear Diatreme where the San Juan River crosses the Comb Ridge monocline, and Church Rock and Agathla Peak near Kayenta, Arizona. In some cases, however, diatremes seem to be randomly scattered without any noticeable basement control, as in the Hopi Buttes, north of Holbrook and Winslow, Arizona.

How to Use This Book

For ease of organization, I have arbitrarily divided the Colorado Plateau into seven "districts," based on generalized geologic similarities. The driving directions for the roadlogs may seem haphazard, but they reflect the cobweb of routes I drove in preparation for writing this book.

The roadside geologic descriptions within each district are recorded by mileage between geographic marker points, such as towns or major highway intersections. Mileage is indicated as **Mile 27: Milepost 210**. The first number is the trip odometer reading, and the milepost number is the corresponding highway marker number. Because vehicle odometers vary considerably, the readings listed in the text should be considered general. Note that mileposts tend to disappear or may not have been replaced after highway construction, and numbers restart at state boundaries.

Figure 12 is a regional index map, but I urge you to use state highway maps for more detailed navigation. They are generally provided free of charge at state tourist offices located near state lines as well as in key cities.

Intersections with minor roads and stream crossings are given as an additional aid in geographic identification and to help those traveling in the opposite direction to the text.

Regrettably, descriptions for each road are provided for only one direction. To describe each journey in both directions would have doubled the size and cost of the book. For travel in the opposite direction of a roadlog, use the fixed milepost numbers and geographic markers. Beyond that, I suggest subtracting the specific mile point from the final mile point given for each section. A pocket calculator will help. Keep in mind that descriptions of features seen "to the left" must be viewed as features to the right, those "to the front" or "ahead" indicate features behind, and so on. In other words, read all descriptions in reverse.

An introductory statement generalizing the geology of each district precedes each part. The more complete text of the roadlogs localizes special

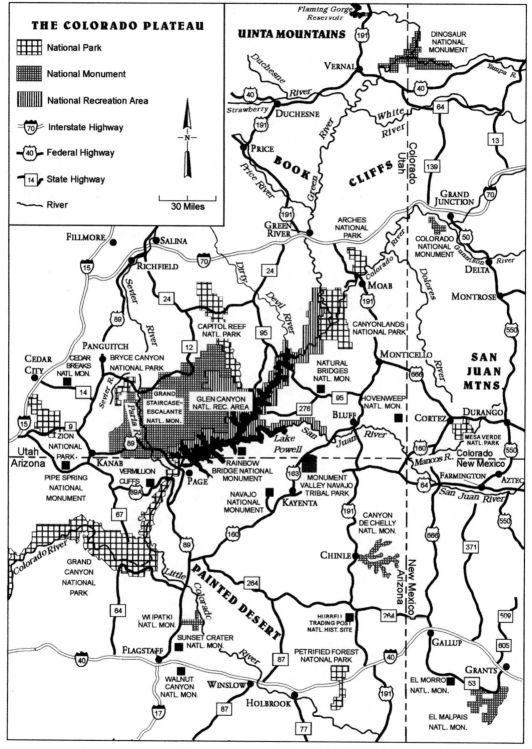

Figure 12. Road map of the Colorado Plateau Province.

geologic features seen along the road, such as the formations that comprise the scenery, formation contacts, faults, folds, and volcanic bodies.

The first group of roadlogs, Part I, is in the "Paradox District," which forms the northeastern quarter of the Colorado Plateau Province. The region is largely that of the Paradox basin and related features in eastern Utah and western Colorado. Moving clockwise around the Plateau, Part II describes the geology along roads in the San Juan Basin of northwestern New Mexico. From there, Part III is to the west, describing roads in and around the Black Mesa Basin of northeastern Arizona. It follows naturally that Part IV includes the highways in the vicinity of Grand Canyon, leading northward to Part V in the west-central Colorado Plateau. Part VI includes the transition zone to the west, the so-called High Plateaus, where Basin and Range–type normal faults of Tertiary age down-fault typical Colorado Plateau geology into the more complexly faulted Basin and Range Province. Finally, the Uinta Basin and Uinta Mountains, including Dinosaur National Monument, comprise Part VII, marking the northern boundary of the Colorado Plateau and closing the great loop around this magnificent province.

I

The Paradox Basin

The northeastern region of the Colorado Plateau Province is distinguished by the presence of a great thickness of salt at depth, deposited in the Paradox basin during Middle Pennsylvanian time. The elongate, down-faulted basin extends in a northwesterly direction from the Four Corners area to about Green River, Utah (Figure 13). The salt, and related gypsum, dolomite, and black shale deposits, formed in some 29 cycles in stagnant, highly saline waters, starting in the deeper center of the basin and spreading laterally with time. Depositional thickness of the salt section is difficult to estimate in the deeper parts of the basin, but something well over 4,000 feet is present east of Moab, Utah. In the eastern part of the basin, salt flowage has created giant walls of salt that penetrate the overlying strata and may have flowed onto the surface as salt glaciers in some parts of geologic time. In the Paradox Valley salt-intruded anticline, where deep drilling penetrated the entire salt section, about 15,000 feet of salt and related evaporite beds form the core of the structure.

Although because of present-day climatic conditions the thick salt is never exposed at the surface, its existence at depth is made obvious not only by the great salt-intruded structures but also by topographic features that were formed by dissolution and collapse of the overlying rocks. Such features distinguish the Paradox basin across Arches and Canyonlands National Parks.

The Precambrian basement beneath the Paradox basin is broken into orthogonal blocks by northwest- and northeast-trending wrench faults that first formed around 1.6 billion years ago. Those fault zones that trend toward the northwest dominate the structural geology of the Paradox basin. A major fault zone borders the Paradox basin on the east, separating the down-faulted basin from an up-faulted block of basement rocks, the Uncompahgre Uplift, on the east. This dominant fault zone was reactivated in Middle Pennsylvanian time to form the salt basin and the closely related

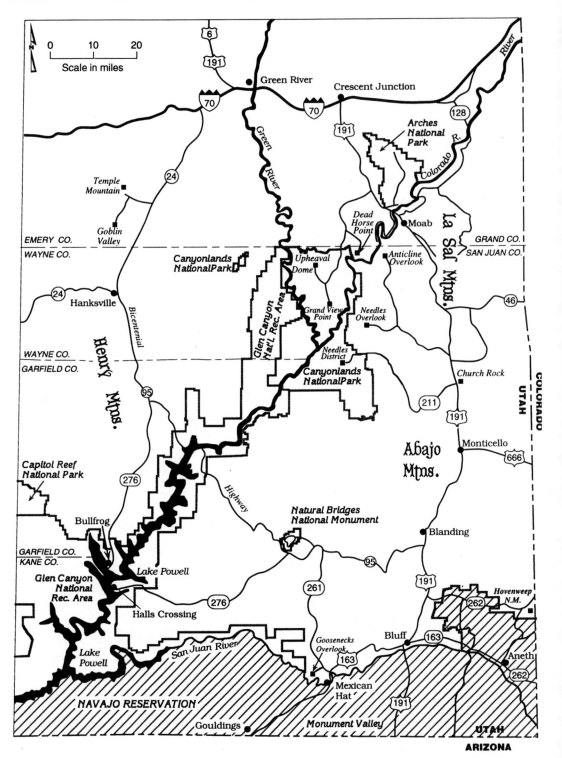

Figure 13. Road map of southeastern Utah.

uplift. Extension of the earth's crust along northeast-trending faults allowed the basin to sag and extend laterally along the northwest-trending fault zone. Rocks of early Paleozoic age are everywhere present beneath the salt, but as they are nowhere exposed at the surface, they are not described here (more information can be found in Baars 1992, 1993, 1995, and 2000).

Sedimentary rocks of Pennsylvanian and Early Permian age are cyclic in nature throughout the world, owing to rising and falling sea level related to glacial cycles in the polar regions. The salt-laden evaporite beds of the Paradox Formation are strongly cyclic in nature, as are more or less contemporaneous limestone deposits along the shallow shelves surrounding the deeper basin. Seawater of normal salinity periodically washed across the shallow shelves of the structurally controlled Paradox basin, coming into the basin mostly from the south through the ancestral San Juan Basin. As seawater filled the deeper basin, extreme aridity caused the surface water to evaporate and the dense brine layers in the water sank into the basin depths. The dense brine could not escape the basin across the shallow shelf thresholds, and salt and gypsum were precipitated in the deeper water. This sequence of events would continue as sea level lowered, only to be repeated again and again as sea level fluctuated over time.

Meanwhile, the adjacent Uncompahgre Uplift was rising, forming mountainous terrain to the east that was naturally attacked by erosion. The metamorphic rocks of the basement uplift weathered into boulders, sand, and mud that washed down from the highlands by streams that deposited the sediments into the quieter waters of the sea. Sand and gravel were intermixed with beds of salt, as the finer mud was washed out to sea. All this resulted in coarse-grained, red and brown sandstone and conglomerate interfingering with evaporite deposits along the eastern margin of the Paradox basin. By Early Permian time, sediments from the Uncompahgre Uplift dominated the scene, flooding almost the entire Colorado Plateau with red beds.

As thick sediments from the Uncompahgre Uplift buried salt deposits in the eastern Paradox basin, the salt began to flow laterally in search of areas of lesser overburden. Upon encountering the large blocks of hard rock along basement faults, the salt had no place to go but upward, and upward it flowed (see Figure 11). The culprit fault blocks are no small items! Deep drilling along the southwest flank of Paradox Valley shows that there is about a mile of structural relief along the bounding fault.

Rapid subsidence of the Paradox basin waned by the end of Middle Pennsylvanian time, but most of the Colorado Plateau was still flooded with seawater. Alternating (cyclic) beds of limestone, sandstone, and shale were deposited across the salt basin to form the later Pennsylvanian Honaker

Trail Formation, seen in the canyons of the San Juan River and in Cataract Canyon on the Colorado River in Canyonlands National Park. The sea finally withdrew toward the northwest in latest Pennsylvanian time, into the Oquirrh Basin of north-central Utah. Deposits similar to the Honaker Trail Formation, the Elephant Canyon Formation, followed the withdrawal of the sea and buried the northwestern corner of the Paradox basin.

Highlands of the Uncompahgre Uplift continued to shed red sand and mud across the Paradox basin in Early Permian time, wandering around the rising salt-intruded anticlines and to the west, spreading out across nearly the entire breadth of the Colorado Plateau. The red sedimentary rocks, known as the Cutler Group (or Formation), gradually become finer in grain size away from the source uplands. In Canyonlands country, coarse red sandstone beds from the Uncompahgre highland interfinger with marine rocks of the Elephant Canyon Formation in Meander Canyon along the Colorado River below Moab, and farther south they become fine-grained mudstone and siltstone of the Halgaito Shale near Monument Valley. As time progressed, light-colored sand was washed in, or blown in, from the northwest into Canyonlands to form the Cedar Mesa Sandstone. Interfingering of the Cedar Mesa Sandstone with red sand of the Cutler Formation produced the banded rocks seen in the buttes and pillars of the Needles District of southern Canyonlands National Park. In turn, the Cedar Mesa Sandstone was buried by red mud and siltstone derived from the Uncompahgre highland to form the Organ Rock Shale, which extends through Monument Valley to Grand Canyon, where it is called the Hermit Shale. Finally, the marine rocks of the Toroweap Formation of Grand Canyon fame spread eastward into western Canyonlands country. Shoreline deposits, there known as the White Rim Sandstone, cap the Permian stratigraphic section and form prominent benches west of the Colorado River in Canyonlands. Offshore bars and barrier bars occur to the west, grading eastward into windblown, nearly white sandstone near the eastern pinchout of the formation.

Rocks of Triassic age, the Moenkopi Formation overlain by the Chinle Formation, unconformably cover the Paleozoic section across the Paradox basin. The Early Triassic Moenkopi Formation consists of dark brown, thin-bedded mudstone and siltstone, deposited in broad coastal mudflats that thicken and grade westward into marine deposits in western Utah. Ripple marks, mud cracks, burrows, and raindrop impressions on bedding surfaces attest to the tidal flat origin. A period of weathering followed in mid-Triassic time, and stream and lake deposits of the Chinle Formation covered the widespread unconformity. Channel deposits at the base of the Chinle consist of pebble conglomerate and sandstone of the Shinarump Member in the south and the Moss Back Member to the north of White Canyon west of

Figure 14. Diagrammatic cross section showing the proposed relationships between the rocks of Middle Jurassic age on the San Rafael Swell, to the west, and the rocks of the Moab and Arches National Park regions, to the east. Drawn from written communication with Fred Peterson, 2001.

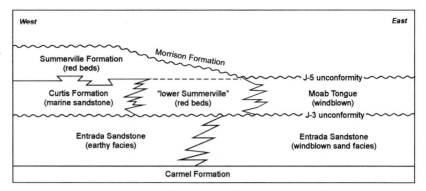

Blanding, Utah. Upper members of the Chinle across the Paradox basin country consist of the Petrified Forest Member, overlain in turn by the Black Ledge and Church Rock Members. All are stream and lake deposits containing considerable amounts of petrified wood and some other fresh-water fossils.

Massive desert sand deposits of the Glen Canyon Group provide formidable reminders of Early Jurassic time. The lower sandstone, the Wingate, forms massive cliffs that stand guard over the interior of Canyonlands, prohibiting ready crossings into the park's inner realms. The Wingate Sandstone is capped by resistant ledges of the stream-deposited Kayenta Formation, which is in turn overlain by the white, highly cross-bedded Navajo Sandstone. These cliff-forming sandstone units are the youngest rocks found in Canyonlands National Park.

Rocks of Middle Jurassic age are best seen in the San Rafael Swell and in Arches National Park. Known as the San Rafael Group, the younger Jurassic rocks of the Paradox basin consist of brown mudstone and siltstone of the Carmel Formation (locally called the Dewey Bridge Member of the Entrada Sandstone), often highly crinkled, which overlies the easily recognizable Navajo Sandstone. The Entrada Sandstone, another windblown desert deposit, forms the massive cliffs above the Carmel (Dewey Bridge) that erode to form myriad natural arches, especially well developed in Arches National Park. The Entrada, as originally defined, consists of the lower Slickrock Member, overlain at a prominent horizontal bedding plane by the Moab Tongue. The Moab Tongue is believed to be an eastern facies of the light-colored marine sandstone of the Curtis Formation that occurs across the San Rafael Swell to the west (Figure 14).

Finally, in latest Jurassic time, the distinctive Morrison Formation was deposited in stream and lake environments. The lower unit, the Tidwell Member, is about 40 feet thick and consists of mudflat, lake, and overbank mudstone; some scarce stream-channel sandstone beds; some thin limestone of freshwater lake origin; and local windblown sandstone and gypsum beds, with a basal bed of grit-bearing sandstone. The Tidwell Member grades upward into the Salt Wash Member, which forms ledgy cliffs of fluvial sandstone that may be seen in structurally low areas across most of the Colorado Plateau. The Salt Wash Member hosts abundant uranium/vanadium deposits in the region, and numerous dinosaur fossils have been recovered at several localities. The upper or Brushy Basin Member, also widespread across the plateau country, consists of gray, light green, and pale red colors of mudstone and shale that formed in overbank floodplains adjacent to stream channels and in lakes. Other members are present locally in other districts.

Stream deposits on an erosional surface at the top of the Morrison Formation are called the Cedar Mountain Formation along the western side of the Colorado Plateau and the Burro Canyon Formation to the east. These cross-bedded, coarse-grained sandstone and conglomerate beds of Early Cretaceous age are difficult at times to distinguish from the overlying Dakota Sandstone of Late Cretaceous age, especially in distant views. Where the roadlogs refer to the Dakota Sandstone, some Burro Canyon or Cedar Mountain strata may have been unintentionally included with the Dakota.

Concluding the geologic legacy of the Paradox Basin District of the Colorado Plateau, the Late Cretaceous dark gray, very thick, marine shale beds of the Mancos Shale are found nearly everywhere a city or main highway has been built. Although easy to bulldoze into road, airport, and house foundations, the clay-laden formation is treacherous when wet. Highways and runways have been known to sink or slide into oblivion, and homes often slither down slopes of the slippery shale. Fortunately, this "gumbo" is exposed only in structurally low, synclinal regions of the district. Various and sundry formations of the Mesaverde Group form cliffs above the Mancos Shale, especially prominent in the Book Cliffs north of the district. Deposited in coastal regions as the Late Cretaceous sea advanced and retreated across the Colorado Plateau, the formations are of different ages in different localities, so they have been given myriad names to match. Only the term *Mesaverde Group* remains in regional usage. Although these thick deposits of Cretaceous age once covered the entire Colorado Plateau, they have been stripped from the uplifted regions by erosion in the last few million years and have been carried away and redeposited by the Colorado River and former river systems.

1. Grand Junction,
Colorado, to
Green River, Utah, via
Interstate 70

Grand Junction is the largest city in western Colorado and on the Colorado Plateau, and it seems appropriate to begin our road descriptions here. The city lies along the Recent floodplain of the Colorado River downstream from the junction with the Gunnison River. The Colorado River was once known as the Grand River from its headwaters to the confluence with the Green River. The city's name is derived from the "junction" of the "Grand River" and the Gunnison, thus *Grand Junction*. The city nearly fills the Grand Valley, with the Uncompahgre Plateau to the south, the Book Cliffs to the north, and Grand Mesa, capped by lava flows of Tertiary age, to the east.

The Uncompahgre Plateau is a faulted segment of the ancient Uncompahgre Uplift, which forms the northeastern boundary of the Paradox basin and Colorado Plateau Province. The ancestral Uncompahgre Uplift was elevated along a Precambrian continental-scale wrench-fault system in Middle Pennsylvanian time (about 300 million years ago) as a segment of the Ancestral Rocky Mountains (Figure 15). Erosion of the ancient mountain range supplied great volumes of sand, gravel, and boulders to the adjacent basins, the Paradox basin to the west and the Eagle basin to the east. It was

Figure 15. The structural geology of southwestern Colorado as it appeared during Precambrian time. The Uncompahgre Uplift was a large segment of the Ancestral Rocky Mountains that was rejuvenated and uplifted in Middle Pennsylvanian time. Each faulted segment of mountain ranges provided vast amounts of coarse-grained sediments to the adjacent basins. Heavy solid lines are the locations of known faults, and the dashed lines indicate the approximate locations of inferred extensions of the major basement fault zones.

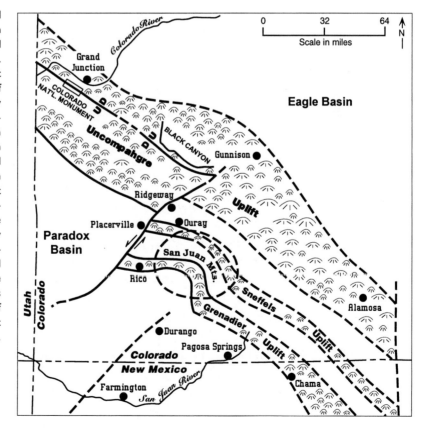

leveled by erosion in Late Triassic time, when the metamorphic roots of the mountains were buried by the Chinle Formation, as seen in Colorado National Monument west of Grand Junction.

Dark gray exposures of the Mancos Shale (Late Cretaceous) may be seen in and around Grand Junction and along the base of the Book Cliffs as far west as Helper, Utah. The organic-rich marine shale is about 4,000 feet thick in this area. As the Mancos sea retreated toward the east, fluctuating deposits of beach sand and offshore bar deposits, interspersed with coastal swamp deposits of coal, comprised the overlying Mesaverde Group, which forms the cliffs above the Mancos slopes in the Book Cliffs to the north of Grand Junction. Above the Mesaverde Group, and set back northward to form the Roan Cliffs, are sandstone and shale deposits of the Wasatch and Green River Formations of Tertiary age. The Book and Roan Cliffs mark the southern boundary of the Uinta Basin, filled with stream and lake deposits of Tertiary age.

From Grand Junction to about the Utah state line, Interstate 70 crosses the northwesterly plunging nose of the Uncompahgre Uplift. From the Colorado-Utah boundary westward, the highway travels along the relatively flat bench formed at or near the top of the Dakota Sandstone (Late Cretaceous) and base of the Mancos Shale. Views to the south are of the La Sal Mountains and salt-intruded anticlines of the Paradox basin; to the north the Book Cliffs are visible the entire distance.

Mile 0.0: I-70 at Exit 26, west side of Grand Junction, Colorado.

I-70 is on the Colorado River floodplain, with low bluffs of Pleistocene terrace deposits to the left, from Grand Junction to Exit 19, Fruita, at **Mile 6.0**. Turn off here for a most scenic tour of Colorado National Monument on the edge of the Uncompahgre Plateau to the left (south; see Roadlog 2). At Exit 15, Loma, at **Mile 10.5** the highway is on the Mancos Shale. (To go north on Colorado 139, see Roadlog 57.) At **Mile 12.0** the Dakota Sandstone hogback is on the left (a hogback is a sharp ridge along an upturned bed). **Mile 14.4**: Exit 11. The Mancos Shale is the dark gray to black, soft-weathering rock that forms the valley and slopes to the right (north). The dark shale weathers to a tawny surface color where the organic-rich rock is oxidized. It is a very thick (4,000 to 5,000 feet) marine shale that was deposited in Late Cretaceous time, as shown by the many marine fossils found in the formation, including ammonites, clams, oysters, shark teeth, and microscopic foraminifera. The Dakota Sandstone, which forms hogbacks to the left (south), comprises nearshore sand, beach, and swamp sedimentary rocks that were deposited along the advancing shoreline of the Late Cretaceous sea. The cliff-forming buff-colored sandstone here dips rather strongly toward the northeast as the formation crosses the major fault and monocline along the margin of the Uncompahgre Plateau to the left.

At **Mile 16.2** the highway climbs the dipslope of Dakota Sandstone and then crosses down through the Dakota (Late Cretaceous) and Burro Canyon (Early Cretaceous) Formations into the Morrison Formation (Late Jurassic) at **Mile 20.0.** Exit 2 at **Mile 23.8** goes to Rabbit Valley, home of numerous dinosaurs, in the Morrison Formation. The quarry there is still actively producing dinosaur remains, and a short hiking trail to displays of dinosaur finds is readily accessible. From Rabbit Valley the highway crosses up through the Dakota Sandstone and onto the lower Mancos Shale at **Mile 24.9.**

Mile 25.8: Colorado-Utah border; the highway is in the Mancos Shale. **Mile 27.2** is Utah **Milepost 228.** Exit 225, to the village of Westwater, is at **Mile 30.2,** with access to the Colorado River at the head of Westwater Canyon. The highway is on top of the Dakota Sandstone; the La Sal Mountains, one of several laccolithic ranges on the Colorado Plateau, are visible on the left at 10:00. At **Mile 32.5: Milepost 223** the southwest flank of the Uncompahgre Uplift is on the left at 9:00, marking the boundary between the Paradox basin and the Uncompahgre Uplift; the La Sal Mountains are at 10:00. **Mile 35.4:** Exit 220, a ranch exit.

Mile 42.9: Exit 212, the first Cisco exit; oilfield pumps and tanks are visible in the area. These wells are part of the "Greater Cisco" oilfield, a conglomeration of five distinctive oilfields discovered in 1924 by Utah Oil Company. Oil and gas have been produced from the Entrada Sandstone, the Morrison Formation, the Burro Canyon Formation, and the Dakota Sandstone in small quantities since the discovery. Production is from small faulted anticlinal folds above the deep-seated Uncompahgre fault zone, which separates the Uncompahgre Uplift from the Paradox basin in this area. As of 1982 the Greater Cisco field had produced more than 320,000 barrels of oil and 10 million cubic feet of gas. It is believed that the oil migrated up the Uncompahgre fault zone from Paleozoic rocks of the Paradox basin and that the gas is from the locally derived decay of organic matter from the Mesozoic rocks.

Mlle 52.7: Exit 202, to Utah 128, is the second Cisco exit and goes to Moab via the Colorado River road (see Roadlog 4). **Mile 64.0:** Exit 190, a ranch exit. **Mile 67.8:** Exit for rest area. **Mile 70.0:** Exit 185, to Thompson Springs. Exit 180, to Crescent Junction, Utah, at **Mile 75.3.** There is a good view to the left (south) of the Salt Valley salt-intruded anticline in Arches National Park. Crescent Junction marks the intersection of I-70 and U.S. 191 to Moab, Arches and Canyonlands National Parks, and points south (see Roadlog 3).

Collapse structures on the right at **Mile 77.7** are in the Mancos and Mesaverde Formations, a result of collapse into the northwestern termination of the Salt Valley anticline. A distant view of the San Rafael Swell is ahead, and the La Sal Mountains can be seen to the left rear. The highway from here to Green River is in the Mancos Shale with Mesaverde sandstones above in the western Book Cliffs to the right (north).

Mile 82.3: Exit 173, a ranch exit. **Mile 83.9**: Good view of the San Rafael Swell, from 10:00 to 2:00 ahead. **Mile 85.8: Milepost 170.**

Mile 93.4: Take Exit 162, the first exit to Green River, Utah. A bridge crosses the Green River at **Mile 95.4.** Entering Green River and Emery County at **Mile 96.2. Mile 98.1.** (Connect with Roadlog 45, Green River to Salina, Utah, via I-70, or Roadlog 52, Green River to Helper, Utah, via U.S. 6 and 191.)

2. Colorado National Monument

Perhaps the least known and least understood area in the National Park System is Colorado National Monument, just outside Grand Junction, Colorado. There colorful rocks of Triassic and Jurassic age are exposed in breathtaking canyons that never fail to amaze visitors. The monument is accessed by either the West Entrance near Fruita, Colorado, or the East Entrance near downtown Grand Junction. A 23-mile road, called Rim Rock Drive, connects the two entrance gates, passing numerous spectacular viewpoints (Figure 16) and the Visitor Center, located 4 miles from the West

Figure 16. Independence Monument, a 500-foot-high monolith in Monument Canyon, stands guard near the west entrance of Colorado National Monument. The prominent landmark consists of cliffs of the Wingate Sandstone, capped by a thin erosional remnant of the Kayenta Formation, both Jurassic in age. Grand Junction is in the valley in the middle distance.

Figure 17. Generalized cross section through the monoclinal fold along the northern flank of Colorado National Monument. The heavy lines are major faults that originated in Precambrian time and were reactivated throughout geologic time. The sedimentary rocks that overlie the metamorphic basement rocks have draped across the faults as later events caused further uplift of the ancient fault block.

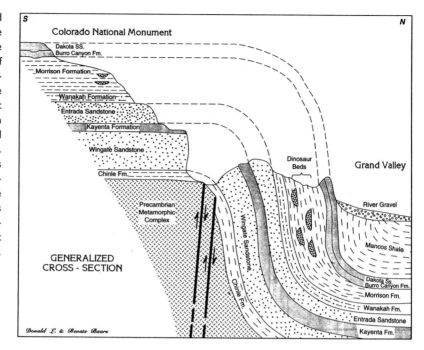

Entrance. Brochures containing a detailed map of the monument are available at either entrance gate or at the Visitor Center.

The paved road entering Colorado National Monument from either entrance winds its way in tortuous paths across a beautifully exposed monoclinal fold that crosses exposed faults in the Precambrian basement. The dark red shale of the Triassic Chinle Formation, the magnificent cliffs of the Jurassic Wingate Sandstone, ledgy stream-deposited sandstone of the Kayenta Formation, capped by the red cliffs of the windblown Entrada Sandstone, are draped across major northwest-trending faults exposed in metamorphic rocks of Precambrian age (Figure 17).

The best-kept secret in the geologic story of Colorado National Monument is that about 1.5 billion years of earth history is missing here. Where did it go? How do we know?

The Uncompahgre Plateau, where the monument is located, is part of an ancient upland, or mountain range, that geologists call the Uncompahgre Uplift. The highland began to rise along major faults in the earth's crust some 300 million years ago (Middle Pennsylvanian time) to form a chain of formidable mountains. As the mountains became gradually higher, weathering attacked the upland, and boulders, pebbles, sand, and mud were washed down from above by streams and were deposited in adjacent lowlands. Erosion soon exposed the ancient metamorphic "basement rocks," which date at about 1.7 billion years old. These, the oldest rocks known to

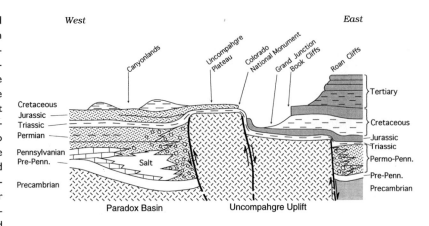

Figure 18. Regional schematic cross section through the Uncompahgre Uplift and the adjacent Paradox basin to the west. Streams from the uplift carried vast amounts of coarse-grained sand and gravel to the eastern margin of the basin in Pennsylvanian and Permian times to interfinger with salt and other marine rocks being simultaneously deposited. Mud and sand eventually covered the marine rocks, extending across Canyonlands to as far west as Grand Canyon. From Baars 1998c.

be present in the region, consist of highly altered gneiss and schist, with some granitic dikes and sills.

Sediments derived from this metamorphic core soon began to fill the adjacent Paradox basin to the west and the Eagle basin to the east. These sediments became sandstone, mudstone, and conglomerate, rich in fragments of the parent metamorphic rocks and with a high feldspar content. Associated iron-rich minerals soon weathered to the red and brown hues so characteristic of the red rocks seen in the Cutler Group of Pennsylvanian and Permian age in the region around Moab, Utah. By Triassic time the moun-

Figure 19. The Coke Ovens in Monument Canyon, Colorado National Monument, have been eroded from cliffs of Wingate Sandstone and capped by thin exposures of the Kayenta Formation, both Jurassic in age. The cliffs in the far canyon wall consist of lower slopes of the Triassic Chinle Formation and massive cliffs of the Wingate Sandstone, overlain by wooded slopes of the Kayenta Formation. The upper, thinner cliffs are in the Entrada Sandstone, capped by hills at the skyline of the Wanakah and Morrison Formations.

tainous region of the Uncompahgre Uplift had worn down to a nub by erosion, and red mudstone and siltstone of the Chinle Formation of Late Triassic age buried the bones of the once-prominent highland. The Chinle Formation rests directly on the ancient metamorphic basement in Colorado National Monument; no rocks of Paleozoic age are to be seen anywhere (see Figure 17).

Thus, a time gap, called an *unconformity* by geologists, representing the entire time interval of about 1.5 billion years of missing rocks, is present where red sedimentary rocks of the Chinle Formation rest directly on dark gray metamorphic rocks. The unconformity may be seen in the deepest canyons near the entrances to Colorado National Monument. Sedimentary rocks of this time span are up to 20,000 feet thick along the eastern margin of the Paradox basin, as shown by deep drilling (Figure 18). The sedimentary rocks thinned and became finer in grain size as the sediments were carried farther away from the upland source area by the ancient streams. The resulting formations constitute the red sedimentary rocks around Arches and Canyonlands, Monument Valley, Canyon de Chelly, and even as far away as the Grand Canyon.

The rocks that tell of this major event in the geologic history of the Colorado Plateau are well exposed and accessible in Colorado National Monument (Figure 19).

3. Crescent Junction to Moab, Utah, via U.S. 191   U.S. 191 from Crescent Junction south to Moab, Utah, is on the Mancos Shale (Late Cretaceous) and follows the southwest flank of the Salt Valley salt-intruded anticline for about 15 miles. It then follows the Moab fault zone down a colorful canyon, descending through the geologic section into rocks of Paleozoic age. From the entrance to Arches National Park in another 12 miles, the Moab Valley salt-intruded anticline is beautifully exposed ahead. Collapse structures along the flanks of the valley are obvious from the park entrance southward through the town of Moab.

Mile 0.0: Crescent Junction, Utah, intersection of Interstate 70 and U.S. 191.

Mile 0.5: Heading south, U.S. 191 is on the Mancos Shale (Late Cretaceous); the Salt Valley salt-intruded anticline is clearly visible on the left. The La Sal Mountains, a laccolithic mountain range with the second highest peak in Utah, form a spectacular backdrop.

A good view down the axis of the Salt Valley anticline is at **Mile 2.5** with the La Sal Mountains in the distance. At **Mile 4.2** the Dakota Formation hogbacks are on the left with the Mancos Shale on the right. The road follows the Dakota-Mancos contact. **Mile 13.5:** The road on the right goes to the Moab Airport.

Figure 20. The Moab fault as seen at the entrance to Arches National Park along U.S. 191. The heavy lines outline the fault zone. The rocks exposed to the left of the fault along the highway are in the uppermost Pennsylvanian Elephant Canyon Formation and contain abundant marine fossils. The rocks to the right are sandstone beds of Jurassic age. The talus-covered slope across the highway contains thin beds of the Cutler Formation (Permian), the Triassic Moenkopi and Chinle Formations, beneath the high, massive cliffs of the Wingate Sandstone, with a thin cap of the Kayenta Formation.

The highway crosses the Dakota/Burro Canyon outcrop at **Mile 15.3.** The bluish-green shale ahead is in the Brushy Basin Member of the Morrison Formation (Late Jurassic). The color comes from traces of iron in the mineral analcime, which formed from volcanic ash that fell into a saline-alkaline lake. The highway is on the Morrison Formation at **Mile 16.6** with the La Sal Mountains directly ahead. The railroad tracks on the right lead to Potash, an underground mine that produces potash salt for fertilizer and explosives by solution mining. The Moab fault lies along the base of the cliffs on the right at **Mile 19.0.** From top to bottom, the cliffs consist of the Kayenta Formation, the ledgy caprock; the Wingate Sandstone, the high massive cliffs; the Chinle Formation, the variable-colored shale slopes below the Wingate cliffs; the Moenkopi Formation, in the lower brown shale slopes; and the Cutler Formation, the lower rubbly brown cliffs. Bulldozer scars are in the Moss Back Member of the Chinle Formation (Late Triassic) from uranium prospecting in the 1950s.

The road on the right at **Mile 20.5** is Utah 313, going to Canyonlands National Park (Island in the Sky District) and Dead Horse Point State Park (see Roadlog 6). Highway 191 starts its descent toward the Colorado River along the Moab fault at **Mile 22.7.** Cliffs on the right expose the section from the Kayenta Formation (Jurassic) down to the Cutler Formation (Permian), the cliff-forming dark reddish brown sandstone, here about 800 feet thick. The

Figure 21. Aerial view looking toward the southeast of Moab Valley. The Colorado River is seen crossing the salt-intruded anticline in which the town of Moab is situated, and the La Sal laccolithic mountains are in the background. The flanks of the northwest-trending valley are collapse blocks of strata of Mesozoic age, having caved into the top of the breached salt structure. The rectangular white area near the center of the photo is the Atlas uranium mill tailings, to be moved or buried in the near future. For a cross section of the geology, see Figure 11.

Morrison Formation (Late Jurassic) is exposed on the left on the downthrown side of the Moab fault.

Limestone beds at the top of the Elephant Canyon Formation (Late Pennsylvanian age), containing numerous marine invertebrate fossils, are on the right and left of the highway at **Mile 26.0.** The Cutler Formation, the dark brown sandstone cliffs exposed on the right, is here only about 400 feet thick, thinning as it laps onto the Moab salt-intruded anticline.

The Moab fault is on the left at **Mile 26.3**; the section exposed beyond the fault to the left is made up of Entrada and Navajo Sandstones (Figure 20). The entrance to Arches National Park is on the left at **Mile 26.5** (Roadlog 5). The jumbled rock pile along the Moab fault on the left is from the collapse of the Navajo Sandstone into the northwest nose of the Moab salt-intruded anticline.

The road to the right at **Mile 27.5**, Utah 279, goes to Potash (Roadlog 7). Tailings from the Atlas Minerals uranium concentration plant ahead are on the right. The mill buildings have been removed from the original site, but as of 2001 the tailings pile, the subject of much environmental controversy, has not been moved and buried elsewhere. A good view of the length of Moab Valley salt structure is ahead. **Mile 28.9:** Bridge over the Colorado River (Figure 21). Good exposures of the Glen Canyon Group are on the left: the Navajo Sandstone forms the upper massive, light-colored cliff; the

Kayenta Formation is the thinner-bedded, ledgy sandstone in the middle; and the lower brown, massive cliff is Wingate Sandstone (Jurassic). At approximately this site the Old Spanish Trail crossed the Colorado River.

The road to the left at **Mile 29.1** is Utah 128, heading up the Colorado River to Cisco (Roadlog 4). From this point through the town of Moab, soft-weathering gray exposures of gypsum and black shale of the Paradox Formation (Middle Pennsylvanian), brought up by salt flowage in the Moab salt-intruded anticline, form the low hills bordering the valley floor. Across the valley careful observation reveals that the Cutler Formation is missing in the section, owing to salt flowage in Cutler time. The Triassic Moenkopi Formation lies directly on upturned beds of Pennsylvanian age.

Mile 30.3: Entering Moab, Utah. The townsite was first settled by Mormon colonists in 1855, and after twenty years of confrontation with the Ute Indians it was resettled in 1876. The name has a biblical origin, although some suggest that the Paiute Indian name for "mosquito," *moapa*, was responsible for the name. Moab was a thriving center for uranium prospecting and mining in the 1950s. The surrounding slickrock country is now a popular site for back-country bicycling and four-wheeling vehicles.

4. Utah 128, West Cisco Exit on Interstate 70, to Moab, Utah, along the Colorado River

The "River Road" from Interstate 70 to Moab begins in the lower Mancos Shale and proceeds down-section to the Entrada Sandstone at Dewey Bridge on the Colorado River. From Dewey Bridge the road follows the river downstream in a beautiful canyon that exposes the stratigraphic section down to the top of the Permian-age Cutler Formation at the mouth of Onion Creek. Salt-intruded anticlines crossed by the river and highway are Onion Creek (Fisher Valley), Cache Valley, Castle Valley, and finally Moab Valley. The lower canyon, carved in the Glen Canyon Group of Jurassic age, provides magnificent scenery.

Mile 0.0: Exit 202 (West Cisco Exit) on I-70; take this exit, which goes to Utah 128 and Moab. (The east exit to Cisco also leads to Utah 128 but is a poor road.)

Mile 0.8: Milepost 44: The road is in the lower Mancos Shale. **Mile 1.8: Milepost 43.**

Intersection at **Mile 3.2**: to the right is Utah 128 to Moab; ahead is the old highway to the abandoned site of Cisco. Turn right. At **Mile 4.6** there is a good view ahead of the La Sal Mountains. To the left at about 2:00 is the western faulted margin of the Uncompahgre Uplift. **Mile 4.8: Milepost 40.**

Mile 6.8: Milepost 38: The view ahead is down the axis of the Sagers Wash syncline, lying between the ancient Uncompahgre Uplift to the left (east) and the Onion Creek–Fisher Valley salt structure to the right (west),

Figure 22. The distribution and names of the individual salt anticlines in the eastern Paradox basin.

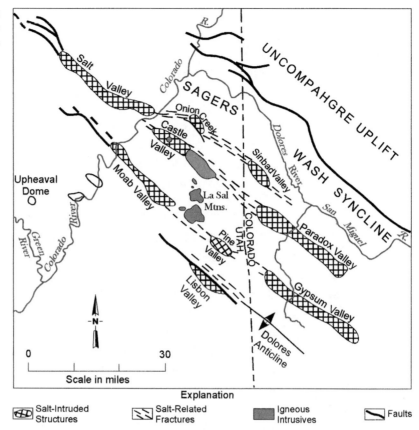

Figure 22. The distribution and names of the individual salt anticlines in the eastern Paradox basin.

with the La Sal Mountains beyond (Figure 22). The road is traveling down the axis of the Sagers Wash syncline in the lower Mancos Shale. **Mile 9.9: Milepost 35:** Utah 128 is designated a Scenic Byway.

Mile 11.0: The road is at the top of the Dakota Sandstone, starting down in the section.

The Colorado River is soon visible on the left. The Burro Canyon Formation (Lower Cretaceous), largely stream-deposited sandstone, is to the right and left at **Mile 11.7** and for the next half-mile. The top of the Brushy Basin Member of the Morrison Formation (Late Jurassic), varicolored floodplain and lake mud deposits, is at **Mile 12.4. Mile 12.9: Milepost 32.** The road is back in the Burro Canyon Formation. At **Mile 13.4** the road is along the right bank of the Colorado River, and the Burro Canyon–Brushy Basin contact is on the left. Exposures of the Salt Wash Member of the Morrison Formation, mainly stream deposits, are to the right at **Mile 13.5.** The Moab Tongue of the Entrada Sandstone is exposed on the right at **Mile 13.7. Mile 13.9: Milepost 31:** Contact of the Moab Tongue and the Slickrock Member of the Entrada Sandstone (windblown sand deposits) is on the right and left. Terrace gravels from previous levels of the Colorado River are along the highway at **Mile 14.3** on the right beneath the Entrada cliffs.

Mile 14.9: Bridge over the Colorado River. The old Dewey Bridge is preserved as a footbridge to the left. The Dewey Bridge Member of the Entrada Sandstone (equivalent to the Carmel Formation) is quite thin here and makes a bench between the Slickrock Member of the Entrada Sandstone and the Navajo Sandstone below. The Navajo Sandstone, a classic wind-blown sand accumulation, forms the lower cliffs ahead. Dewey Bridge Recreation Area, with a picnic area and boat ramp, is on the right at **Mile 15.2**. The top of the Kayenta Formation is exposed on the left.

The top of the Wingate Sandstone, another windblown (eolian) deposit, rises to road level at **Mile 15.7**. **Mile 16.0: Milepost 29**: Cliffs to the right and left are the Wingate Sandstone. The contact between the Wingate Sandstone and the underlying Chinle Formation is at road level at **Mile 16.2**. The Chinle slopes here are covered by fallen blocks of Wingate Sandstone. The caprock is the lower Kayenta Formation. **Mile 17.0: Milepost 28**. Good exposures of the Chinle Formation, largely fine-grained stream and lake sediments, are ahead on the left.

Mile 18.4: A thick section of the Chinle and Moenkopi Formations is visible ahead. The Shinarump Member of the Chinle Formation is not evident here, and the section is difficult to subdivide. The view ahead at **Mile 18.8** is of Fisher Towers and the La Sal Mountains in the background. At **Mile 19.2** a thin white marker bed is present high in the red bed slopes, marking the contact between the Chinle Formation and the abnormally thick Moenkopi Formation. **Mile 20.1**: There is another nice view downstream of Fisher Towers and the La Sal Mountains. The road is now well down into the Moenkopi Formation, but the contact with the Chinle Formation is not clear.

Mile 20.5: The view ahead is of the Priest and Nuns and Castle Rock on the east flank of Castle Valley. The low cliffs ahead are in the top of the Cutler Formation (Permian), more than 12,000 feet thick in this area near the Uncompahgre Uplift, the source of the coarse arkosic sediments. **Mile 21.0: Milepost 24**. Note that the upper Cutler sandstone beds in the low cliffs are very coarse-grained and conglomeratic, indicating a nearby source for the sediments. Hittle Bottom Campground is on the right at **Mile 21.7**.

A view of Fisher Towers and Onion Creek is on the left at **Mile 21.9**. Onion Creek has carved a deep, spectacular canyon into the crest of the Fisher Valley salt-intruded anticline. **Mile 22.0: Milepost 23**. Looking back across the river at **Mile 22.8**, we see an angular unconformity at the top of the Cutler Formation (purplish brown) and base of the Moenkopi Formation (dark brown). The nature of the unconformity indicates that an anticline formed here and was beveled by erosion before Moenkopi time, but geophysical studies suggest the presence of little or no salt at depth and certainly no salt thickening due to flowage. On the left is an excellent view of Fisher Towers and the mouth of Onion Creek. **Mile 23.2**: Exposures of the top of the Cutler Formation are present to the right and left. The Priest and Nuns and Castle Rock are ahead on the left at **Mile 23.6.** The Fisher Valley salt-intruded structure is on the left.

Mile 24.0: Milepost 21: The road to the left goes to a picnic area at the base of Fisher Towers and the old drill site of the Harry Hubbard, Onion Creek Number 1 well. The hole was drilled to a depth of about 16,000 feet and was still drilling in the Cutler Formation. Fisher Towers, capped by thin beds of Moenkopi Formation, was the site of the original Chevrolet convertible television commercial. The road here is very near the top of the Cutler Formation. The road to the left at **Mile 24.9** goes up Onion Creek to Fisher Valley. Fisher Valley Ranch is 13 miles away over a slow, sinuous, and often rough road. At last count the road crossed Onion Creek 32 times.

Mile 25.0: Milepost 20: The open area to the left is known as Professor Valley or the Richardson Amphitheater, both named for Dr. Sylvester Richardson, the postmaster and teacher in the valley around the turn of the century.

Mile 25.4: The base of the cliffs across the valley is the purplish sandstone of the Cutler Formation, overlain above an unconformity by the thick, dark brown Moenkopi Formation. About two-thirds of the way up the shale slope is a white ash bed that marks the contact with the Chinle Formation, which comprises the upper one-third of the shale slope. The brown cliffs above are in the Wingate Sandstone, capped by basal beds of the Kayenta Formation. The road to the left at **Mile 25.9** goes to the Taylor Creek Ranch.

At **Mile 26.7** the road enters a highly collapsed zone that connects the Cache Valley collapse across the river with the Onion Creek structure. The fault zone crosses the river in an S pattern, indicating that there has been left-lateral fault movement along a basement fault zone that generally follows the course of the Colorado River. **Mile 27.1: Milepost 18.** The Cache Valley collapse structure, the southern terminus of the Salt Valley salt-intruded anticline, is beautifully exposed across the river to the right at **Mile 27.4.** Roads to the left and right at **Mile 27.8** go to the Sorrel Ranch luxury bed and breakfast establishment.

Mile 28.8: There is a good view of the north-plunging nose of the Castle Valley salt-intruded anticline ahead. Note that the salt structure is terminated at the Colorado River. The road is in the lower Moenkopi Formation. The road to the left at **Mile 29.5** goes to Castle Valley. A 2-mile drive takes

Figure 23. An angular unconformity is apparent across the Colorado River from Utah 128, better known as the River Road, to Moab, Utah. Beds of the lower Chinle Formation, just above river level, are dipping rather strongly up-river, toward the camera. At a plane about one-third of the way up the lower slope across the river, the dipping beds are truncated at an erosional surface, and nearly horizontal beds of the upper part of the Late Triassic Chinle Formation overlie the surface. The unconformity was apparently caused by salt removal from a syncline that is better exposed to the north in Chinle time. The upper massive cliffs are in the Lower Jurassic Wingate Sandstone, capped by the Kayenta Formation.

one into the well-developed salt-intruded and collapsed Castle Valley structure. The White Rim Sandstone pinches out against the northerly nose of the salt structure.

Mile 30.1: Milepost 15: The low cliffs on the left are in the uppermost Cutler Formation in the north-plunging nose of the Castle Valley salt structure.

Mile 31.1: Milepost 14: Crossing Castle Creek. The ranch gate on the right was for years a false-front fortress gate, built for an army post for an old John Wayne–John Ford western. Numerous western movies have been photographed in this vicinity. The little rapid in the river behind the ranch is where the John Wesley Powell boat, the *Noname*, was broken up for the Disney movie *Ten Who Dared*. According to the late Kenny Ross of Bluff, Utah, an old-time river runner, he was the stand-in for river shots. The director didn't believe Ross could hit the rock squarely to break the boat in half, so cables were attached at both the bow and stern to direct the boat onto the rock. When the "break-away" boat hit and broke up, the bow and stern swung in neat arcs to the left and right banks in unison. Needless to say, another break-away boat had to be built in Hollywood and the scene rephotographed. The historical event actually occurred in Disaster Falls in Lodore Canyon on the Green River in 1869, but Disney decided to shoot the scene here where the canyon is more picturesque.

Mile 32.1: Milepost 13: The light-colored rock layers left and right mark the contact between the Moenkopi and Chinle Formations. The road to the right at **Mile 32.9** goes to a boat-launching ramp. The highway is at about the top of the Moenkopi Formation here. A turnout is on the right at **Mile 34.3.** An excellent view of an angular unconformity within the Chinle Formation is across the river (Figure 23). Big Bend Campground is on the right at **Mile 37.4.** The road is in the Moenkopi Formation. **Mile 38.0:** Oak Grove Campground is on the right. **Mile 38.2:** Hal Canyon Campground is on the right.

Mile 39.2: Milepost 6: The road climbs slowly up-section as the rocks dip gently westward into the Courthouse Wash syncline, a structural low caused by salt withdrawal between salt-intruded anticlines. **Mile 40.2: Milepost 5:** The Wingate cliffs are dipping more strongly westward into the syncline. **Mile 41.2: Milepost 4:** The cliffs to the right and left are in the Wingate Sandstone.

The road is near the top of the Wingate and base of the Kayenta Formation at **Mile 41.5.** The white, rounded cliffs ahead are in the Navajo Sandstone. **Mile 42.3:** The top of the Kayenta Formation and base of the Navajo Sandstone is at road level. The Goose Island Campground is on the right at **Mile 43.8.** The rocks ahead are dipping strongly toward the east into the Courthouse Wash syncline along the eastern flank of the Moab salt-intruded anticline. The road is in the Kayenta Formation with the white, massive cliffs above in the Navajo Sandstone. Sharp, brown cliffs of the Wingate Sandstone are visible downstream. **Mile 44.2: Milepost 1.** The road crosses the Kayenta-Wingate contact at **Mile 44.3.** The low cliffs here are in the Wingate Sandstone, rising sharply to the Moab anticline. Across the river to the right at **Mile 44.6** is a beautifully exposed section of the entire Glen Canyon Group of Early Jurassic age. **Mile 45.0:** Wedding Bell Spring to the left at the base of the Navajo Sandstone provides excellent drinkable water.

Mile 45.2: Milepost 0: Intersection of U.S. 191 and Utah 128. Go left to Moab; go right to Crescent Junction (Roadlog 3).

5. Arches National Park

The road into Arches National Park first crosses the Moab fault and then climbs up the steeply dipping top of the Navajo Sandstone, heading east out of the north end of the Moab salt-intruded anticline. Ahead lies the Courthouse Wash syncline, formed by salt withdrawal at depth to provide thickened salt for the adjacent Moab anticline on the west and the Salt Valley salt-intruded anticline on the east. The road then climbs the west flank of the Salt Valley anticline and follows along the top of the Navajo Sandstone, providing beautiful views of exposures of the Entrada Sandstone on the left. Crossing the collapsed crest of the Salt Valley anticline, the road reaches the highly fractured east limb of Salt Valley, where most of the park's natural arches are located.

Mile 0.0: U.S. 191 at the turnoff to Arches National Park. The park entrance pay station is at **Mile 0.1.** The Visitor Center is to the right. The road is along the Moab fault. To the right is the heavily fractured Navajo Sandstone, which has collapsed into the northern nose of the Moab salt-intruded anticline, toward the Moab fault. To the left, or west, is the up-thrown block of the Moab fault, where rocks of the latest Pennsylvanian Elephant Canyon Formation are exposed at road level and reddish brown, coarse-grained sandstone beds of the Permian Cutler Formation, here only about 400 feet thick, are in the low cliffs above. Overlying the Cutler sandstone beds are exposures of the Moenkopi brown mudstone beds, the varicolored mudstone of the Chinle Formation beneath brown cliffs of the Wingate Sandstone, capped by thin remnants of the Kayenta Formation at the skyline. Immediately ahead are exposures of the Entrada Sandstone draping down to the left, or west, into the fault.

In Arches National Park the Entrada Sandstone is subdivided into the lower crinkled red beds of the Dewey Bridge Member (elsewhere called the Carmel Formation), the middle cliffs of the Slickrock Member, and the upper cliff of the Moab Tongue of the Entrada Sandstone. To simplify geologic mapping in this area, USGS geologists changed the status and name of the Carmel Formation to the Dewey Bridge Member of the Entrada.

Mile 1.0: The road winds its way up the folded and collapsed top of the light-colored windblown deposits of the Navajo Sandstone. Here is a good view to the southeast of the full length of the Moab Valley salt-intruded anticline, with the La Sal Mountains in the distance. A turnout at **Mile 1.3** overlooks the Moab fault. The highway below leads from Crescent Junction to the north and Moab to the south. The railroad visible above the highway is a spur from Potash, down the Colorado River below Moab, to Crescent Junction, used to transport the potash salt extracted from the Pennsylvanian Paradox Formation.

The road at **Mile 1.7** is at the top of the Navajo Sandstone, exposed to the right; the three members of the Entrada Sandstone form the cliffs to the left. The rocks here dip eastward away from the Moab salt anticline toward the Courthouse Wash syncline, a salt-withdrawal syncline. At **Mile 1.9** there is a good view of the La Sal Mountains to the south.

A turnout and a trailhead for Park Avenue are to the left at **Mile 2.3.** The road is in the brown crinkly beds of the lower Dewey Bridge Member with pink, massive cliffs above of the Slickrock Member and the white upper cliff of the Moab Tongue of the Entrada Sandstone. The parking turnout for

Figure 24. Balanced Rock in Arches National Park. The upper, more massive balanced bed is the Slickrock Member, and the lower crinkly beds are in the Dewey Bridge Member of the Entrada Sandstone, Middle Jurassic in age.

Courthouse Towers is at **Mile 3.7.** The road descends into the Courthouse Wash syncline, formed as salt was being withdrawn to flow into the Moab salt anticline. **Mile 4.6**: The bridge over Courthouse Wash is in the axis of the Courthouse Wash syncline. The contact of the Navajo Sandstone and Dewey Bridge–Carmel is in roadcuts to the right and left at **Mile 5.3.**

The road is at the top of the Navajo Sandstone at **Mile 5.6**, and crinkled beds of the Dewey Bridge Member are obvious to the left. Crinkling of the bedding in the Dewey Bridge here, and the Carmel elsewhere, is due to slumping prior to lithification (making the sediment into rock) of the red beds during deposition of the Slickrock Member of the Entrada Sandstone. In this location it may be the result of salt flowage from the Courthouse Wash syncline into the Moab salt anticline to the west and the Salt Valley salt-intruded structure to the east. It has been suggested that the crinkling resulted from a meteorite impact at Upheaval Dome in northern Canyonlands National Park, but the origin of Upheaval Dome is still being debated. The proposed meteorite impact is believed to have occurred, in various estimates, anytime between Middle Jurassic and "a few million years ago." Anyway, crinkling in the Carmel Formation, however formed, is almost ubiquitous across the Colorado Plateau.

Mile 6.1: Turnout for viewing the Petrified Dunes. The term *petrified* is obviously used incorrectly, but the viewpoint is located to emphasize the cross bedding in the Navajo Sandstone, ancient *cemented* sand dunes. There

Figure 25. Double Arch in the Windows Section of Arches National Park. The arches are formed in the Slickrock Member, above crinkly beds of the Dewey Bridge Member in the Entrada Sandstone. Arches usually form where springs emerge at the Slickrock–Dewey Bridge contact, causing spalling of the overlying massive sandstone.

is obvious crinkling in the Dewey Bridge Member to the left at **Mile 7.0. Mile 7.4**: Turnout for viewing interesting erosional features on the left. A view of the Windows Section is to the right at **Mile 8.5**, where a few natural arches may be seen. **Mile 8.7**: Elevation 5,000 feet.

A turnout for Balanced Rock is on the right at **Mile 9.0.** A rounded erosional remnant of the Slickrock Member is poised above a pedestal of the Dewey Bridge Member of the Entrada Sandstone, rising above the top of the Navajo Sandstone (Figure 24). **Mile 9.3**: Turnoff to the Windows Section; turn right. After the turn there is a good view of the south end of the Salt Valley salt-intruded anticline to the left. The road is near the top of the Navajo Sandstone.

The Garden of Eden viewpoint is to the left at **Mile 10.4.** There are beautiful exposures to the left at **Mile 10.7** of large-scale cross bedding in the Navajo Sandstone. The back side of Double Arch is high on the left at **Mile 10.9.** Note the deep alcoves (incipient arches) in the Dewey Bridge Member on the left at **Mile 11.1.** The parking area for the upper Windows Section is on the right at **Mile 11.3.** The parking area for Double Arch is at **Mile 11.9** (Figure 25). Return to the main road. The intersection with the main road is at **Mile 14.2**: the Visitor Center is to the left, and Devil's Garden is to the right. Turn right.

Mile 14.4: Good view to the left (northwest) into the Salt Valley salt-intruded anticline. The road descends into the core of the collapsed salt structure. The road to the right at **Mile 15.2** goes to Panorama Point. There is a good view to the left at **Mile 15.5** of the Fins on the eastern flank of Salt Valley, where most of the natural arches are located. Below and to the left severe collapse of the salt structure is evident, with exposures of Entrada, Morrison, Dakota, and Mancos Formations all dropped down into the crest of the structure. A view into the south end of Salt Valley, also called Cache Valley, is at **Mile 15.9.** The Cache Valley overlook is on the right at **Mile 16.4.**

The road to the right at **Mile 16.6** goes to Delicate Arch. Turn right. At **Mile 16.8** the road winds its way through huge collapsed blocks of dark gray Mancos Shale. Collapsed blocks of Entrada Sandstone are obvious on the left at **Mile 17.5.** A hogback of Dakota Sandstone is on the left at **Mile 17.7.** What a mess! **Mile 17.8**: Trailhead for Delicate Arch. A hogback of Dakota Sandstone and Burro Canyon Formation overlying the Morrison Formation is at **Mile 17.9** on the left. At **Mile 18.1** a hogback of Dakota Sandstone and a collapsed block of Mancos Shale are on the right. The Dakota hogback is on the left at **Mile 18.6.** The rocks are dipping steeply into the valley. The parking area for the Delicate Arch view is at **Mile 18.8.** A short walk permits a distant view up and to the left of Delicate Arch. **Mile 19.0**: Exit the parking area and return to the main road.

Mile 21.3: Intersection with the main road; turn right to Devil's Garden. The road leads up the eastern flank of the Salt Valley structure through collapsed blocks of everything up to the Mancos Shale, dipping steeply at all angles and all directions as they settle into the salt structure.

Collapsed blocks of Entrada Sandstone are in the roadcuts at **Mile 23.5.** The Salt Valley overlook is on the right at **Mile 23.6.**

The Fiery Furnace overlook road is to the right at **Mile 23.8.** At **Mile 24.0** there is a good view ahead of the Fins, remnant walls of sandstone between enlarged vertical fractures in the Entrada Sandstone along the eastern flank of the Salt Valley salt anticline. The Fins are well developed to the right and ahead at **Mile 24.7.** The road here is in the Dewey Bridge Member of the Entrada Sandstone. Trailhead parking on the right is for Sand Dune Arch. **Mile 26.2**: Skyline Arch is ahead; trailhead parking is at **Mile 26.4.**

Entering Devil's Garden and parking area at **Mile 27.0.** The road to the Devil's Garden campground is on the right (make reservations well ahead). **Mile 27.4**: Trailhead parking for the numerous arches that are scattered along the Fins for miles. **Mile 27.8**: Leaving the Devil's Garden parking loop. Return to the park entrance.

6. Utah 313 from U.S. 191 to Dead Horse Point State Park and the Island in the Sky District of Canyonlands National Park

Utah 313 first crosses the Moab fault and climbs up through the stratigraphic section from the Cutler Formation (Permian) through canyons carved into rocks of Triassic and Jurassic age to the top of the Big Flat dome. There a road to the left, in the Kayenta Formation, goes to Dead Horse Point State Park. Dead Horse Point provides spectacular vistas of the natural chaos of northern Canyonlands country from a breathtaking viewpoint above cliffs of the massive Wingate Sandstone across terrains of exposed rocks of Permian age. The laccolithic ranges of the La Sal and Abajo Mountains dominate the eastern and southern skylines. Heading west from the intersection, the road across the plateau on the Kayenta Formation leads to the Island in the Sky District of Canyonlands National Park, where the Grand View and Green River overlooks offer magnificent views of northern and western Canyonlands country. The excursion is climaxed by an examination of Upheaval Dome: is it a spectacular example of a meteorite impact structure or a pinched-off salt dome? You decide.

Mile 0.0: Intersection of U.S. 191 and Utah 313. Head west on 313.

The road crosses the Potash railroad spur and the approximate location of the Moab fault at **Mile 0.2**. The section ahead consists of Wingate Sandstone cliffs above slopes of the Chinle and Moenkopi Formations. Note the bulldozer scars made for uranium exploration in the 1950s at the level of the Moss Back Member of the Chinle. **Mile 1.5: Milepost 21.** A campground is on the left, and the Wingate-Chinle Formation contact is low on the right. The road is in the Wingate Sandstone (Lower Jurassic) at **Mile 2.2**. The contact between the Kayenta and Wingate Formations is to the right and left at **Mile 3.1** in the Kings Bottom syncline. The road winds its way up through the ledgy cliffs of the Kayenta Formation. Well-developed cross bedding in roadcuts in the Kayenta were formed by stream deposition; the deeply down-curved bedding represents stream channels that have been filled with ancient channel sand deposits. Such medium-scale cut-and-fill cross bedding is typical of the formation wherever it is exposed.

Mile 4.5: Scenic turnout. Across the canyon are the two monoliths named the Monitor and the Merrimac, after the iron-clad gunboats in the Civil War, capped by cliffs of the Moab Tongue and Slickrock Members of the Entrada Sandstone above a dark brown rubbly break of the Carmel Formation (Dewey Bridge Member) and the lower white cliffs of the Navajo Sandstone (all Jurassic in age). The turnout is at the Navajo-Kayenta contact. The rounded white cliffs above are in the Navajo Sandstone. Beautiful desert sand-dune cross bedding in the Navajo Sandstone is exposed on the right at **Mile 44.7**.

Mile 5.7: Milepost 17: The road for several miles is in the lower Navajo

Sandstone with the view to the north (right) of the Entrada Sandstone cliffs and a thin red slope of the Wanakah Formation, capped by the Morrison Formation (Late Jurassic in age). Take the turnout to the right at **Mile 6.4** to view the Entrada cliffs. The light-colored cliffs at the top are the Moab Tongue, which makes a sharp contact with the lower, reddish cliffs of the Slickrock Member of the Entrada. The dark brown notch-forming unit at the base of the cliffs is the Carmel Formation (Dewey Bridge Member of the Entrada Sandstone in Arches National Park). **Mile 8.7: Milepost 14.**

The Navajo-Kayenta contact is exposed on the left at **Mile 10.4.** The road is climbing up-dip from the Kings Bottom syncline toward the Big Flat structure. **Mile 10.7: Milepost 12.**

Mile 11.7: Milepost 11: Oil tanks and well sites are visible on the left. A good view of the La Sal Mountains is on the left at **Mile 12.0.** The unimproved road on the right at **Mile 12.5** leads 15 miles to Mineral Bottom and a boat launching ramp on the Green River, about 50 miles upstream from the confluence of the Colorado and the Green. The road is now on the Kayenta Formation. The road on the left at **Mile 13.4** goes to a trailhead to Gemini Bridges. **Mile 13.9: Milepost 9.**

Mile 14.9: Utah 313 turns left to Dead Horse Point State Park. After the turn the broad grassy area to the right is Big Flat, a broad structural dome and the site of the Pure Oil Company discovery of the 1950s that produced oil from the Mississippian Leadville (Redwall) Limestone. The field was soon abandoned because of mechanical problems in the Paradox salt section. **Mile 15.9: Milepost 7.**

The dirt road straight ahead at **Mile 16.6** goes to the Long Canyon oilfield, which produces from fractured shale (Cane Creek zone) in the Pennsylvanian Paradox Formation. **Mile 16.9: Milepost 6.** The road is in the Kayenta Formation across the Big Flat area. The Abajo (Blue) Mountains are visible in the far distance in a half-mile with the La Sal Mountains to the left (east).

Mile 19.2: Entering Dead Horse Point State Park; 3 miles to Dead Horse Point. **Mile 19.9: Milepost 3.** Entry pay station to Dead Horse Point and Visitor Center at **Mile 21.5.** The road to the campground is on the right at **Mile 21.7.** The turnout for the Dead Horse Point overlook is on the right at **Mile 22.5.** A stick fence across a narrows replicates the historical(?) gate used by cowboys to capture wild horses; the weaker ones were left behind, trapped to die of thirst. According to modern historians, such events never occurred—but they still make a good story. A picnic area is to the right at **Mile 22.7.**

Mile 22.9: Milepost 0: Turnaround parking area for the Dead Horse Point overlook. From the covered overlook the Abajo Mountains are ahead

Figure 26. The view from Dead Horse Point is straight down to the Colorado River as it passes through Shafer Dome, capped by the Elephant Canyon Formation. The strata exposed in the canyon wall are upward from the Cutler Group to the Navajo Sandstone on the horizon.

to the south, and the gently curved skyline to the right is the crest of the Monument Upwarp. In the middle distance is the Needles District of Canyonlands National Park, and the Colorado River and Shafer Dome are immediately below the overlook, which is atop the sheer Wingate cliffs, capped by the lower Kayenta Formation (Figure 26). To the east (left) are the La Sal Mountains above the western flank of the Moab salt-intruded anticline.

A short walk to the left of the covered overlook offers a view of the bright blue evaporation ponds of the Potash mining operation. River water is pumped some 2,000 feet down into the salt beds of the Paradox Formation, allowed to dissolve the sodium and potash salts, and then pumped up to the evaporative ponds to re-precipitate the salts. The blue color of the ponds is due to chemicals added to the brine to increase the rate of evaporation. The salts are harvested with bulldozers and trucked to the large green buildings near the river for separation. Potash is used to make fertilizer and explosives.

A short stroll to the right of the covered structure provides a view of the stratigraphic section of the Elephant Canyon Formation, the Cutler Formation, and the White Rim Sandstone of Permian age in the lower cliffs. Above the obvious White Rim are the Moenkopi, Chinle, Wingate, Kayenta, and Navajo Formations. A careful look reveals parts of Shafer Trail, which winds its way up through the entire section to the Island in the Sky District of Canyonlands National Park.

Spider Rock, located at the junction of Monument Canyon and Canyon de Chelly, is a tower of DeChelly Sandstone of Permian age 825 feet thick. Lower Permian red beds form the low slopes beneath the towers. The white rocks capping Spider Rock are, in Navajo mythology, the bones of children who disobeyed their elders and were carried there by Spider Woman.

The White House Ruin Trail descends the southern cliffs in Canyon de Chelly National Monument. The red, cross-bedded cliffs are in the DeChelly Sandstone, deposited in a vast wind-blown desert setting in Permian time.

Pedestal pillars, capped by erosional remnants of the White Rim Sandstone, guard Monument Basin in the Island in the Sky District of Canyonlands National Park. Softer red beds of the underlying Cutler Group erode more readily along fractures, leaving isolated caps of the more resistant sandstone. The Abajo Mountains form the distant skyline.

Exposures of Navajo Sandstone (Lower Jurassic) near the east entrance to Zion National Park in southern Utah. Cross bedding in the cliffs indicates that the deposits are of a windblown origin.

Magnificent colors and cliffs typify the scenery along Shafer Trail in the Island in the Sky District of Canyonlands National Park. The prominent white cliffs that extend across Shafer Canyon in mid-photo are in the White Rim Sandstone of Middle Permian age, above reddish brown rocks of the softer Cutler Group. Slopes above the White Rim are in the Triassic Moenkopi and Chinle Formations, capped by massive cliffs of the Lower Jurassic Wingate Sandstone with its omnipresent thin cap of the Kayenta Formation. Note the Jeep rounding a bend on Shafer Trail for scale.

The Totem Pole and Yeibichi (*ye'ii bicheii*, or "masked dancers") in southern Monument Valley Navajo Tribal Park in northern Arizona. Pillars and massive cliffs are the DeChelly Sandstone overlying slopes of the Organ Rock Shale, both of Middle Permian age.

Woodenshoe Arch forms the middle skyline butte in the Needles District of Canyonlands National Park. White sandstone beds of the Cedar Mesa Sandstone are here interspersed with red sandstone beds of the Cutler Group, Early Permian age.

Kachina Bridge in Natural Bridges National Monument, eroded from the Permian Cedar Mesa Sandstone. Natural bridges, unlike arches, are formed where meandering stream channels undermine the narrow goosenecks to shorten the course of the stream.

The Narrows along the North Fork of the Virgin River in Zion National Park. The high, white, distant cliffs are in the Navajo Sandstone, lying above brown cliffs of the Kayenta Formation, both of Early Jurassic age.

Colorful cliffs of rocks of Triassic and Early Jurassic age mark the upper end of Long Canyon along Burr Trail in the Grand Staircase–Escalante National Monument in southern Utah. The lower slopes were eroded from the Moenkopi and Chinle Formations beneath upper cliffs and crags of the Wingate Sandstone.

Massive, colorful sandstone cliffs form the southern boundary of the Colorado Plateau Province near Sedona, Arizona. The upper light-colored cliffs are in the Coconino Sandstone, and the lower brown cliffs are in the Schnebly Hill Formation, both of Middle Permian age.

From Harpers Corner in Dinosaur National Monument there is a magnificent view of the Mitten Park fault where it crosses the Green River at the head of Whirlpool Canyon. Drag folds are in the Pennsylvanian Morgan Formation, with white upturned beds of the Weber Sandstone on the right.

The Palisade rises majestically above the town of Gateway, Colorado, marking the western boundary of the ancient ancestral Uncompahgre Uplift and the eastern limits of the Paradox basin. The lowermost rocks exposed here are conglomerate beds of the Permian Cutler Formation, overlain by shale slopes of the Triassic Moenkopi and Chinle Formations. The upper massive brown cliffs are in the Lower Jurassic Wingate Sandstone, overlain by ledgy slopes of the Kayenta Formation, capped by white beds of the Middle Jurassic Entrada Sandstone. The highest point on the ridge is in the Salt Wash Member of the Morrison Formation of Late Jurassic age. Rocks of the ancestral Uncompahgre Uplift form the dark skyline in the right distance.

The Comb Ridge monocline forms the boundary between the Blanding Basin to the east (right) and the Monument Upwarp, here seen to the left. The view is toward the north along the San Juan River from the Mule Ear diatreme. The red ridge in the lower right of the photograph is an exposure of the nearly vertical DeChelly Sandstone lying stratigraphically above a low swale to the left of dark brown Organ Rock Shale; pink hills in the lower left are in the Cedar Mesa Sandstone, all of Permian age. The dark brown beds dipping steeply toward the right off the Lime Ridge anticline are in the uppermost Pennsylvanian Halgaito Shale, with gray beds of the Pennsylvanian Honaker Trail Formation visible along the left skyline.

Independence Monument, consisting of the Lower Jurassic Wingate Sandstone above lower red slopes of the Triassic Chinle Formation, rises from Monument Canyon in Colorado National Monument near Grand Junction, Colorado. Colorado National Monument lies on a segment of the ancestral Uncompahgre Uplift that was a prominent mountain range in Pennsylvanian and Permian time. Grand Junction, situated in the Grand Valley, is in the distance beneath cliffs of Cretaceous-age strata in the Book Cliffs.

Aerial view to the northwest of Upheaval Dome, the most controversial structure in Canyonlands National Park. The outer white ridge that nearly surrounds the dome is the Navajo Sandstone, the darker-colored "racetrack valley" is in the Kayenta Formation, and the inner massive red cliffs are in the Wingate Sandstone, all of Early Jurassic age. The lower red and brown slopes within the "crater" are in the Triassic Chinle and Moenkopi Formations, pierced by pinnacles of sandstone dikes that have squeezed upward from the underlying White Rim Sandstone of Permian age. Is this the throat of a pinched-off salt dome or the deep-seated rebound structure of a meteorite impact?

Retrace the route back to the Canyonlands road intersection at **Mile 30.0.** Turn left to Canyonlands National Park, 4 miles. After the turn the road crosses the northern Big Flat structure on about the top of the Kayenta Formation. **Mile 34.7**: Entering San Juan County from Grand County.

Mile 35.5: Entering Canyonlands National Park, marked by another symbolic stick fence.

Mile 36.6: Fee station for Canyonlands National Park, Island in the Sky District. The road to the left at **Mile 36.8** is Shafer Trail, not recommended for the faint of heart or those without high-clearance and/or four-wheel-drive vehicles. **Mile 37.8**: The Visitor Center is on the right.

Leaving the Visitor Center at **Mile 38.0**, the road wanders around through exposures of the Navajo Sandstone above the ledgy Kayenta Formation, seen to the left. The turnout on the left at **Mile 38.3** gives a breath-taking view down the myriad switchbacks of Shafer Trail as it winds down the cliffs to the White Rim bench far below. The road ahead now is in the Navajo Sandstone. The Lathrop Canyon trailhead is on the left at **Mile 39.9.** The Mesa Arch trailhead is on the left at **Mile 43.9.**

Mile 44.2: Intersection: the road to the right goes to Upheaval Dome, and the road to the left goes to Grand View Point. Turn left. The road is in the lower Navajo Sandstone. An overlook toward the canyon of the Green River is to the right at **Mile 45.1.** The broad white bench below is the White Rim Sandstone, overlain by slopes of the Moenkopi and Chinle Formations, capped by cliffs of the Wingate and Kayenta Sandstones. The butte to the left across the Green River is Elaterite Butte, towering above Elaterite Basin. The road to the right at **Mile 46.6** goes to the Murphy trailhead. The Buck Canyon overlook is on the left at **Mile 47.3.** The road left at **Mile 48.2** goes to a picnic ground. Another scenic overlook is on the right at **Mile 49.7.**

Mile 50.1: Parking loop at Grand View overlook, elevation 6,080 feet. Junction Butte is the prominent butte to the right, and Cataract Canyon is just to the left in the distance. The Needles District is visible to the south, below the gently curved skyline of the Monument Upwarp, and the Abajo Mountains are to the left, or southeast. Directly below the viewpoint is Monument Basin, where erosion has penetrated beneath the highly frac-tured White Rim Sandstone into the Cutler Formation (Organ Rock Shale) below, skirted by the White Rim Trail road. The La Sal Mountains are visi-ble to the east. The viewpoint is in the Kayenta Formation above massive cliffs of the Wingate Sandstone.

Retrace the road back to the intersection with the Upheaval Dome road.

Mile 56.0: Intersection with the Upheaval Dome road. Turn left. The road to the left at **Mile 56.3** goes to the Green River overlook and camp-ground, 1.2 miles. The road ahead goes to Upheaval Dome, 5 miles; proceed

Figure 27. Aerial view of Upheaval Dome, looking toward the northwest. Light-colored rocks exposed in the deep center of the structure are in the Moenkopi Formation, with spines of sandstone dikes penetrating upward from the White Rim Sandstone. The lower massive cliffs are in the Wingate Sandstone, overlain and encircled by the Kayenta and Navajo Formations. The Green River is visible in the upper center of the photograph.

ahead. An overlook and the Aztec Butte trailhead are to the right at **Mile 56.5.** The road here is in the Navajo Sandstone.

The road at **Mile 57.8** descends the dipslope of the Kayenta Formation into the rim syncline adjacent to Upheaval Dome; Whale Rock in the Navajo Sandstone is visible ahead. The Navajo is dipping strongly back eastward away from the core of Upheaval Dome.

Mile 61.0: Parking area, picnic ground, and trailhead for Upheaval Dome in the Kayenta Formation. The hike is a 20-minute climb up 200 vertical feet of rocky trail to the rim of Upheaval Dome in the Kayenta Formation. The view down into the crater-like structure is of upturned, jagged spires of sandstone dikes, squeezed upward from the White Rim Sandstone at shallow depth. The surrounding gray and brown slopes are in the Moenkopi and Chinle Formations below massive cliffs of Wingate and Kayenta Sandstones. An outer rim of Navajo Sandstone surrounds the upturned core of the dome (Figure 27).

The origin of this spectacular structure has been debated for decades. Suggested theories include a cryptovolcanic origin (a volcano that tried to pierce the rocks and didn't make it), a salt-piercement dome, and a meteorite impact crater. The salt dome theory prevailed for years because there is known salt at depth in the vicinity, and obvious salt-piercement structures, such as Moab Valley, lie a short distance to the east. The well-developed rim syncline, normally formed by salt withdrawal into the dome, encircles the uplift. Then, in the 1990s, the late Gene Shoemaker, a brilliant research geologist who formed the Lunar Geology Laboratory for the U.S. Geological Survey in Flagstaff, Arizona, was looking for large meteorite impact structures on the Colorado Plateau and pronounced that Upheaval Dome was formed by meteorite impact some 65 million years ago. The age of the proposed impact is now estimated to be anytime between Middle Jurassic and "a few million years ago," according to various authors. Petroleum geologists familiar with Gulf Coast and European salt domes could not believe the interpretation. A technical article by M. P. A. Jackson and others which appeared in the *Bulletin of the Geological Society of America* in late 1998 provided ample measurements and details to make a very convincing story that Upheaval Dome is a pinched-off salt dome. Another technical publication by Shoemaker and others, published in late 1999 in the *Journal of Geophysical Research*, provided convincing evidence that a meteorite impact formed these deep-seated impact features. So which story do you prefer?

Return to the intersection with U.S. 191 at **Mile 93.7**.

7. Utah 279 to Potash and Beyond to Shafer Trail in Canyonlands National Park: A Scenic Byway

This is the most scenic and geologically rewarding drive in northern Canyonlands country. Utah 279 exits the Moab Valley salt-intruded anticline through the Portal and then crosses the Kings Bottom syncline through spectacular exposures of the Glen Canyon Group (Early Jurassic in age). Potash, site of a solution mine for potash salts, lies on the eastern flank of the salt-thickened Cane Creek anticline. The pavement ends just beyond at the boat-launching ramp for river trips through Cataract Canyon. From here the primitive road crosses the Cane Creek anticline on the top limestone of the Elephant Canyon Formation (Late Pennsylvanian) and proceeds to cross another salt-thickened anticline at Shafer Dome. The route then follows Shafer Wash up through the stratigraphic section from the Lower Cutler arkosic sandstone to the top of the White Rim Sandstone (Permian) at the foot of Shafer Trail. Shafer Trail ascends the mighty cliffs surrounding Canyonlands National Park up-section from the top of the White Rim Sandstone through red beds of Triassic age, the Moenkopi and Chinle Formations, and inches upward across cliffs of the Wingate Sandstone, capped here by the Kayenta Formation. The top of Shafer Trail is near the entrance to the Island in the Sky District of Canyonlands National Park.

Mile 0.0: Intersection of U.S. 191 and Utah 279, north of the bridge across the Colorado River. After turning west onto 279, note the railroad spur to Potash entering a tunnel high on the right. The road is on the toe of talus of the Moenkopi, Chinle, and Wingate Formations, slumped from the cliffs above to the right. To the left is the Atlas Minerals uranium tailings pile, controversial in 2001 as to whether it should be moved or buried. To the left is an excellent view of the Colorado River floodplain and Moab Valley with the La Sal Mountains in the distance.

Exposures on the right at **Mile 1.2** are of the Moenkopi Formation (Triassic), resting with angular unconformity on the Honaker Trail Formation (Pennsylvanian), with no intervening Cutler Formation (Permian).

Mile 2.3: An angular unconformity within the Chinle Formation is visible ahead across the river, indicating salt movement into the salt structure during Chinle time (Late Triassic), overlain by the Wingate, Kayenta, and Navajo Formations. At **Mile 2.5** the road enters the Portal, the canyon through which the Colorado River exits Moab Valley toward the southwest. Exposures of the Chinle Formation are on the right at **Mile 2.7**, capped by the sheer cliffs of the Wingate Sandstone. **Mile 3.0**: The road crosses the contact between the varicolored shale slopes of the Chinle Formation and the cliff-forming Wingate Sandstone above, both dipping strongly toward the west away from the Moab salt-intruded anticline. At **Mile 3.3** the road climbs up-section into the Kayenta Formation. The Kayenta stream deposits are quite thick here in the Kings Bottom syncline.

The road crosses up-section through the ledgy beds of the Kayenta Formation into the overlying cliff-forming Navajo Sandstone at **Mile 4.2.** The massive cliffs ahead are in the Navajo. Well-preserved Indian petroglyphs are scattered along the desert varnish–coated Navajo cliffs for a couple of miles on the right. Some of the best petroglyphs are on the right at **Mile 5.0.** The road is back down-section into the Kayenta Formation at **Mile 5.5**, as the strata are rising toward the west onto the flank of the Cane Creek anticline. Dinosaur tracks are visible in the Kayenta Formation, high and to the right at **Mile 5.9.**

Mile 6.3: The road level is back down to the top of the Wingate Sandstone. The apparent reversal of dips in the canyon is due to the meandering of the canyon and road in the Kings Bottom syncline. The low cliffs now are in the Wingate Sandstone. The Colorado River bottom to the left is heavily overgrown with tamarisk bushes, making it difficult to see the river.

Mile 7.4: Milepost 8: The road is back up into the very thick Kayenta Formation.

The contact between the Kayenta and Navajo Formations is at road level again at **Mile 8.8.** The cliffs along the canyon walls are in the light-colored Navajo Sandstone. At **Mile 9.5** the road is back down into the stream-deposited Kayenta Formation. The railroad exits its tunnel high on the right at **Mile 9.7.** The main exit on the left to the Gold Bar Campground is at **Mile 10.0.** The road here is in the Kayenta Formation, and the high cliffs are in the Navajo Sandstone. The railroad spur is obvious at the base of the cliffs to the right.

The cliffs on the right at **Mile 10.4** are the Wingate Sandstone, capped by the Kayenta Formation. The Wingate Sandstone cliffs are heavily stained with desert varnish throughout the canyon. The railroad tracks are nearly down to road level at **Mile 10.8**. At **Mile 11.7** the Wingate Sandstone–Chinle Formation contact is on the right just above the railroad tracks.

Jug Handle Arch is on the right at **Mile 13.5**, with a parking area just beyond. The rocks are now dipping strongly toward the east, away from the Cane Creek anticline. The road is well down into the Chinle Formation, with the Black Ledge Member prominently exposed above the road. The prominent green structure ahead at **Mile 14.1** is the mill for the potash mine. The top of the Cutler Formation (Permian) is exposed to the right and left at **Mile 14.5**. It forms the low, dark brown, rough-weathering sandstone at the base of the section.

Mile 15.4: Milepost 0: Potash: Entrance to the PCS potash operation, with its carefully camouflaged bright green buildings, product warehouses for the potash mine. Underground mining operations began in 1964, when a shaft was dug 3,200 feet down into the flank of the Cane Creek salt-thickened anticline, and a 2,500-foot-long tunnel was extended under the crest of the structure. The potash-rich salt body in the Paradox Formation was originally mined with heavy equipment, but high temperatures, gas emissions, and contorted bedding in the salt made mining operations difficult and hazardous. An underground explosion in the mine killed 18 workers.

The operation was converted to solution mining in 1970, when river water was pumped into the mine, allowed to become saturated with salt for 300 days, and then pumped through pipelines to the 23 evaporation ponds nearby. The re-precipitated potash and halite salts are harvested, separated in a flotation process, and shipped from the mine by rail. Potash is mainly used for fertilizer and explosives, and halite is used for highway de-icing.

The rocks at **Mile 15.6** dip sharply to the east from the Cane Creek anticline, most notably to the left and ahead. The road enters a tamarisk thicket. An abandoned, plugged well site is on the left at **Mile 16.1**. Rocks of the upper Cutler Formation are beautifully exposed at **Mile 16.5**.

The boat ramp for launching Cataract Canyon trips is on the left at **Mile 16.8**. The topmost limestone bed of the Elephant Canyon Formation is ahead on the right. The unimproved road ahead, which leads to the foot of Shafer Trail and the White Rim Trail, is for high-clearance or four-wheel-drive vehicles only.

Reset the trip odometer to **Mile 0.0:** Boat ramp on the left. The unimproved road ahead climbs the east flank of the Cane Creek anticline on the

Figure 28. View across the Colorado River of the crest of Shafer Dome near the River Road. The cliffs along the river are in the latest Pennsylvanian Elephant Canyon Formation, here mostly coarse-grained arkosic sandstone, but with several beds of fossiliferous limestone. The high cliffs on the left are in the Cutler Formation of Permian age.

topmost limestone bed of the Elephant Canyon Formation, the informally named "Shafer lime."

At **Mile 1.0** plugged wells of the MGM (Monty G. Mason) Cane Creek operation are on the left. According to Bob Norman, a Moab geologist, the wells were drilled on the crest of the Cane Creek anticline between 1953 and 1956 and encountered some oil at depths of about 7,500 feet in the Paradox Formation. The field never produced oil in commercial quantities. Oil and gas flows found during tests were filmed and used to promote the development of the field. Monty Mason was convicted of overselling stock and died in a California prison in 1965.

Mile 2.0: The top of the Elephant Canyon Formation, the highest gray, marine limestone bed, and base of the brown, coarse-grained cliffs of the Cutler Formation are on the right at road level. The road crosses the brine pipelines for the potash mine operation at **Mile 2.2.** The first of several evaporation ponds is on the left at **Mile 3.1.** The crest of the hill overlooking the many evaporation ponds is at **Mile 4.0.** The gate guarding the west entrance to the evaporation pond farm is at **Mile 5.0.**

The road is again at the top of the "Shafer lime" (top of the Elephant Canyon Formation) at **Mile 6.0**, with a view of Shafer Dome ahead and to the left. The road crosses a wash at **Mile 7.0** and begins the steep climb up the northeast flank of Shafer Dome on top of the "Shafer lime."

Mile 8.7: Viewpoint at the crest of Shafer dome, locally known as the Thelma and Louise overlook, named after the popular movie, is directly beneath Dead Horse Point, looking down into the canyon of the Colorado River. The "Shafer lime" caps the bench, with the Elephant Canyon Formation comprising the cliffs of the inner gorge (Figure 28).

Figure 29. The stratigraphic section as viewed from near the junction of the River Road and the Shafer Trail. The road in the lower right of the photo is in the Organ Rock Formation of the Cutler Group (Permian), and the prominent ledge above is in the White Rim Sandstone. The slopes include the Moenkopi and Chinle Formations, Triassic in age, and the massive cliffs of the Wingate Sandstone and Kayenta Formation, both of Early Jurassic age.

Shafer Dome is an odd anticline, as the axis curves from a northeast-trending structure to a northwesterly nose. Salt bulged the structure, much like at the Cane Creek anticline, but did not penetrate overlying sedimentary rocks. An old well drilled on the axis of the anticline at river level penetrated about 5,000 feet of salt section, somewhat more than the regional thickness. The road descends the northwest flank of Shafer Dome (Figure 29) from the viewpoint on top of the "Shafer lime," where fossil clamshells, along with brachiopods and a rare trilobite or two, may be found weathered out on and near the road.

Mile 11.3: Entering Canyonlands National Park. The road is in the Permian-age Lower Cutler Formation, following Shafer Wash. There is a

Figure 30. The stratigraphic section in the vicinity of the Green and Colorado Rivers, Canyonlands National Park.

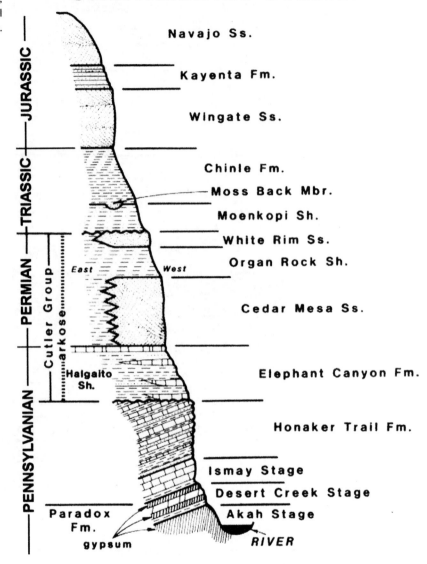

GREEN & COLORADO RIVERS STRATIGRAPHIC COLUMN

Navajo Ss.

Kayenta Fm.

Wingate Ss.

Chinle Fm.

Moss Back Mbr.

Moenkopi Sh.

White Rim Ss.

Organ Rock Sh.

Cedar Mesa Ss.

Elephant Canyon Fm.

Honaker Trail Fm.

Ismay Stage

Desert Creek Stage

Akah Stage

RIVER

JURASSIC

TRIASSIC

PERMIAN

PENNSYLVANIAN

Cutler Group
arkose

Halgaito Sh.

East   West

Paradox Fm.
gypsum

turnout and a restroom facility in Shafer Wash at **Mile 12.0.** The road crosses upward through the contact of the Lower Cutler and Organ Rock Shale at **Mile 13.6.** At **Mile 14.2** the road crosses upward through the White Rim Sandstone. Note the crinkly, horizontal bedding in the lower half of the formation (tidal flat deposition?) and the highly cross-bedded nature of the upper sandstone (windblown, or eolian).

Mile 15.3: Intersection with Shafer Trail. The road to the left is the White

Rim Trail; to the right are Shafer Trail and the Visitor Center in the Island in the Sky District of Canyonlands National Park. Turn right.

The intersection is in the lower Moenkopi Formation. The road crosses upward through the gentle slopes of the Moenkopi Formation to the prominent light-colored sandstone of the Moss Back Member of the Chinle Formation. The slope steepens through the Chinle Formation, with prominent dark brown beds of the "Black Ledge member" at about midsection. The road then tackles the massive cliffs of the Wingate Sandstone and ascends through most of the ledgy Kayenta Formation. At the top of the steep grade the road traverses a bench in the upper Kayenta Formation, and light-colored cliffs of the Navajo Sandstone are exposed above to the left (see Figure 30).

Mile 18.9: Intersection with the Canyonlands National Park paved road. The intersection is at about the top of the Navajo Sandstone. Turn left to the Visitor Center or right to Moab.

8. Moab to Monticello, Utah, via U.S. 191

From Moab to Monticello, U.S. 191 follows the full length of the Moab Valley salt-intruded anticline, where obvious collapse features can be seen along the valley walls. The salt structure lies along deep-seated basement faults that controlled the location and northwesterly orientation of salt flowage in Late Pennsylvanian through Jurassic time. The salt flowed into the structure until it was depleted, for deep drilling has revealed that no salt is present beneath the flanks of the structure while some 12,000 to 15,000 feet of salt is present beneath the valley floor. Groundwater leached the salt from the top of the flowed mass during the past few million years, leaving a leached cap of gypsum and black shale some 800 feet thick above the salt. The removal of near-surface salt left a void in the structure, causing collapse of the overlying formations that can still be seen at either end of the valley, along with collapsed valley walls (see Figure 11).

The highway climbs the southwest flank of Moab Valley about 8 miles south of Moab and then follows exposures of the Navajo-Carmel-Entrada Formations for about 30 miles. There the highway climbs back up through the Morrison Formation to the top of the Dakota Sandstone to the town of Monticello, Utah.

Mile 0.0: The southern city limits of Moab. Entering San Juan County, Utah, at **Mile 2.4**. On the left at **Mile 3.6** is the La Sal Mountains road, which provides access to the La Sal Mountains via a circular route, ending at La Sal Junction.

Collapsed blocks of Mancos Shale are exposed on the left at **Mile 6.6**. The highway begins its ascent from Moab Valley onto the southwest flank of the

Moab anticline at **Mile 7.8.** The highway on top is on the Morrison Formation (Late Jurassic). The clearing to the right at **Mile 9.0** is the buried Pacific Northwest gas pipeline. The highway descends into Kane Springs through the reddish-colored cliffs of the Entrada Sandstone (Jurassic). The obvious fault on the left at **Mile 9.7** drops the Salt Wash Member of the Morrison Formation into juxtaposition with the Entrada Sandstone. Kane Springs rest area is on the left at **Mile 10.2.** The cliffs surrounding the oasis are in the Entrada Sandstone.

The highway crosses the Entrada-Carmel contact at Mile **10.7**, and road-cuts at **Mile 10.9** are in brown mudstone and siltstone of the Carmel Formation (Middle Jurassic). At **Mile 11.3** the highway crosses down through the Carmel–Navajo Sandstone contact but is back up into the Carmel Formation at **Mile 11.9.** A Pacific Northwest gas pipeline pump station is on the right at **Mile 15.4.** The highway is still on the Carmel Formation bench.

Mile 17.3: La Sal Junction. The road to the left, Utah 46 (Roadlog 13), goes to the La Sal Mountain Loop road junction, Paradox Valley salt-intruded anticline, and on to Naturita, Colorado, on Colorado 90. Wilson Arch is on the left at **Mile 20.4** in the Entrada Sandstone. The Abajo Mountains, a laccolithic range, also known as the Blue Mountains, are visible ahead. At **Mile 23.7** the highway enters a series of washes down into the nearly white, massive beds of the Navajo Sandstone and again ascends into the Carmel Formation for the next 2 miles. The road to the left at **Mile 25.5** goes to Lisbon Valley Industrial Center. The Lisbon Oilfield, discovered in 1960 by Pure Oil Company, lies along the southwestern flank of the Lisbon Valley salt-bulged anticline. Production of high-gravity oil and natural gas is from the Mississippian Leadville Formation (Redwall Limestone of Grand Canyon usage) and McCracken Member of the Elbert Formation (Late Devonian) at depths of about 8,500 feet. Because gas pressure was sufficient to lift the oil to the surface, pumps were never needed. Oil recovery from the field is estimated at about 1.5 million barrels. The area was also the site of the Big Indian uranium district, one of the largest on the Colorado Plateau, where Charlie Steen discovered his legendary Mi Vida Mine in the Chinle Formation in the 1950s.

The highway crosses the Navajo-Carmel contact at **Mile 26.8.** The road on the right at **Mile 27.8,** Utah 133, goes to the Canyon Rims Overlooks (Roadlog 9). The nearest viewpoint looks down onto the Needles District of Canyonlands National Park, a spectacular vista. Another viewpoint, much farther north, overlooks the Cane Creek anticline west of Moab. The road left at **Mile 30.1** goes to Lisbon Valley Oilfield and uranium and copper mines. The southwestern limb of the Lisbon Valley salt structure is visible on the left, as are oil tanks and other facilities of the oilfield.

Mile 35.2: The road to the right, Utah 211, goes to Newspaper Rock State Park and the Needles District of Canyonlands National Park (see Roadlog 10). Church Rock, also called the Lampstand, eroded from the Entrada Sandstone, is on the left.

The highway ascends into the Entrada Sandstone at **Mile 36.3** and continues to climb through the Entrada, Wanakah, and Morrison Formations toward the Monticello plateau region. At **Mile 41.7** the highway crosses the Morrison–Dakota–Burro Canyon contact, with exposures of Dakota Sandstone, a cliff-forming light-colored sandstone of Cretaceous age, ahead to the right and left. The highway is on top of the Dakota Sandstone at **Mile 43.7**, where it caps the vast beanfield plateau once known as the Great Sage Plain. The Monticello Airport is on the right at **Mile 45.2**, and the Abajo (Blue) Mountains are beyond.

Mile 48.5: Entering Monticello, Utah. The Mormon Church called five families from Bluff to settle the village in 1887, and they chose to name it Monticello. **Mile 49.6**: Junction with U.S. 666, heading to the left to Cortez, Colorado; U.S. 191 goes straight ahead to Blanding and Bluff, Utah. (For U.S. 666, see Roadlog 18. For Blanding and Bluff, see Roadlog 12.)

9. Utah 133 to Canyon Rims Overlooks from U.S. 191

Utah 133 joins U.S. 191 27.8 miles south of Moab, Utah. Take Utah 133 to the Canyon Rims overlooks and campgrounds: Windwhistle Campground, 6 miles; Needles Overlook, 22 miles; Hatch Point Campground, 25 miles; and Anticline Overlook, 32 miles. The road is largely on grassy plains on the Carmel Formation with some scattered buttes and mesas of Entrada Sandstone. The overlooks are magnificent.

10. Utah 211 to the Needles District of Canyonlands National Park from U.S. 191

Mile 0.0: Intersection of Utah 211 and U.S. 191. Turn west on Utah 211. The road is on sagebrush-covered flats on the Carmel Formation with buttes and mesas of Entrada Sandstone in all directions. The high plateau to the left is capped by the Dakota Sandstone, Wanakah, and Morrison Formations above the Entrada cliffs. Occasional glimpses of the Abajo Mountains are in the left distance.

Mile 0.5: Milepost 19: Remnants of Marie Ogden's "Home of Truth" religious commune are scattered along the foothills to the left of the road for the next mile. Exposures of the light-colored cliffs of the Entrada Sandstone are to the right and left at **Mile 2.5**, capped by thin Wanakah Formation (Middle Jurassic) and the Tidwell Member of the Morrison Formation (Late Jurassic), a reddish brown mudstone and siltstone unit at the base of the Morrison Formation. At **Mile 3.2** the road crosses a divide in exposures of the Entrada Sandstone and then is back down onto brown mudstone and siltstone of the Carmel Formation. **Mile 3.4**: View of the Abajo Mountains

Figure 31. A small portion of the petroglyph panel at Newspaper Rock. The petroglyphs, some about nine hundred years old, are etched into desert varnish in an alcove in the base of the Wingate Sandstone, mostly by Anasazi, Ute, and Navajo Native Americans.

ahead. The Spanish explorers in the 1700s named these the Abajo ("lower") Mountains, in contrast with the higher La Sal Mountains to the north. **Mile 4.4: Milepost 15.**

The top of the light-colored, massive Navajo Sandstone is exposed to the right and left at **Mile 5.5.** The road leaves a valley in the Carmel Formation at **Mile 7.2** and climbs up the dipslope on top of the Navajo Sandstone. **Mile 9.4: Milepost 10.** A new road to Harts Draw and Monticello is to the left at **Mile 9.6** (see Roadlog 11). The road crosses the southern fault of the Shay graben, a northeast-trending down-faulted block, at **Mile 10.1;** the Navajo Sandstone is high on the left but low on the right. The Navajo Sandstone is high on the left and right at **Mile 10.6,** and exposures of the stream-deposited Kayenta Formation are ahead in the Shay graben. Exposures and roadcuts of Navajo Sandstone are on the right and the Kayenta Formation is on the left at **Mile 10.9.** The Navajo Sandstone is on the right and the Kayenta Formation is on the left at **Mile 11.4,** and then the Navajo-Kayenta contact is on the right. At **Mile 11.6** the road is down into the Kayenta Formation, with its typical cut-and-fill cross bedding. The road then goes down into the heavily cross-bedded Wingate Sandstone. The Wingate is here nearly white, rather than its usual brownish color, and is easily confused with the Navajo Sandstone. The peak to the left at **Mile 12.2** is composed of the Navajo-Kayenta-Wingate Formations, in descending order.

Mile 12.4: Newspaper Rock, now a BLM Recreation Site, is on the right. The petroglyphs are carved in heavy desert varnish in a spalled-off cove in the base of the Wingate Sandstone. They date from nearly the present back some nine hundred years (Figure 31).

The road follows a sharp, narrow canyon in the Wingate cliffs at **Mile 12.5**. At **Mile 14.5: Milepost 5** the road is at about the top of the Chinle Formation. Good exposures of the varicolored shale slopes of the Chinle Formation (Triassic) at **Mile 15.1** are below the Wingate cliffs ahead. There are more good exposures of the Chinle Formation at **Mile 15.2** to the right at road level. The valley widens at **Mile 16.0** with grassy river bottomland to the left. There are good exposures of the Chinle Formation below the Wingate cliffs. Beautifully developed desert varnish may be seen on the Wingate Sandstone cliffs ahead at **Mile 17.1**. The valley is now wide and grassy at **Mile 17.6: Milepost 2**, surrounded by spectacular cliffs of the Chinle through the Navajo Formations. The Sixshooter Buttes are visible ahead and slightly to the left at **Mile 18.8**.

Mile 19.1: Cattle guard and **Milepost 0**. "End of state maintenance." Beautiful grassy meadows and Dugout Ranch, an important cattle ranch for more than a century, are visible to the left. At **Mile 20.0** the low cliffs in the valley bottom are in the Shinarump Member of the Chinle Formation (Triassic), which floors the flat meadowlands upstream. The Shinarump is the basal member of the Chinle Formation and consists of stream-channel sandstone and conglomerate deposited on the Chinle-Moenkopi unconformity. It is typically resistant to erosion and forms low cliffs and benches where it is developed. The road to the left at **Mile 20.3** goes to Beef Basin, Elk Ridge, and eventually Natural Bridges National Monument.

At **Mile 20.6** the road is just above the top of the Shinarump Member of the Chinle Formation with exposure of the Shinarump on the left. **Mile 21.4**: Ahead and to the left there is a good view of Sixshooter Buttes, capped by the Wingate Sandstone, with exposures in the valley of the Shinarump Member and Moenkopi Formation (Early Triassic) (Figure 32).

The road descends down-section through the Shinarump cliffs into the top of the brown mudstone and siltstone of the Moenkopi Formation at **Mile 21.9**. The broad valley ahead at **Mile 22.5** is underlain by brownish red sandstone at the top of the Cutler Formation (Permian). There is a good view back to the left, up a tributary canyon, of Elk Ridge. The road to the left at **Mile 23.2** goes to Lavender Canyon via Davis Canyon.

Across the broad valley floor, against the low cliffs, was the proposed location of a high-level nuclear waste disposal site that will not be built. A turnout to the right at **Mile 23.7** provides a good view of the Sixshooter Buttes and the proposed nuclear waste site. The dirt road to the left at **Mile**

Figure 32. The Sixshooter Buttes are Wingate Sandstone pinnacles above slopes of the Chinle Formation. The nearly horizontal bench is the Shinarump Member of the Chinle Formation on the lower slopes of the Moenkopi Formation.

25.0 went to the site, which is now abandoned. Public outcry was against it, and the work of peer-review panels indicated that the site was not practical. High-level nuclear waste would have been buried at a depth of about 3,000 feet in salt of the Paradox Formation of Pennsylvanian age, but fractures provide groundwater communication with the surface in nearby Salt Creek. Close proximity to the entrance to Canyonlands National Park and the necessary road and railroad construction to the site provided other complications.

The road crosses the top of the Cutler Formation at **Mile 25.7** and descends into the formation. **Mile 26.5**: Road to Davis Canyon, 9 miles, and Lavender Canyon, 12 miles. The road and valley are in stream-deposited brown sandstone of the Cutler Formation. A thin bed of white Cedar Mesa Sandstone rims the valley to the right at **Mile 27.4**. This is a far eastern tongue of the nearly white, cross-bedded Cedar Mesa Sandstone (Permian). **Mile 28.9**: The road is here on top of a light-colored Cedar Mesa Sandstone tongue. The road to the right at **Mile 29.2**, County Road 122, goes to Lockhart Basin, 15 miles, and Hurrah Pass, 48 miles (four-wheel-drive vehicles only).

Mile 29.7: The intertonguing relationships of the light-colored Cedar Mesa Sandstone and the brown beds of the Cutler Formation are becoming more obvious. The light-colored sandstone beds were derived from the northwest and deposited in coastal environments; the brown arkosic beds were derived from the Uncompahgre Uplift to the northeast and deposited by streams. Several tongues of Cedar Mesa Sandstone are exposed in the valley walls at **Mile 30.9**.

Turnouts to the right and left at **Mile 31.7** provide a good view of the Needles District of Canyonlands National Park. The brown and white

Figure 33. Woodenshoe Arch, on the right-center of the skyline ridge, is in the Needles District of Canyonlands National Park. Alternating beds of light and dark sandstone resulted from the interfingering of light-colored tongues of Cedar Mesa Sandstone with dark-colored Cutler beds, all of Early Permian age.

ledges are of the interfingering Cedar Mesa and Cutler sandstone beds. The Needles are ahead at about 2:00.

Entering Canyonlands National Park at **Mile 32.1.** There is a good view of the Needles ahead. **Mile 32.7**: Canyonlands National Park entrance sign; good view of the banding in the Cedar Mesa Sandstone ahead in the Needles. At **Mile 33.2** the Needles are visible at about 10:00; at about 1:00 is the Sewing Machine Butte across the Colorado River; and at about 2:00 is Elaterite Butte in the vicinity of the Maze. The Needles Outpost road is to the right at **Mile 33.6,** leading to a small general store and the dirt runway of the Canyonlands Airport. The service road to the left with the closed gate goes up Salt Creek. Salt Creek Wash is at **Mile 34.1. Mile 34.5**: Entrance fee station. The road to the right just ahead goes to the Visitor Center.

Mile 0.0: Leave the Visitor Center and head through the park toward the Squaw Flats Campground. Numerous white tongues of the Cedar Mesa Sandstone are visible around the Visitor Center. Note that the cross bedding in the Cedar Mesa, believed to have a nearshore marine and windblown dune origin, is quite different from the cross bedding in the Navajo Sandstone.

A roadside ruin is on the left at **Mile 0.3**, and Sixshooter Butte is on the skyline. A road to Salt Creek is on the left at **Mile 0.6.** Squaw Flat Campground is straight ahead. The turnoff to the left at **Mile 2.0** goes to Wood-

Figure 34. Interbeds of light-colored Cedar Mesa Sandstone and dark-colored Cutler beds have eroded into myriad forms along the Chesler Park road in the Needles District of Canyonlands National Park.

enshoe Arch. There is obvious red and white banding visible in the Cedar Mesa ahead (Figure 33). The road left at **Mile 2.4** goes to Salt Creek and ranger residences. A good view of the Needles across Squaw Flat is at **Mile 2.6**.

The road to the left at **Mile 2.8** goes to the Squaw Flat Campground and Elephant Hill. The road straight ahead at **Mile 3.0** is to Campground A; turn right to Campground B. The road to the left at **Mile 3.3** goes to Campground B, and the gravel road to the right goes to Elephant Hill. Turn right.

The road wanders through little pedestal rocks, ridges, and strange landforms of Cedar Mesa and interbedded Cutler sandstone beds and descends into a steep little canyon at the head of Elephant Canyon at **Mile 4.5**. The caprock of the little canyon to the right at **Mile 4.7** is the topmost limestone bed of the Elephant Canyon Formation. The road crosses a wash at **Mile 4.8** on the topmost limestone bed of the Elephant Canyon Formation. Pedestal rocks of the white Cedar Mesa on the red Cutler sandstone are ahead on the left at **Mile 5.0** (Figure 34). The top of the Elephant Canyon Formation is on the right in the wash.

Mile 6.0: Picnic area and turnaround at the Elephant Hill trailhead. The trail, for four-wheel-drive vehicles and hikers only, goes to Chesler Park and the Grabens, where large fault blocks of Cedar Mesa Sandstone dominate the landscape. A graben is a down-faulted block of rock, and a horst is an up-faulted block. In actuality, the Grabens is a complex series of both horsts

Figure 35. The Grabens. Key: Ml = Mississippian Redwall Limestone, IPps = Pennsylvanian Paradox Formation salt (shaded), IPht = Honaker Trail Formation, Pec = Elephant Canyon Formation (latest Pennsylvanian to Permian), and Pcm = Cedar Mesa Sandstone.

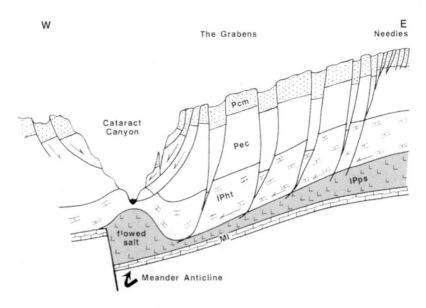

and grabens, caused by the slumping of surface strata by gravity down the western flank of the Monument Upwarp toward Cataract Canyon of the Colorado River, gliding on salt beds of the Paradox Formation found at shallow depth (Figure 35). Retrace the road back to Squaw Flat. We are back to the pavement at Campground B at **Mile 8.8.** Enter the main loop road out of the campgrounds at **Mile 9.0** and head toward the Visitor Center. Intersection in Squaw Flat at **Mile 9.3.** The road to the left goes to the Confluence Overlook trailhead; the road to the right goes to the Visitor Center. Turn right to the Visitor Center at **Mile 12.1.**

Mile 14.7: Leaving Canyonlands National Park, heading back toward Newspaper Rock, BLM Recreation Area.

Mile 37.1: Intersection. County Road 136, on the right, goes to Monticello, Utah (18 miles) (see Roadlog 11).

11. Needles District to Monticello, Utah, via County Road 136

Mile 0.0: Intersection at County Road 136, 9.6 miles from U.S. 191. The road climbs to the south toward the northeast side of the Abajo Mountains, which are visible at about 1:00. At the top of the first climb the road is on the Entrada Sandstone with a beautiful view of the Abajos ahead. The valley wall to the east (left) at **Mile 1.7** is made up of the Entrada, Wanakah, and Morrison Formations, capped by the Dakota Sandstone. The Entrada Sandstone is exposed along the road as well.

Entering the Manti–La Sal National Forest at **Mile 2.7.** The mountain to the right in the Abajos at **Mile 3.6** is an intrusive igneous stock that pierced upward through the Entrada Sandstone and caused the obvious steep dips. Nearly flat-lying exposures of the Entrada Sandstone are ahead. The mountainside is heavily covered with talus at **Mile 4.5**, but seen from across the valley, the road appears to be at about the top of the Entrada here, going up-section into the Salt Wash Member of the Morrison Formation. The mountainside to the left at **Mile 5.0** is the Entrada Sandstone, overlain by the brush-covered Morrison Formation and capped by the Dakota Sandstone.

There is a good view to the left at **Mile 5.2** of the La Sal Mountains. The road climbs steeply into the Abajo Mountains, heavily covered in trees and brush. Poor exposures of the Brushy Basin Member of the Morrison Formation are on the right at **Mile 6.1.** Above the road ahead and on the right at **Mile 6.4** are pretty good exposures of the Dakota–Burro Canyon Formations. Exposures of the Dakota Sandstone are above the road to the left at **Mile 7.5.** At **Mile 8.0** tree cuts for a ski hill are visible high ahead.

Mile 8.5: Stop sign at a T intersection with County Road 101; Torrey Lake is 2 miles to the right, and Monticello is 10 miles to the left. Turn left toward Monticello. The road is at about the Morrison–Burro Canyon contact at **Mile 8.9**; conglomeratic exposures of the Dakota–Burro Canyon are in the cliffs ahead and to the right. **Mile 9.3: Milepost 15:** The mountainside is heavily covered here, but the road appears to be at the top of the Dakota Sandstone.

Mile 9.8: The road crosses through a scrub oak thicket and below groves of aspen trees. Monticello Lake and a picnic area are on the left at **Mile 10.7.** A small exposure of the Dakota Sandstone is across the valley to the left at **Mile 11.0**, and the road is now in a ponderosa pine–aspen forest. At **Mile 11.9** a poor exposure of Dakota Sandstone is in the roadcut on the left. Leaving Manti–La Sal National Forest. At the crest of the hill the road to the right goes to the Buckboard Canyon Campground. The road now skirts the northeast flank of the Abajo Mountains.

Mile 12.4, County Road 2432, Milepost 10: Triangulation markers are painted in the road along here for several miles. The road heads down a steep mountainside.

Cross Creek is at **Mile 12.9.** The dirt road to the right leads to Indian Creek (9 miles), Nazoni Campground (18 miles), and Blanding (32 miles). The road to the right at **Mile 13.5** goes to a picnic area and campground. The road to the right at **Mile 13.7** goes to Dalton Springs.

Mile 14.2: Leaving National Forest land; deer are all around us. Entering National Forest land again at **Mile 14.9.**

Mile 15.5: Milepost 5: The San Juan Mountains are visible in the far distance ahead along with the San Miguel Range. Lone Cone and the Wilson Peaks, as well as the La Plata Mountains, are discernible ahead. The La Sal Mountains can be seen to the left and the Ute Mountains to the southeast. Monticello is visible low ahead. Leaving Manti–La Sal National Forest at

Mile 16.3 with a good view of Monticello down below. The road to the right at **Mile 17.7** leads to Lloyds Lake.

Mile 17.9: Entering Monticello. **Mile 18.6**: Intersection with U.S. 191 in Monticello, across the road from the post office and state liquor store, the only liquor store in San Juan County (no street name visible).

12. Monticello to Bluff, Utah, via U.S. 191

U.S. 191, from Monticello to Blanding, Utah, skirts the eastern base of the laccolithic Abajo (Blue) Mountains. The range was named by the early Spanish explorers in the region, *abajo* meaning "lower" in Spanish, relative to the higher La Sal ("salt") Mountains to the north. The highway is generally on the Dakota Sandstone, dropping into the Morrison Formation in several gullies. From Blanding to Bluff the highway descends from the Dakota-capped plateau of White Mesa onto Bluff Bench in the Morrison Formation and on down into the San Rafael Group at Bluff.

Mile 0.0: Intersection of U.S. 666 and U.S. 191. Head south on 191 toward Blanding, Utah.

Mile 0.7: Leaving Monticello. At **Mile 2.1** the Ute Mountains, a laccolithic mountain range west of Cortez, Colorado, is visible to the left at 10:00. The highway is at or near the top of the Dakota Sandstone, a basal Late Cretaceous formation of coastal origin, at **Mile 4.1**, with good views of the great beanfield plateau, formerly the Great Sage Plains, to the left. **Mile 5.9**: The highway descends through the Dakota Sandstone into varicolored mudstone and siltstone of the Morrison Formation (Late Jurassic) in washes and then climbs back up to the Dakota Sandstone bench for the next 5.6 miles. The Devil's Canyon Forest Service campground is to the right at **Mile 12.0**. The Bears Ears on the crest of the Monument Uplift, capped by the Wingate Sandstone, are visible on the right at **Mile 12.6**.

At **Mile 15.5** the highway descends through the Dakota Sandstone into the Morrison Formation. The highway crosses Recapture Dam, with its reservoir to the right, at **Mile 16.9**. The highway again ascends through the Dakota Sandstone at **Mile 17.5** and is on the dark gray Mancos Shale at **Mile 17.9**. The top of the canyon is at **Mile 18.0** with the Bears Ears visible ahead.

Mile 20: Entering Blanding, Utah, which was settled in 1887 by Mormons from Bluff seeking better farming conditions.

Mile 20.5: Stoplight. Turn left on U.S. 191. Leaving Blanding at **Mile 22.1**. The Blanding Airport is to the right at **Mile 23.8**.

The road to the right at **Mile 24.8** is Utah 95, which heads west across the Monument Upwarp to Natural Bridges National Monument, then to Hite on Lake Powell, and on up North Wash to Hanksville, Utah (see Roadlog 16).

Mile 25.6: The Carrizo Mountains in northeastern Arizona, another laccolithic range in the Four Corners region, are visible at 11:00. The road to the right at **Mile 27.6** goes to the White Mesa uranium concentration mill, now on standby until uranium/vanadium prices improve. At **Mile 32.2** is White Mesa village, a small reservation for Ute, Paiute, and some Navajo Indians.

The highway begins its descent through the Dakota, Burro Canyon, and Morrison Formations to Bluff Bench at **Mile 34.7. Mile 35.5:** Junction with Utah 262, which goes east to Aneth and Hovenweep National Monument.

The very gentle Bluff anticline is visible on the skyline to the left at **Mile 37.7.** It is the site of the rather small Bluff Oilfield, discovered by Shell Oil Company in 1956. Comb Ridge is visible to the right at **Mile 38.7.** It consists of jagged peaks eroded from the upturned Navajo Sandstone along the Comb Ridge monocline, which forms the eastern flank of the Monument Upwarp.

At **Mile 44.9** the highway descends into Cow Canyon through cliffs of the Bluff Sandstone and red slopes of the San Rafael Group (Wanakah, Entrada, and Carmel Formations).

Mile 45.7: Entering Bluff, Utah. On the right are the Navajo Twins, two towers of Wanakah Formation red beds capped by the Bluff Sandstone of the Morrison Formation. The townsite is on the floodplain of the San Juan River, developed in the Carmel Formation. The village, settled in 1880 by members of the Mormon Hole-in-the-Rock expedition, was named for the surrounding sandstone bluffs that later became the type section for the Bluff Sandstone Member of the Morrison Formation (Late Jurassic). The village suffered in the early years from Indian attacks and flooding from the San Juan River, forcing many residents to safer, higher ground at Blanding.

The road to the left at **Mile 46.0** goes to Aneth, Utah, site of the huge Aneth Oilfield, along the San Juan River.

Continue south on U.S. 191 to Chinle, Arizona (Roadlog 23), or follow U.S. 163 to Kayenta, Arizona (Roadlog 36).

13. La Sal Junction, Utah, through Paradox Valley to Naturita, Colorado, via Utah 46/ Colorado 90

The road from La Sal Junction, Utah, to Naturita, Colorado, crosses the eastern Paradox basin and salt anticline region from west to east, crossing three salt-intruded anticlines. At first the route follows the southern base of the laccolithic La Sal Mountains, passing just north of the Lisbon Valley salt structure. The road then crosses the much smaller and poorly exposed Pine Valley salt anticline, south of the La Sals, and next crosses the syncline between the Pine Valley and Paradox Valley salt anticlines. After crossing the west flank of Paradox Valley, the road follows the full length of Paradox Valley to emerge from the southeastern flank of the salt-intruded anticline to near Naturita in the Sagers Wash syncline, adjacent to the west flank of the Uncompahgre Uplift.

Mile 0.0: Intersection of U.S. 191 and Utah 46. Rock exposures to the left are cliffs of the Entrada Sandstone, overlain by ledgy slopes of the Wanakah Formation and Tidwell Member of the Morrison Formation, succeeded by cliffs of the Salt Wash Member of the Morrison Formation.

The valley walls at **Mile 0.5** consist of low cliffs of the Entrada Sandstone, capped by the Wanakah, Tidwell, and ledgy cliffs of the Salt Wash Member of the Morrison (all Jurassic in age). **Mile 1.0: Milepost 1:** Exposures to the left and right are of the massive windblown Navajo Sandstone (Jurassic). The countryside is heavily covered at **Mile 2.3**, but the road is in the Brushy Basin Member of the Morrison Formation; the La Sal Mountains are ahead. Exposures of the Entrada Sandstone are on the left at **Mile 2.6.** A nice view of the La Sal Mountains is ahead. **Mile 2.9: Milepost 3.** The area is heavily covered at **Mile 3.4**, but the bedrock is in the Brushy Basin Member. **Mile 4.9: Milepost 5.**

The headframe on the right at **Mile 5.5** is for an abandoned uranium mine in the Salt Wash Member which is at shallow depths underground. The road is now on high plains, covered with grass and brush, on the south slope of the La Sal Mountains. **Mile 5.9: Milepost 6:** The Lisbon Valley salt structure is visible at about 2:30 on the right, with the La Sal Mountains on the left at 9:00. The road to the right at **Mile 6.5** goes to the Lisbon Valley mining district (copper and uranium) and the Lisbon Valley Oilfield. **Mile 8.7:** Lone Cone and the San Miguel Mountains are visible ahead and to the right at about 2:00. The road is still on the wide open, grass-covered plateau on the south flank of the La Sal Mountains.

Mile 8.8: Entering the village of La Sal. Leaving La Sal at **Mile 9.3.** On the left are uranium tailings from an abandoned mine in the Salt Wash Member of the Morrison Formation (Late Jurassic in age).

Mile 9.9: Milepost 10: The road climbs up the west limb of the Pine Valley salt anticline on dipslopes of the Salt Wash Member. The hills are heavily covered in scrub piñon pines and juniper. Entering Manti–La Sal National Forest at **Mile 11.6.** Some ponderosa pine is in the dense forest here. The road crosses a divide at **Mile 12.6** and starts its descent into the Pine Valley salt structure in rocks of the Salt Wash Member. **Mile 13.0: Milepost 13.** Entering "Old La Sal" at **Mile 15.1.** At **Mile 15.2** the San Juan Mountains are visible in the distance ahead through a forest of scrub oak.

Pine Valley is an open valley at **Mile 16.4**, narrowing down into a canyon ahead. The Salt Wash Member is poorly exposed on either side of the narrow valley at **Mile 17.4.** Exposures at **Mile 18.0: Milepost 18** in the Salt Wash

Member of the Morrison Formation are to the right and left. The road descends the east flank of the Pine Valley salt structure toward the withdrawal syncline between the Pine Valley and Paradox Valley salt anticlines.

Mile 18.7: Abandoned uranium mining operations on the left in the Salt Wash Member. **Mile 19.0: Milepost 19:** Sharp hairpin curve—slow down to 15 mph. Numerous uranium mining operations are visible. **Mile 20.0: Milepost 20**: More sharp hairpin curves; the road heads down into a sharp canyon. Entrada Sandstone cliffs are visible low in the canyon walls at **Mile 20.3**, capped by the Salt Wash Member of the Morrison Formation, the host rock of most uranium mines in the area.

Entrada Sandstone cliffs are prominent low in the canyon walls at **Mile 20.6.** Brown mudstone and siltstone of the Carmel Formation is exposed in the roadcut on the left at **Mile 21.2.** The lower canyon walls are now Entrada cliffs. The lower cliffs near the valley floor at **Mile 21.6** are in the Navajo Sandstone.

Mile 21.7: Welcome to Colorado. Entering Montrose County; the road is now Colorado 90. Naturita is 35 miles from **Mile 21.9.** Exposures of the light-colored windblown Navajo Sandstone are on the left. The road is near the bottom of the syncline at **Mile 22.5.**

The Salt Wash Member is exposed high on the canyon walls above a middle cliff of Entrada Sandstone and with slopes of the Carmel Formation between the Entrada and Navajo Sandstone cliffs below. **Mile 22.7: Milepost 1** (Colorado). Exposures of the Carmel Formation are visible at **Mile 23.0** between cliffs of the Entrada Sandstone above and the Navajo Sandstone below. The Salt Wash Member still caps the section along the canyon rim. Massive low cliffs to the right and left at **Mile 23.5** are in the Navajo Sandstone. The Carmel Formation is well exposed at **Mile 24.0** between the Entrada and Navajo cliffs. Rare exposures and roadcuts on the left at **Mile 25.1** are in the Kayenta Formation, here red and thin-bedded. Good exposures of the Kayenta Formation are to the right and left at **Mile 25.5.**

Mile 26.1: The road begins its ascent of the canyon wall in the Kayenta Formation; exposures down through the Wingate Sandstone are in the deep canyon to the right. There is a good view of the section down into the canyon on the right at **Mile 26.4.** The rocks here dip gently westward along the flank of the Paradox Valley salt-intruded anticline, becoming steeper at **Mile 26.6. Mile 26.8: Milepost 5.** The road is approaching the western rim of the Paradox Valley salt structure at **Mile 27.0.** The Navajo Sandstone thins dramatically and pinches out against the flank of the salt structure ahead. The road is still in the Kayenta Formation. **Mile 27.8: Milepost 6:** The road is in the top of the Wingate Sandstone beneath ledges of the

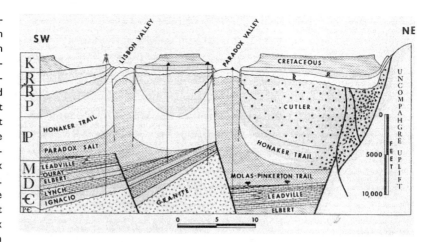

Kayenta Formation. **Mile 28.2**: Cliffs of the Wingate Sandstone are on the right and left. Rocks of the Kayenta Formation are a reddish color, and the Wingate is here almost white.

Mile 28.6: The road crosses the high divide on the flank of the Paradox Valley salt structure. There is a good view of Paradox Valley to the right and the Wingate Sandstone cliffs to the left. The Mesozoic section is seen to collapse into the northwestern termination of the salt structure around the curve to the right at **Mile 28.7: Milepost 7**. The Chinle Formation, green and red mudstone, is exposed on the left. The road begins its descent into Paradox Valley across immense collapsed blocks of any and all of the formations above the Chinle Formation. Collapsed Kayenta(?) blocks are in roadcuts on the left at **Mile 29.0**. There is a good view ahead at **Mile 29.5** down the full length of Paradox Valley. The San Juan Mountains are visible in the distance ahead. **Mile 29.7: Milepost 8**: The road is steep and crooked ahead as it descends the western flank of Paradox Valley through collapsed blocks of the various sedimentary rock formations (Figure 36).

Mile 31.4: The road to the left goes to Paradox, 1 mile. The main road ahead goes to Naturita, 30 miles. The road now skirts the main valley floor, which is on the left. The cliffs across the valley to the left consist of the upper Salt Wash Member, Tidwell Member, and Wanakah Formation; upper red cliffs of the Entrada Sandstone above slopes of the Carmel Formation; lower cliffs of thin Navajo Sandstone and Kayenta Formation; and the Wingate Sandstone. The lower slopes are of the Chinle and Moenkopi Formations. The escarpment consists of clean, uncollapsed exposures of the section, although the cliffs to the right are badly collapsed.

Mile 33.7: Milepost 12: The road is heading directly to the southeast along the floor of Paradox Valley. The sharp canyon visible on the left at **Mile 35.2** marks the course of the Dolores River as it leaves Paradox Valley toward the northeast. **Mile 36.3**: Entering Bedrock, Colorado. A T-shirt reads: "Where in the hell is Bedrock?" and on the back: "Across from Paradox." The road to the right at **Mile 36.6** goes into the Dolores River canyon and boat launching site. **Mile 36.7: Milepost 15.**

Mile 37.0: Bridge over the Dolores River. Unlike most respectable rivers, the Dolores does not flow along the major valley but crosses it at right angles. Henry Gannett, a mapmaker for the U.S. Geological and Geographical Survey of the Territories, better known as the Hayden Survey, recognized the incongruity in 1875 and named the valley Paradox Valley. Many years later it was realized that the valley was a salt-intruded anticline that had been buried deeply by younger sedimentary rocks. The course of the Dolores River was well established at that time, as erosion was only beginning to expose the harder rocks below. The river was thoroughly entrenched into its own canyon when it eroded downward into the older rocks of the salt structure below and had no recourse but to continue down-cutting across the upturned strata of the salt anticline below. Thus, the canyon of the Dolores River was necessarily established at right angles to the later-to-be-collapsed salt structure that lay athwart its path.

Mile 38.1: Entering the BLM Stock Driveway. Watch for cattle for the next 16 miles. The road to the left at **Mile 38.4** goes down the Dolores River Canyon through the Roc Creek salt dome. **Mile 38.7: Milepost 17.**

Mile 39.3: Near here, to the left in the middle of the valley, Conoco drilled its Scorup Number 1 well, which penetrated some 15,000 feet of Paradox (Middle Pennsylvanian) salt and interbedded rocks to bottom a few feet into the Mississippian Leadville Limestone. Behind the high cliffs to the right, Shell Oil Company drilled its Wray Mesa Number 1 well, which encountered *no* salt but drilled from post-salt limestones into the Devonian Elbert Formation, which directly underlies the Leadville Limestone. Displacement along the major fault that underlies the western flank of Paradox Valley is thus about 5,000 feet, and salt deposited along the high side of the fault had flowed entirely into the salt-intruded structure (Figure 36). **Mile 39.7: Milepost 18.**

The low hills in the bottom of the valley at **Mile 40.0** are erosional remnants of the gypsum cap on top of the salt-flowage structure. The low canyon in the gypsum cap in the valley floor at **Mile 40.5** is the type section of the Paradox Formation. When USGS geologists mapped Paradox Valley, they realized that the contorted gypsum, black shale, and dolomite beds had flowed upward from depth and were here out of place. At the time, the existence of the salt basin was unknown, but they believed the gypsum to be of Pennsylvanian age because of the presence of a few fossil brachiopods and named it the Paradox Formation for the location in Paradox Valley. When it became known through deep drilling that a large salt basin underlay this entire region, it was named the Paradox basin for the original name of the formation.

Mile 41.7: Milepost 20: The road here climbs up onto a low hill of partially exposed gypsum caprock. The road is at the top of the hill of gypsum caprock at **Mile 42.7: Milepost 21.** The roadcuts, especially on the right, contain large masses of gypsum. **Mile 43.3**: The San Miguel Range of the San Juan Mountains is exposed ahead and to the right. To the right at about 2:00 is a large abandoned open-pit uranium mine of the Uravan Mineral Belt.

The open-pit uranium mine to the right at **Mile 44.4** appears to have been mining a collapsed block of the Salt Wash Member of the Morrison Formation. **Mile 46.7: Milepost 25.** The dirt road to the right at **Mile 47.2** goes to the uranium mine. **Mile 50.1**: The road wanders through hills in a collapsed block of red rock that may be the Kayenta Formation. The collapsed walls of the eastern and western limbs of the salt structure are visible on the left and right at **Mile 50.5** with views ahead of the San Juan Mountains in the distance. **Mile 50.8: Milepost 29.**

The Wilson Peaks and Lone Cone in the San Miguel Range can be seen ahead at **Mile 51.5**, and a view of the San Juan Mountains, Sneffels Range, is to the left at about 2:00. The rocks exposed here are in the Salt Wash Member of the Morrison Formation, collapsed down into the south plunge of the Paradox Valley salt structure. The anticlinal fold on the west rim of the salt structure at **Mile 52.5** is a collapsed block of Salt Wash but was named the Coke Oven anticline; it was drilled for oil some years ago. **Mile 52.8: Milepost 31.** The road to the right at **Mile 53.4** goes to the Coke Oven Ranch.

Mile 53.6: The road enters a sharp canyon cut across the southeast rim of the salt structure. The rocks in the Salt Wash Member of the Morrison Formation are dipping toward the valley in a collapse structure. At **Mile 54.4** the Salt Wash Member is dipping eastward away from the salt structure toward the Sagers Wash syncline. The Uncompahgre Plateau is visible on the skyline to the east. This is the ancient Uncompahgre Uplift of Pennsylvanian age.

Mile 55.7: Intersection with Colorado 141, the Unaweep-Tabeguache Scenic and Historic Byway. Uravan is to the left, and Naturita is to the right. Join Roadlog 14 at **Mile 92.2.**

This is an important segment of highway, as it first crosses the Uncompahgre Plateau, the western fault block of the Uncompahgre Uplift, at right angles and then parallels the frontal, southwestern fault zone along the eastern margin of the Paradox basin from northwest to southeast. Colorado 141 crosses a basement fault zone and its associated monocline just west of Whitewater, where rocks of Cretaceous and Jurassic age are draped across the deep-seated fault(s). Exposures of the Triassic Chinle Formation, resting directly on the Precambrian basement metamorphic complex, make it apparent that this was a major faulted mountain range in late Paleozoic time. The Middle Pennsylvanian to Early Permian uplift supplied abundant coarse-grained sediments to the basins on either side of the ancient mountain range. The highway crosses the western fault zone, marginal to the Paradox basin, in spectacular style at Gateway, Colorado. The great faulted uplift is a segment of the late Paleozoic Ancestral Rocky Mountains. Perhaps as much as 18,000 to 20,000 feet of rocks of Paleozoic age underlie the course of the highway from Gateway to Placerville, as suggested by deep drilling for petroleum products.

Mile 0.0: Intersection of U.S. 50 and Colorado 141, Whitewater, Colorado. The intersection is on the Upper Cretaceous Mancos Shale. Colorado 141 crosses the Union Pacific Railroad tracks and the bridge over the Gunnison River at **Mile 0.4,** heading west toward Gateway. The Uncompahgre Plateau and monocline are visible ahead. The Gunnison River, known as the Tomichi by the Ute Indians and Río Javier by the Spanish, was named for Captain John Williams Gunnison, who made a reconnaissance survey for a railroad route in 1853.

Exposures of the Dakota Sandstone, beach and nearshore deposits of an advancing Cretaceous seaway, may be seen on the right at **Mile 0.7.** The beds are dipping gently northward across the monocline. The formation of Late Cretaceous age is well exposed from **Mile 1.0** to **1.6.** Exposures of the Lower Cretaceous Burro Canyon Formation, very thick at this location, are on the right at **Mile 1.7.** Beds of the Burro Canyon were deposited largely by rivers, and channel-filled, coarse-grained sand and gravel comprise much of the formation. Beneath the Burro Canyon lie heavily covered slopes of the Morrison Formation at **Mile 2.7.** The Brushy Basin Member of the Morrison Formation, primarily lake and stream mudstone and siltstone of variable colors, is exposed to the left and right at **Mile 3.2,** and the canyon widens for the next mile in the soft Morrison exposures, seen capped here by the Burro Canyon sandstone beds.

Mile 8.1: Grand Valley Overlook. The highway flattens as it enters Unaweep Canyon (Figure 37). To this point the road has been gradually climb-

Figure 37. The entrance to Unaweep Canyon, looking toward the south. The massive cliffs are metamorphic rocks of Precambrian age (about 1.75 billion years old), capped by a high slope of the Triassic Chinle Formation; the very high cliff in the upper right of the photo is the Wingate Sandstone, of Jurassic age. The course of tiny East Creek is in the right middleground.

ing dipslopes of the Morrison Formation while crossing the monocline. It is now crossing the Uncompahgre Plateau, an ancient highland that stretches 25 to 30 miles wide and extends southeastward some 100 miles to the vicinity of Ridgway, Colorado. This highway was once known as the Uranium Road, as it provided access from the mines in the vicinity of Gateway to the mill in Grand Junction.

The eastern part of Unaweep Canyon is drained by the tiny stream of East Creek; west of a divide little West Creek drains the western flank of the plateau. The two drainages are obviously too small to have carved the broad canyon ahead. It has been speculated for decades that Unaweep Canyon was carved by a much larger river, but opinions are divided as to whether it was the ancestral Gunnison or the ancestral Colorado River. One plausible interpretation has it that the canyon was originally carved by the Gunnison, which was later captured by headward erosion by the Colorado River and diverted to its present course.

The reddish-colored cliffs of the Entrada Sandstone (Jurassic) are exposed to the left and right at **Mile 8.8.** The road to the left at **Mile 9.0: Milepost 145.1** goes to Cactus Park. About a mile up the road is a gravel pit containing pebbles and cobbles believed to have been deposited by the ancestral Gunnison River about 1 to 5 million years ago when Unaweep Canyon was carved. This road provides access to the top of the Uncompahgre Plateau, used mainly for recreational purposes and Forest Service access. Exposures of the Wingate Sandstone, a brownish-colored sandstone of windblown origin, are to the right and left. The highway crosses a wide fault zone comprising the Cactus Park fault at **Mile 9.2** and then crosses Cactus Park Creek at **Mile 9.4.**

The sequence of sedimentary rocks here begins with the Precambrian basement, which is directly overlain by red mudstone and siltstone of the Triassic Chinle Formation, followed by the Wingate Sandstone and the Kayenta Formation (both Lower Jurassic) and the Entrada Sandstone (Middle Jurassic). The Uncompahgre Plateau is an ancient faulted uplift formed in Middle Pennsylvanian time only to be eroded to a relatively smooth plain and buried by sediments of the Upper Triassic Chinle Formation. The Wingate-Chinle contact is exposed on the left at **Mile 9.7** below good exposures of the Wingate, which is here about 300 feet thick. The top of the Precambrian basement may be seen beneath the Chinle Formation at **Mile 10.9**. Exposures of the metamorphic rocks, 1.3 to 1.7 billion years old, are to the right and left. There are good exposures of the red Chinle Formation ahead.

The abandoned townsites of Copper City and Pearl City, occupied from 1897 to 1914, are in the meadow to the left at **Mile 11.2**. The tent cities were home to 150 to 200 miners who came here during the copper and silver rush of the late 1890s.

At **Mile 11.3** good exposures of Precambrian granite are visible on the right; they have Rb-Sr (rubidium-strontium) and K-Ar (potassium-argon) dates of 1.3 to 1.37 billion years before the present. The road to the right at **Mile 13.9** goes to mines in Nancy Hanks Gulch. Exposures of a pegmatite dike in Precambrian quartz monzonite are visible in the rock quarry on the right at **Mile 15.5**. Exposures of the quartz monzonites in Unaweep Canyon date from 1.48 to 1.42 billion years, intruding metamorphic rocks that date to about 1.7 billion years. The valley opens up ahead into a typical Unaweep Canyon profile. The road to the left at **Mile 16.3** goes to the plateau above.

The drainage divide (elevation 7,048 feet) between East Creek and West Creek is at **Mile 18.0**. The drainage is now going toward the west. A nice view of Unaweep Canyon is ahead (see Figure 37) at **Mile 21.2**. *Unaweep* is a Ute Indian word for "valley with two mouths" or "dividing waters."

The ruin of Driggs Mansion is on the left at **Mile 24.7**. It was built by a wealthy New Yorker between 1914 and 1918 of locally cut sandstone from Mayflower Canyon. At **Mile 26.1** the canyon narrows ahead, cut deeply into the Precambrian rocks. **Mile 31.2: Milepost 123.** The canyon again narrows abruptly ahead at **Mile 33.8**.

An interpretive sign for Unaweep Seep is on the left at **Mile 34.5**. This area has been named an Area of Critical Environmental Concern by the BLM because it is a rare habitat for the Nokomis Fritillary butterfly.

At **Mile 36.7** the highway crosses a complex fault zone that bounds the southwestern margin of the Uncompahgre Uplift. Sagers Wash syncline and the eastern Paradox basin are ahead. The La Sal Mountains are visible ahead at **Mile 38.8.** The entire stratigraphic section from the Permian Cutler Formation up to sandstone beds in the Salt Wash Member of the Morrison Formation is visible ahead in the Palisade. Thick deposits of outwash gravels and boulders from the Uncompahgre Uplift occur to the left and right. **Mile 40.2: Milepost 114.** Exposures of Cutler Formation are seen on the right at **Mile 40.4,** capped by thick outwash deposits. The reddish-brown Cutler Formation is very conglomeratic, having been derived from the Uncompahgre highland in Permian time. The Cutler conglomerates, overlain by coarse outwash deposits, suggest a continuation of Permian depositional conditions from Permian time to the Recent. Ahead and to the left at **Mile 41.2,** the lower cliffs are in the Cutler Formation, overlain by thick brown shale beds of the Moenkopi and Chinle Formations, capped by the light brown, windblown Wingate Sandstone and a few feet of the stream-deposited Kayenta Formation. The pink cliffs of the Entrada Sandstone are seen to overlie the Kayenta Formation in the upper cliffs at **Mile 41.5.** More boulder deposits in outwash from the Uncompahgre highland are obvious at **Mile 42.0.**

Mile 42.7: Entering Gateway, Colorado, the "gateway" to the spectacular slickrock country of the Dolores River. Exposures of the Cutler through Kayenta Formations are visible on the right. From here to Placerville the highway follows south along the Sagers Wash syncline adjacent to the Uncompahgre Uplift to the east. There is a bridge over the Dolores River at **Mile 43.4.** The road to the right at **Mile 43.8** goes to Moab via John Brown Canyon and the La Sal Mountains loop. An angular unconformity at the base of the Chinle cuts out the Moenkopi Formation toward the east to directly overlie the Cutler and eventually Precambrian rocks onto the Uncompahgre Uplift in the Palisade to the east beyond Gateway. The Palisade has been designated the Palisade Outstanding Natural Area by the BLM. The highway here is in the Cutler Formation (Permian) (Figure 38).

Mile 44.3: Milepost 110. Good exposures of the Cutler Formation are to the left and right at **Mile 45.0.** Deep wells drilled in the vicinity penetrated about 18,000 feet of Cutler and bottomed in the Cutler. The top of the Cutler Formation is exposed at road level at **Mile 47.4.**

The highway is in brown siltstone of the basal Moenkopi Formation at **Mile 48.2.** The Triassic section at around **Mile 51.0** should be called the Moenkopi-Chinle, as the two cannot be easily separated. Perhaps the name *Dolores Formation*, used in the San Juan Mountains for the Moenkopi-Chinle strata, would be more appropriate.

The road to the right at **Mile 53.0** is 7.2 Road, heading along Salt Creek

Figure 38. The Palisade, looking toward the northeast across the Dolores River and the village of Gateway, Colorado, not visible in this view. The long lower slopes in the mesa are in the Moenkopi and Chinle Formations of Triassic age. The Moenkopi pinches out against the flank of the Uncompahgre Uplift, seen in the right distance, leaving the Chinle to rest directly on the Precambrian basement on the plateau. The massive cliff is the Wingate Sandstone, with the Kayenta Formation capping the cliff, beneath a higher, light-colored cliff of the Entrada Sandstone and the Salt Wash Member of the Morrison Formation, all of Jurassic age.

into Sinbad Valley, a small, salt-intruded anticline. **Mile 57.3: Milepost 97.** Entering Montrose County at **Mile 58.6.** The highway is on the upper Chinle (Dolores) Formation. The Chinle-Wingate contact is visible on the right at **Mile 60.2** with exposures of the Kayenta, Entrada, and Morrison Formations ahead. The Chinle Formation is again exposed beneath Wingate cliffs at **Mile 60.8** and again at **Mile 64.3.**

A good view of Wingate–Entrada–Salt Wash strata lies ahead at **Mile 64.8.** The highway crosses up-section from the Wingate Sandstone lower cliffs through the ledgy Kayenta Formation and into the pink cliffs of the Entrada Sandstone between here and the bridge over the Dolores River at **Mile 65.0.** The Morrison Formation is ahead above the Entrada Sandstone at **Mile 66.5.** The Brushy Basin Member of the Morrison Formation here has a strong greenish blue cast and is capped by the cliff-forming Burro Canyon Formation.

C. H. Turner and N. S. Fishman (1991) explain: "The greenish-blue, or turquoise, cast of the Brushy Basin Member is the result of volcanic ash that fell into a large saline-alkaline lake, called Lake T'oo'dichi', in Late Jurassic time. The ash was altered in the highly alkaline waters to analcime and potassium feldspar, although the turquoise probably comes from small quantities of iron compounds incorporated in the analcime. The lake was quite large, stretching southeastward almost to Albuquerque and northward almost to the Book Cliffs."

An abandoned uranium mining operation in the Salt Wash Member of the Morrison Formation, mainly stream-deposited sandstone, may be seen above to the left at **Mile 67.3.** River terrace gravels may be seen in roadcuts on the left at **Mile 68.1.** At **Mile 68.5** the road to the left is P-5. A bridge over

Figure 39. The Hanging Flume is perched high above the San Miguel River. The lower massive, smooth cliff is the Wingate Sandstone, and remnants of the flume can still be seen near the top of the formation. The ledgy cliffs above are in the Kayenta Formation, and the highway is on its upper bench. The higher smooth cliffs are in the Entrada Sandstone, capped by the Morrison Formation at the distant skyline.

Mesa Creek is at **Mile 68.9.** The Entrada Sandstone forms the prominent cliffs to the left and right. Another mining operation is in the lower Salt Wash Member at **Mile 69.8.**

The Kayenta-Entrada contact is along the highway on the left at **Mile 71.3.** A deep canyon cut well into the Kayenta Formation is on the right. The highway is near the top of the Kayenta at **Mile 72.4** with cliffs of Entrada to the right and left.

Hanging Flume overlook is on the right at **Mile 72.8.** The flume was built in 1889-90 to deliver water from the San Miguel River to the Lone Tree placer mine, just 40 feet above the Dolores River. The placer mine was not economical and was abandoned soon after construction of the flume, which is about 100 to 150 feet above the river near the top of the Wingate Sandstone (Figure 39).

The Uncompahgre Plateau is visible ahead on the skyline at **Mile 74.0.** The Kayenta-Entrada contact is on the left at **Mile 75.8.** Channels at the Kayenta-Entrada contact, visible at **Mile 76.3**, indicate that a disconformity is present ahead and on the left. Evaporation ponds for contaminated groundwater that is being pumped up from the underlying alluvium can be seen on the right across the river. The disposal cell for mill tailings and other radioactive waste is on Club Mesa, not visible from the highway. A uranium mining operation is on the left at **Mile 76.6** in the Salt Wash Member of the Morrison. A bridge and Road T-117 to the tailings operations is at **Mile 76.8**. Exposures of the Entrada Sandstone are obvious on the left and right. Umetco Minerals' reclamation operations are on the right at **Mile 78.0.**

A Point of Interest sign for Uravan is on the right at **Mile 78.3.** Uravan was established in 1936 when the U.S. Vanadium Corporation moved its plant here. It was abandoned in 1994, when the market for uranium and vanadium was exhausted. At one time, more than 800 people lived here. Between 1936 and its closure in 1994, the mill produced 42 million pounds of uranium and 220 million pounds of vanadium. Since 1983 Umetco Minerals Corporation has been involved in the $70 million reclamation program. Abandoned milling operations are evident on the hill across the river. The private road to the right crosses a bridge to the reclamation operations at **Mile 79.0.** Colorado 141 is in the Salt Wash Member of the Morrison Formation.

Mile 80.3: Bridge over the San Miguel River. The highway now follows the San Miguel River. **Mile 82.5: Milepost 72.** The top of the Salt Wash Member and base of the Brushy Basin Member of the Morrison, capped by the Burro Canyon and Dakota, are seen on the skyline at **Mile 84.4.** The top of the Salt Wash Member is in the roadcut on the left; the highway is in the Brushy Basin Member, capped by the Burro Canyon Formation at **Mile 88.3.** There are old river terrace gravels in the roadcut on the right at **Mile 89.4.**

Mile 90.2: Bridge over the San Miguel River to the left**. Mile 92.2**: Junction of Colorado 141 and Colorado 90, which goes to Paradox Valley and eventually to La Sal Junction (see Roadlog 13). **Mile 92.7**: Another road to the right. The highway is now in the Burro Canyon Formation. Beyond the junction with Colorado 90 there is another massive tailings reclamation operation.

Mile 94.0: Milepost 61.0: Entering Naturita, Colorado, elevation 5,431 feet, in the Burro Canyon Formation. Naturita, established in 1882, was named by "Grandma" Blake, a founding resident, because of its beautiful setting beside the river. The town was developed primarily as a business center for nearby ranching and mining interests. **Mile 94.6**: The road to left is Colorado 97, which goes to the airport and Nucla, Colorado. **Mile 94.9**: Leaving Naturita. The highway crosses Naturita Creek at **Mile 95.9.** Mudstone of the Brushy Basin Member of the Morrison Formation is at road level, overlain by the Burro Canyon and Dakota Formations.

The highway crosses up through the Burro Canyon and Dakota Formations at **Mile 96.6** with the Uncompahgre Plateau visible in the distance ahead and to the left. A view of the Wilson Peaks and Lone Cone is on the right at **Mile 97.1.**

Mile 99.1: **Milepost 56.8**: Intersection of Colorado 141 and Colorado 145. Colorado 141 goes right to Gypsum Valley and beyond. Turn left on 145 to Norwood and Placerville. The contact between green mudstone of the Burro Canyon Formation and conglomerate at the base of the Dakota Sandstone is on the left at **Mile 99.5.** At **Mile 100: Milepost 116** thin coal seams in the Dakota are exposed in roadcuts on the left.

At **Mile 104.3** the road is in the San Miguel syncline; the Paradox Valley salt-intruded structure is to the west, and the Uncompahgre Plateau is to the east. The Dakota Sandstone is exposed in the center of the structural depression.

The Redvale Post Office is at **Mile 106.0: Milepost 110** on the Dakota-capped plateau. The townsite was named Redvale in 1909 for the soil in the area. There is a good view of the southeast end of the Paradox Valley salt-intruded anticline to the west. The Montrose County line is at **Mile 112.6.** The highway from here to Norwood is on young stream and windblown deposits that lie on the Dakota Sandstone.

The Norwood Post Office is at **Mile 115.1.** Established in 1887, Norwood was named by its settlers for Norwood, Missouri. The flat plain and farmland from here to the top of Norwood Hill (**Mile 118.0**) is on the top of the Dakota Sandstone of Late Cretaceous age. At the top of Norwood Hill several mountain ranges are visible to the southeast, including Mount Wilson (elevation 14,246 feet) and Lone Cone (elevation 12,613 feet). To the west, flanking beds are visible at the Gypsum Valley and Paradox Valley salt-intruded anticlines, and the La Sal Mountains are in the distance.

The road down Norwood Hill crosses down-section through the Dakota and Burro Canyon Formations into the Brushy Basin and Salt Wash Members of the Morrison Formation. The bridge across the San Miguel River is in the basal Morrison Formation at **Mile 120.2.** Rocks of the Morrison Formation, capped by the Dakota Sandstone, form the canyon walls to **Mile 123.7,** where the massive, salmon-colored cliffs of the Entrada Sandstone appear in the lower canyon walls. Tree-covered slopes of the underlying, stream-deposited Wanakah Formation appear near road level ahead. The dark red slopes of the Triassic Dolores Formation (approximately equivalent to the Chinle and Moenkopi Formations) are seen 1 mile ahead.

The mouth of Saltado Creek is at **Mile 125.8**, where the Entrada Sandstone is nearly white, overlying a bed of red sandstone in the upper Dolores Formation. This red sandstone is believed by some geologists to be equivalent to the Wingate Sandstone. At **Mile 126.4** the road crosses the Black King fault, an east-west-trending normal fault that down-drops the Salt Wash Member of the Morrison Formation against red beds of the Dolores Formation. The valley floor ahead to **Mile 127.9** is lined with Pleistocene river gravel terraces that were once the sites of placer mining operations.

Dark red cliffs of the Dolores Formation line the canyon walls ahead until the characteristic purplish red conglomerate of the Cutler Formation (Permian) appears at a bridge to the right to Specie Creek at **Mile 129.3**. Red and purple arkosic sandstone and conglomerate beds of the Cutler Formation form the lower canyon walls from here to Placerville; coarse conglomerate beds, 5 to 25 feet thick, are impressive along the highway. The sediments and gravel in the Cutler Formation were washed down by streams from the Uncompahgre Uplift from the east and may be hundreds or even thousands of feet thick in the subsurface here. The formation was drilled by Shell Oil Company in a deep well near Nucla that penetrated more than 16,000 feet of the Cutler conglomerate.

Between **Mileposts 77 and 78** a dark greenish brown igneous dike is exposed on the left. Terrace gravels from here to Placerville hosted numerous placer gold mines in past years.

Placerville, named for the gold placer mining operations nearby, is near the intersection of Colorado 62 and Colorado 145 at **Mile 132.2**. Colorado 62 goes east to Ridgway, and 145 goes ahead to Telluride.

Here our journey connects with the San Juan Skyway loop, described in my book *The American Alps* (Baars 1992).

15. Utah 24 from Interstate 70 to Hanksville, Utah (including Goblin Valley State Park)

Utah 24 descends rapidly down-section into the San Rafael Group, consisting of the Summerville, Curtis, Entrada, Page, and Carmel Formations. The formations and the group are named for exposures found on the San Rafael Swell, and all are Middle Jurassic in age. The highway weaves its way up and down through exposures of these formations, and the San Rafael Swell and monocline ("reef") are beautifully exposed to the right (west). The extensive anticlinal structure is bordered on the east by a major basement fault system at depth, over which the sedimentary rocks of the monocline are draped. Uplift on the San Rafael Swell began in Paleozoic time, as rocks of Permian age rest directly on strata of Mississippian age across the uplift. All strata of Paleozoic age thicken dramatically down the west flank of the structure into the Basin and Range Province of western Utah and Nevada.

As we see it today, the uplift and folding of the swell occurred in Late Cretaceous and Early Tertiary time during the Laramide Orogeny, but like most structures on the Colorado Plateau, the uplift may have occurred along reactivated major fracture systems of Precambrian age.

Mile 0.0: Intersection of Interstate 70 and Utah 24, about 11 miles west of Green River, Utah. Head south to Hanksville. The highway is on top of the Salt Wash Member of the Morrison Formation, light-colored, stream-deposited, ledgy sandstone beds. The variably color-banded slopes above the Salt Wash are in the Brushy Basin Member of the Morrison, capped by sandstone and conglomerate beds in the Lower Cretaceous Cedar Mountain Formation.

Mile 4.7: The base of the Morrison Formation is at road level. Brown, thin-bedded bluffs on the right are in the Summerville Formation in a short distance. The Summerville Formation is a dark brown series of thinly bedded mudstone and siltstone, once intertidal mudflats along the edge of a seaway that lay to the west. At **Mile 4.9** the San Rafael Swell and monocline are obvious to the right.

The highway is on the San Rafael Desert at **Mile 6.3**, probably about on top of the Curtis Sandstone. The countryside is heavily covered with wind-blown sand; no rock exposures are to be found. Exposures of the Curtis Sandstone are to the right and left at **Mile 8.0.** The Curtis is a light-colored to pale green sandstone of shallow marine origin. The pale green color is due to the presence of glauconite, an iron-silicate mineral that forms only in shallow marine conditions. The road goes back on the sand-covered plain immediately. **Mile 9.7**: The contact of the Summerville and Curtis Formations is exposed in the hill to the left. Entrada Sandstone exposures lie ahead. Good exposures of the top of the Curtis and base of the Summerville are to the right and left of the highway at **Mile 12.0.** The road is back on the sand-covered San Rafael Desert at **Mile 12.6.** A view of the Henry Mountains (Mount Ellen) at 10:00 to 11:00 is in the distance at **Mile 13.5.**

Mile 14.2: Milepost 145: The Nequoia Arch, a deep-seated, northwest-trending fault block, passes through here at a depth of about a mile below the surface but is not evident at the surface. The sedimentary rocks overlying the faulted block have buried the structure and dampened the relief. The road to the right at **Mile 22.4** goes to Goblin Valley State Park. Roadlog mileage continues into Goblin Valley and back to Utah 24.

Goblin Valley State Park: The road heads west toward the San Rafael Swell at **Mile 23.1**; the craggy peaks on the right are Temple Mountain, a blowout vent on the monocline. The Summerville Formation on the Curtis Sandstone is ahead at **Mile 26.7**. There is too much cover to locate our

Figure 40. View downstream along Temple Wash where it crosses the San Rafael monocline. The lower ledgy rocks along the wash are the Chinle Formation (Triassic), overlain by the massive cliffs of the Wingate Sandstone and thin caprock of the Kayenta Formation, both Jurassic in age.

stratigraphic position at **Mile 27.1.** Upturned beds of the Navajo Sandstone form the hogbacks ahead on the San Rafael monocline.

The road to the left at **Mile 27.6** goes to Goblin Valley; the road ahead crosses the San Rafael monocline to Temple Mountain. We will take the Goblin Valley fork on our return. Go straight toward Temple Mountain. **Mile 27.8**: The top of the Navajo Sandstone is exposed on both sides of the road; the Navajo hogback is ahead. Note the well-developed cross bedding in the windblown Navajo Sandstone. At **Mile 28.2** the Navajo Sandstone dips strongly toward the east in the San Rafael monocline. The Navajo-Kayenta contact is exposed on the right and left at **Mile 28.5.** At **Mile 28.7** the road crosses Temple Wash. The Kayenta-Wingate contact is exposed to the right and left at **Mile 28.8.** The massive cliffs ahead are in the Wingate Sandstone (Figure 40).

The pavement ends at **Mile 28.9**; use caution and watch for washouts in the road and crazy drivers. Do not travel here if thunderstorms are threatening. The top of the Chinle Formation is to the right and left at **Mile 29.1**; the road is now on the Chinle varicolored mudstone. **Mile 29.3**: The Shinarump Member at the base of the Chinle Formation, a stream-deposited sandstone unit that fills channels in the top of the underlying Moenkopi Formation, is on the right and ahead. Exposures of the Shinarump are on the right and ahead at the dirt road to the right at **Mile 29.4.** There is a good view of Temple Mountain, showing features of the blowout, on the right (Figure 41). Exposures of the Shinarump-Moenkopi contact are to the right and left at **Mile 30.5**; the road is on dark brown mudstone of the Moenkopi Formation (Lower Triassic). **Mile 30.7**: Photo stop for Temple Mountain

Figure 41. Temple Mountain, a gaseous blowout structure along the San Rafael monocline and the former site of the Temple Mountain Uranium District. The lower slopes are in the Moenkopi and Chinle Formations, both Triassic in age, and the upper cliffs are in the Wingate Sandstone of Jurassic age. Light-colored rocks in the upper cliffs and distant slopes are bleached by gases emitted during the blowout event.

and the abandoned uranium mining operation. Backtrack to the Goblin Valley turnoff.

Mile 32.8: Turnoff to Goblin Valley State Park. Turn right (south) toward Goblin Valley. There is a good view to the right at **Mile 34.7** of the San Rafael Swell with the Page and Navajo Sandstones forming hogbacks and the Carmel Formation in the lowlands. A thin windblown sandstone unit called the Page Sandstone unconformably lies on the much thicker Navajo Sandstone. Because the two formations are similar in rock type and depositional environment, they may be difficult to distinguish. The road seems to be in the Carmel Formation. At **Mile 35.5** the road is in the Entrada Sandstone with hills capped by the light-colored Curtis Formation. **Mile 36.5**: A good view of the southern plunging nose of the San Rafael Swell is on the right at about 1:00 to 2:00. The low country to the right at **Mile 36.7** is in the Carmel Formation with the Navajo Sandstone on the skyline.

The Waterpocket Fold is visible in the distance to the right at **Mile 37.6**. The visible hogbacks are in the Page and Navajo Sandstones. The Henry Mountains (Mount Ellen) are ahead on the skyline at **Mile 37.8**. The buttes on the left and right at **Mile 37.9** are in the Entrada Sandstone, capped by small remnants of the light-colored Curtis Sandstone and dark brown Summerville Formation. The bigger butte to the right and ahead (Wild Horse Mesa) is capped by the Morrison Formation, above cliffs of the San Rafael Group.

Mile 38.7: Entering Goblin Valley State Park. The road to the right goes to Wild Horse Mesa.

The pay station is at **Mile 39.3**, where the pavement begins. A campground is on the right at **Mile 39.5**, surrounded by Entrada "goblins."

Figure 42. Goblin Valley Utah State Park. The "goblins" are being eroded from the soft-weathering Entrada Sandstone of Middle Jurassic age. Here the Entrada is a water-deposited siltstone, although the formation is a massive windblown sandstone in other parts of the Colorado Plateau. The light-colored cliff in the capping butte in the distance is in the Curtis Formation, a marine sandstone unit of Middle Jurassic age.

Although the Entrada Sandstone is a massive, highly cross-bedded, wind-blown sandstone to the east, it is here a more massive water-deposited siltstone unit that weathers to rounded bluffs and comical figurines. Entrada goblins are on the right at **Mile 40.1**, capped by the Curtis Formation. **Mile 40.6:** The end of the road, a turnaround, and picnic shelters in the Valley of the Goblins (Figure 42). Return to Utah 24.

Mile 48.4: Intersection with the road to Temple Mountain; the La Sal Mountains are visible in the far east. The Henry Mountains are visible to the south on the skyline.

Mile 49.7: Intersection with Utah 24. Continue south toward Hanksville. The road left at **Mile 50.2** goes to Flint Trail, Robbers Roost, and the Maze District of Canyonlands National Park. Active sand dunes are on the right at **Mile 51.8**; buttes of soft-weathering siltstone of the Entrada Sandstone are in the distance. **Mile 53.6: Milepost 133.** The Henry Mountains are ahead at **Mile 54.9**, Boulder Mountain is on the right at about 2:00, and Capitol Reef is at 3:00 to the right.

Entering Wayne County at **Mile 60.0.** Originally part of Piute County, Wayne County was designated in 1892 by state legislator Willis E. Robinson in memory of his deceased son.

Exposures of the Entrada Sandstone are on the left at **Mile 62.6: Milepost 124** and again at **Mile 63.4**, where badland topography is on the right. Hanksville emergency airport is on the right at **Mile 65.7**, and an omni air navigational transmitter is at about 2:00 (round, white, flat-topped building with a spike-like antenna on top).

Mile 67.0: Bridge over the Fremont (Dirty Devil) River. Upstream the river is called Fremont; downstream it is called the Dirty Devil River. In 1875 a party led by Andrew Jackson Allred named the Fremont River for General John C. Frémont. When the river was traced downstream, it was realized that the same river had been named Dirty Devil by Major John Wesley Powell in 1869.

Mile 68.9: Welcome to Hanksville. The town of Hanksville, originally known as Graves Valley, was named for Ebenezer Hanks in 1880. Utah 24 goes right to Capitol Reef National Park (Roadlog 37), and Utah 95 goes left toward Blanding and Bullfrog and Hite on Lake Powell (Roadlogs 16 and 17).

16. Hanksville South toward Hite and Blanding, Utah, on Utah 95

From Hanksville, Utah, south to the intersection with Utah 276, Utah 95 winds through exposures of the Entrada Sandstone and the underlying Carmel Formation, both of Middle Jurassic age. South of the Dirty Devil River, the high plateau country is named the Burr Desert, for John Atlantic Burr. There are good views of the Henry Mountains (Mount Ellen) toward the west along the way. The intersection of Utah 95 and 276 is at the head of North Wash, a sinuous, entrenched canyon that descends rapidly down the stratigraphic section from the Page Sandstone (Middle Jurassic) to the Cedar Mesa Sandstone (Early Permian) at Hite on Lake Powell. The man-made reservoir of Lake Powell in Glen Canyon of the Colorado River behind the Glen Canyon Dam was filled between 1962 and about 1968. The extensive lake lies here along the western, rather gentle flank of the Monument Upwarp. From Hite, Highway 95 climbs up and across the western slopes of the upwarp on top of the Permian Cedar Mesa Sandstone all the way to the eastern marginal monocline that lies along Comb Ridge. The highway then climbs back up-section to the Dakota Sandstone at the eastern end of Utah 95 just south of Blanding.

Mile 0.0: Turn left on Utah 95 in beautiful downtown Hanksville. Exposures of the Entrada Sandstone are to the right and left at **Mile 1.6. Mile 2.1: Milepost 2** on Utah 95. Exposures are all around of the Entrada Sandstone at **Mile 3.7.** There are good views of the Henry Mountains to the west. The butte on the right at **Mile 6.6** is the reddish Entrada Sandstone, capped by the nearly white Curtis Sandstone (both Jurassic in age). Scattered exposures to the right at **Mile 10.5** are in the Entrada Sandstone, capped by the Curtis, dipping to the west into the Henry Basin. **Mile 11.1: Milepost 11.**

Mile 16.3: Entering Garfield County, subdivided from Iron County in 1882 and named for President James A. Garfield. Exposures are of the

Entrada Sandstone, capped with pediment gravels of Quaternary age. A *pediment* is an erosional slope formed in past erosional cycles that slopes gently away from a mountain mass. Pediments are usually covered with gravel or boulders washed down from the high country and in this case are being incised by modern weathering processes. Roadcuts at **Mile 18.3** are in pediment gravel derived from the Henry Mountains to the west. The highway crosses back down through the pediment gravels at **Mile 19.2.** Roadcuts and exposures are in the Entrada Sandstone at **Mile 22.3.** The road passes through small-scale folds at **Mile 22.9**, in and out of the Carmel Formation. Pediment gravels cap the mesas to the right. The contact between the Entrada Sandstone and the Carmel Formation is on the left at **Mile 24.9.** Roadcuts and hills at **Mile 26.1** are in red siltstone of the Carmel Formation.

Mile 26.3: Utah 95 goes ahead down North Wash to Hite; Utah 276 goes right to Bullfrog (Roadlog 17). Ticaboo is 28 miles on Utah 276. The Middle Jurassic Page Sandstone is here about 100 feet thick, forming a separate cliff above the Lower Jurassic Navajo Sandstone. Keep left on Utah 95, a designated Scenic Byway. Active sand dunes are present just past the intersection.

Mile 28.1: Milepost 28 on Utah 95, heading into North Wash. The highway is in the Page Sandstone, then the Navajo Sandstone. Note the well-developed cross bedding in the Navajo that typifies wind-deposited sandstone. The Navajo-Kayenta contact is at road level at **Mile 29.1.** Hogg Spring is on the right at **Mile 33.5**; *do not drink the spring water* (note that the restrooms above have been carefully engineered to drain toward the spring). The spring is at the Wingate-Chinle contact; the road descends into the Chinle Formation, a varicolored, slope-forming mudstone/siltstone unit.

Lots of fallen rocks from the overlying Wingate Sandstone lie on the Chinle slopes at **Mile 35.7.** The top of the Moss Back Member of the Chinle Formation is at road level at **Mile 40.1.** The Monitor Butte Member of the Chinle Formation lies beneath the Moss Back Member at **Mile 40.4.** In North Wash and White Canyon both the stream-deposited Shinarump and the Moss Back Members of the lower Chinle Formation are locally present, separated by shaly interbeds called the Monitor Butte Member.

Mile 40.6: Entering Glen Canyon National Recreational Area. The road is in the Moenkopi Formation at **Mile 42.3.** The Shinarump Member of the Chinle Formation is not noticeable along the road, as it thickens and thins along erosional channels, but it is visible ahead on the left below the Monitor Butte Member. The Moss Back and the Shinarump are present together only here and in White Canyon.

The road leaves North Wash at **Mile 43.6** and climbs onto Moenkopi benches toward the canyon of the Dirty Devil River arm of Lake Powell. Turn right at **Mile 45.5** to the Hite Scenic Overlook.

Figure 43. The view from Hite Overlook, between North Wash and Hite Marina. Lake Powell is in the foreground with Straight Canyon in the middle distance to the left. The broad expanse of light-colored slickrock is the upper part of the Cedar Mesa Sandstone (Pcm), here about 1,200 feet thick, capped by thin remnants of the dark-colored Organ Rock Shale (Por), both Permian in age. The small buttes in the distance are capped by the White Rim Sandstone (Pwr) and Moenkopi Formation.

Mile 45.8: Hite Overlook of Lake Powell, a good place for photos of Hite Marina and the Colorado River canyon "upstream" to the left (Figure 43). The top of the Cedar Mesa Sandstone, a nearly white, very thick shoreline sandstone, is just above lake level. The Cedar Mesa bench is overlain by brownish red shale and siltstone of the Organ Rock Shale and the white capping cliff of the White Rim Sandstone, all formations of the Cutler Group of Permian age. Reenter Utah 95 at **Mile 46.2**, heading east toward Hite.

Limestone pebble conglomerate at the base of the Moenkopi Formation is visible in sharp roadcuts at **Mile 46.6.** The pebbles were derived from weathering in Late Permian and earliest Triassic time of the Kaibab Limestone, which is present in the subsurface of the Henry Basin to the west. The conglomerate lies on the White Rim Sandstone, a shoreline equivalent of the marine Toroweap Formation farther to the west. **Mile 46.7**: The contact of the White Rim Sandstone and the Organ Rock Shale (both Permian) is in roadcuts. The road crosses an unnamed "stray sand" in the Organ Rock Shale at **Mile 46.9.**

Figure 44. Bridge over the Colorado River arm of Lake Powell. Formations visible are the Cedar Mesa Sandstone (Pcm), Organ Rock Shale (Por), and White Rim Sandstone (Pwr). There is a thin cap of the Triassic Moenkopi Formation (Trm) at the top of the butte.

The canyon of the Dirty Devil River (an arm of Lake Powell) is visible on the right at **Mile 48.3**, entrenched into the Cedar Mesa Sandstone (Permian). The highway is in the Organ Rock Shale. The Organ Rock Shale–Cedar Mesa Sandstone contact is at road level at **Mile 48.9**. The highway here is on the Cedar Mesa Sandstone.

Mile 50.3: Bridge over the Dirty Devil River arm of Lake Powell. **Mile 51.9: Milepost 47**. Hite "International Airport" is on the right at **Mile 52.0**. The highway descends well into the Cedar Mesa Sandstone, which is here about 1,000 feet thick. **Mile 52.4**: Bridge over the Colorado River arm of Lake Powell (Figure 44).

The road to the right at **Mile 53.6** goes to Hite Marina; a road sign says this is the Bicentennial Highway. **Mile 53.9: Milepost 49**. The road is on the red Organ Rock Shale at **Mile 54.3** with exposures of the Cedar Mesa Sandstone to the left. Another glimpse of Lake Powell is at **Mile 55.1**; the Henry Mountains are at 2:00. Note the highly weathered fracture system in the Glen Canyon Group across the lake from Hite (Figure 45). These surface

Figure 45. The view toward the west from Hite Overlook. Siltstone and mudstone of the Moenkopi Formation (TRm) form the lower rough slopes, overlain by the slope-forming Chinle Formation (TRc); both are Triassic in age. The higher vertical cliff is in the Jurassic Wingate Sandstone (Jw) below the ledgy Kayenta Formation (Jk), with the massive, light-colored, rounded exposures of the Navajo Sandstone (Jn) on the skyline. Note the deeply weathered fractures forming fins in the Navajo Sandstone that extend down into the Wingate Sandstone. This is the surface expression of the deep-seated Four Corners basement lineament.

fractures occur along a northwest-trending basement fracture zone that also forms northwesterly sandstone dikes in the Organ Rock in this vicinity.

Mile 56.6: The road is at the top of the Cedar Mesa Sandstone. From here to Comb Wash, the highway is at or near the Organ Rock Shale–Cedar Mesa Sandstone contact. The Organ Rock is the dark reddish brown slope-forming rock above, and the Cedar Mesa is the hard white sandstone below. The Henry Mountains are visible to the right.

The road to the right at **Mile 58.0** leads to Farley Canyon, which has access to Lake Powell. The highway is on the Cedar Mesa Sandstone at **Mile 58.3**. On the right is White Canyon, which the highway follows to Natural Bridges National Monument. **Mile 58.8**: Bridge over White Canyon wash. Generally speaking, White Canyon lies along the northwest-southeast basement fracture system of the Four Corners lineament. The highway is on the Organ Rock–Cedar Mesa contact at **Mile 59.4.** At **Mile 59.6** the road to the right provides lake access via White Canyon.

Mile 59.7: Leaving Glen Canyon National Recreation Area. The cliffs to the right at **Mile 60.3** expose the entire stratigraphic section up to the Kayenta Formation. Buttes to the left at **Mile 61.2** are capped by the White Rim Sandstone, here thin-bedded and flat-bedded, suggesting that they are beach deposits. This is one of the few areas where the White Rim is known to be present east of the Colorado River. The highway is climbing up the gradual westerly dip of the Monument Upwarp, with strata dipping westward into the Henry Basin.

Mile 64.3: High and to the right at 1:30 is the abandoned Happy Jack uranium mine in the Shinarump Member of the Chinle Formation. The road to the right at **Mile 64.5** goes to the mine. The easternmost exposures of the White Rim Sandstone are on the mesas to the left at **Mile 65.3.** The White Rim Sandstone can be seen to change rock type into red beds ahead from **Mile 65.8.**

In the cliffs ahead and buttes to the left at **Mile 67.0** the White Rim Sandstone interfingers with red beds of the upper Cutler Group. Older geologic maps of this area confuse the red bed section with the overlying Hoskinnini Member of the Cutler Group.

Mile 67.6: The high mesa to the left, behind the high crag, is capped by the Wingate Sandstone. **Mile 68.0: Milepost 63.** The high crag to the left at **Mile 69.0** is Jacobs Chair. At **Mile 70.2** the White Rim Sandstone is entirely replaced by red beds of the upper Cutler Group. The low cliff ahead at **Mile 70.6** is capped by the Shinarump Member of the Chinle Formation (Triassic). **Mile 73.7**: The low buttes to the left and the cliff ahead are capped by the Shinarump. The prominent rounded butte ahead at **Mile 75.4** is capped by the Wingate and Kayenta Formations.

Fry Canyon Lodge is on the right at **Mile 76.6**, and the road is on top of the Cedar Mesa Sandstone (Permian). **Mile 77.0: Milepost 72.** Cheese Box Butte is on the left at **Mile 80.2**, capped by thin remnants of the White Rim red marker beds. On the right at 2:00 are more uranium mines in the Moss Back Member. The Bears Ears on the skyline ahead at **Mile 84.4** are capped by the Wingate Sandstone.

Mile 86.2: The highway emerges from White Canyon onto the broad expanse of Cedar Mesa, capped by the Cedar Mesa Sandstone, with exposures of the Organ Rock Shale on the right. The road to the right at **Mile 88.8** goes to Halls Crossing and Clay Hills Crossing on Utah 276. The highway is on the Cedar Mesa Sandstone. **Mile 90.0: Milepost 85.** The buttes on the skyline to the left at **Mile 91.6** are the Bears Ears (Figure 46). There is a good

Figure 46. The Bears Ears, capped by the Wingate Sandstone, mark the crest of the Monument Upwarp near Natural Bridges National Monument. The highway is here at the top of the Permian Cedar Mesa Sandstone.

view of the Bears Ears ahead at **Mile 94.3. Mile 96.5**: The road to the left, Utah 275, goes to Natural Bridges National Monument. Natural bridges, unlike arches, form where meandering stream courses undercut the thin part of a gooseneck, leaving more resistant rock across the new, shortened course of the stream. The road to the right at **Mile 98.3,** Utah 261, leads to Mexican Hat.

The highway reaches the crest of the Cedar Mesa anticline at **Mile 102.3** and begins its long, gradual descent into Comb Wash. Cedar Mesa is obvious to the right, and the Ute Mountains are ahead across Comb Ridge. The highway is still on top of the Cedar Mesa Sandstone. **Mile 106.1: Milepost 99.** The road left at **Mile 106.9** goes to the Mule Canyon Indian ruins.

Mile 107.4: The Abajo Mountains are to the left at 9:00. **Mile 109.9**: The low cliffs to the left, ahead, and to the right are the northern extension of Comb Ridge. The highway is still on the Cedar Mesa Sandstone. Comb Ridge, ahead from **Mile 111.1,** is capped by the Kayenta Formation, overlying the massive cliffs of the Wingate Sandstone and the lower red slopes of the Chinle Formation. The well-developed fracture pattern in the Wingate cliffs ahead is on a deep-seated basement fracture zone that trends northwest across the Monument Upwarp, down White Canyon to the cliffs across from Hite, and up North Wash.

The highway crosses Comb Wash at **Mile 112.5.** The Organ Rock Shale (Permian) and the Moenkopi Formation (Triassic) are covered in the floor of Comb Wash. At **Mile 113.9** the highway crosses spectacular roadcuts in the Wingate Sandstone. Roadcuts cross upward into the Kayenta Formation at **Mile 114.1.** The white sandstone to the left and right at **Mile 114.2** is the Navajo Sandstone in Comb Ridge. Orange-colored cliffs ahead are in the Entrada Sandstone and Wanakah Formation, capped by the Morrison Formation.

Mile 114.7: The red mudstone of the Carmel Formation is in the valley and lower slopes to the right, and the higher pinkish-colored cliffs are the Entrada Sandstone, with the Salt Wash Member of the Morrison Formation forming the higher ledgy cliffs on the skyline. At **Mile 115.8** the sedimentary rocks are dipping rather gently to the right, or east, from the Monument Upwarp into the Blanding Basin. The highway is in Butler Wash at **Mile 116.4**; the road to the left goes to the Butler Wash Indian ruins. At **Mile 116.9** the highway is in the Carmel Formation with low cliffs to the right in the Entrada Sandstone. The highway crosses upward through the Entrada Sandstone at **Mile 117.8** and into the Salt Wash Member of the Morrison Formation at **Mile 118.1,** overlain by the Recapture Shale Member and Westwater Canyon Sandstone Member of the Morrison Formation. The highway is in the Brushy Basin Member of the Morrison Formation at **Mile 118.6.** The Abajo Mountains are obvious on the left at about 10:00 from **Mile 118.9**, where the highway is going down the dipslope on the Salt Wash Member.

Mile 120.1: The varicolored slopes on the left are in the Brushy Basin Member, capped by the Cedar Mountain Formation and the Dakota Sandstone. The highway crosses upward through the Brushy Basin Member of the Morrison Formation at **Mile 121.9** and through the Morrison–Cedar Mountain–Dakota contact at **Mile 122.8.** A cattle guard at **Mile 123.2** is in outwash gravels from the Abajo Mountains.

Mile 123.6: The highway is on top of the Dakota Sandstone on the Great Sage Plains. The Abajo Mountains are on the left at 9:00, and the Ute Mountains are on the right at 2:00. The highway descends through the Dakota Sandstone to the Brushy Basin Member at **Mile 125.3** and climbs back up through the Dakota Sandstone at **Mile 125.9** to the top of the Dakota at **Mile 126.2: Milepost 121.**

Mile 127.0: End of Utah 95 and junction with U.S. 191. Turn left to Blanding or right to Bluff. (Join Roadlog 12 at **Mile 24.8**.)

17. Utah 276 from Utah 95 to Bullfrog, Utah, on Lake Powell

Utah 276 begins at the intersection with Utah 95 in the Page Sandstone at the head of North Canyon. It wends its way through exposures and roadcuts of the San Rafael Group (Middle Jurassic), capped by extensive deposits of pediment gravels derived from the Henry Mountains, for most of the distance to Bullfrog, Utah, on Lake Powell. Magnificent views of the laccolithic Henry Mountains highlight the drive. First passing along the west flank of the northern Henrys, the highway crosses a saddle and flanks the western margin of the southern intrusive igneous mountains into Bullfrog Basin. The Henry Mountains were studied intensively, first by G. K. Gilbert in 1877 and later by C. B. Hunt and others (1953). Gilbert coined the term *laccolith* for igneous intrusive bodies that are shaped much like a mushroom. From the highway numerous examples may be seen of igneous dikes, sills, and stocks intruded into the red sedimentary rocks that comprise much of the countryside. And at the end of this very scenic drive is Lake Powell!

Mile 0.0: Intersection of Utah 276 and 95. Keep right on Utah 276 to Ticaboo and Bullfrog.

Mile 0.3: The highway crosses upward through the Entrada-Carmel contact and then is back in the red Entrada Sandstone. The soft-weathering Entrada of Goblin Valley has here changed to a red cross-bedded sandstone of apparent windblown origin. Exposure of the Entrada Sandstone is on the left at **Mile 2.7. Mile 4.0: Milepost 4.** Roadcuts at **Mile 4.4** are in pediment gravels previously washed down from the Henry Mountains, probably in Pleistocene time. Two peaks of the Henry Mountains are visible ahead: Mount Pennell, elevation 11,320 feet, in the distance, and Mount Hillers, elevation 10,650 feet, nearer. The highway at **Mile 4.8** crosses a large wash in pediment gravels. **Mile 5.1: Milepost 5:** A dark-colored igneous sill caps the hill on the right. The road passes back up through pediment gravels at **Mile 5.5.** The igneous sill is well exposed on the right.

Mile 6: Milepost 6: A roadcut in the pediment gravels rests on the red Entrada Sandstone. **Mile 6.4**: A large intrusive igneous sill is exposed on the right at about 1:00 to 2:00. The Henry Mountains are the classic laccolithic mountains. A *laccolith* is a mushroom-shaped intrusive igneous body with a central circular stock that fed magma upward and then outward into sills along bedding planes within the host sedimentary strata. The lower contacts of the sills are typically parallel to the bedding and horizontal. The tops of the sills are bulged upward, forming dome-shaped rises on the overlying sedimentary layers. Of course, there are no perfect laccoliths in nature, but the intrusive bodies seen in the Henry Mountains generally fit the original description. What are seen in exposures here are bits and pieces of the stocks and sills, not an entire laccolith. Feeders to the classic laccolithic shapes have numerous variations that were given specific names by Charlie Hunt in his 1953 classic study of the Henry Mountains.

An igneous sill caps the Entrada Sandstone ahead at **Mile 7.3**. The highway is in red rock country at **Mile 7.6**, mostly in the Entrada Sandstone. The highway is down into the Carmel Formation on the left at **Mile 8.4. Mile 9.2: Milepost 9**: The Entrada Sandstone ahead dips steeply away from the igneous stock of Mount Holmes, elevation 7,930 feet, of the Henry Mountains; the road is now in brown siltstone of the Summerville Formation. **Mile 11.0**: The highway is in the Entrada Sandstone in a sharp syncline between Mount Holmes and Mount Hillers. The sharply domed nature of the

Figure 47. Mount Ellsworth in the Henry Mountains north of Lake Powell. Light-colored beds of the Navajo and Entrada Sandstones are sharply upturned around the flanks of the igneous intrusive stock, one of several in these laccolithic mountains.

Entrada and Navajo Sandstones by the intrusive igneous stock is very apparent at **Mile 11.7.** The dominant igneous rock type of the Henry Mountains is gray diorite porphyry, which has been dated by potassium-argon methods at 44 to 48 million years before the present (about Middle Tertiary).

Mile 12.2: The highway crosses a deep gully cut into the Entrada and Carmel Formations. The Entrada and Carmel are dipping steeply away from the Mount Holmes intrusive stock at **Mile 13.5.** Roadcuts ahead at **Mile 14.4** are in the Summerville Formation. The Summerville-Entrada contact is on the right at **Mile 15.1. Mile 17.2: Milepost 17**: The road to the right goes to Starr Springs Campground.

Widespread pediment gravels cap the surface at **Mile 18.3.** A roadcut in the Summerville Formation is at **Mile 19.1.** Mount Ellsworth, elevation 8,150 feet, of the Henry Mountains is ahead at 10:00 to 11:00 (Figure 47). Pediment gravels are everywhere. Roadcuts at **Mile 19.8** are in the nearly white Entrada Sandstone with brown Summerville caps. **Mile 20.3: Milepost 20.** Roadcuts are in the Summerville Formation at **Mile 20.7.** The Entrada Sandstone on the left dips strongly away from the Mount Ellsworth stock.

Mile 21.6: Another laccolithic mountain, Navajo Mountain, elevation 10,388 feet, is ahead. A view of Lake Powell in the Bullfrog embayment is below. Pediment gravels in a broad expanse at **Mile 22.2** look like basalt boulders but are intrusive igneous and sandstone boulders heavily coated with desert varnish. At **Mile 23.1** there are good exposures of the Entrada Sandstone as it dips strongly to the west away from the Mount Ellsworth intrusive stock. **Mile 23.3: Milepost 23.** The Entrada-Carmel contact is exposed to the left at **Mile 24.9.**

Mile 26.3: Milepost 26: The countryside broadens into the Bullfrog Basin in the Rock Spring syncline, the deepest downfold in the Henry structural basin. Ticaboo Lodge and Boat Storage are on the right at **Mile 27.7.** There is another boat storage on the left at **Mile 31.0.** The rolling hills around **Mile 34.0** are in the Entrada Sandstone.

Mile 35.9: The road to the right leads to Burr Trail, Capitol Reef National Park, and on to Escalante in the heart of Grand Staircase–Escalante National Monument (Roadlog 39). The Waterpocket Fold is obvious to the right at **Mile 37.4.** Navajo Mountain is on the skyline at 2:00 at **Mile 38.6.**

Mile 39.2: Pay station at the entrance to Glen Canyon National Recreational Area. Entrada Sandstone exposures are to the right. **Mile 41.2**: Boat ramp for Bullfrog Marina on Lake Powell. Bullfrog Creek drains from the northwest flank of Mount Pennell into Lake Powell, along which there live many bullfrogs, hence the name. One may take the ferry to Halls Crossing on the east bank of Lake Powell, where Utah 276 continues to the eastern intersection with Utah 95 near Natural Bridges National Monument.

II

The San Juan Basin

The San Juan Basin lies generally south of the Blanding Basin in northwestern New Mexico and extends a short distance into southwestern Colorado. It is a structural basin, that is, unlike the Paradox basin, evident at the surface. The oldest rocks exposed in the basin proper are of Late Cretaceous age, although basement uplifts pretty much surround and define the large structural depression. Sedimentary rocks of Late Cretaceous age are widely exposed around the basin perimeter, and Tertiary-age strata are preserved over much of the central basin.

Bounding structural features include the northeast-trending Hogback monocline, which extends from near Shiprock, New Mexico, to about Durango, Colorado, and forms the northwest corner of the basin. The broad, gentle high structure to the northwest of the Hogback monocline is the Four Corners platform, also the southern limit of the subsurface Paradox basin. Another abrupt monocline extends from Durango southeastward to about Chama, New Mexico, forming the northeastern boundary. The Dakota Sandstone of Late Cretaceous age lies directly on Precambrian quartzite on this limb of the basin. The eastern flank of the basin is marked by the north-south-trending San Pedro–Nacimiento Mountains faulted uplift that runs south from about Cuba to near San Ysidro, New Mexico. Rocks of Mississippian through Permian age onlap this ancient highland from south to north, and the Permian-age Cutler Formation (called the Abo Formation in New Mexico) rests directly on Precambrian granite on San Pedro Mountain. The southern limits of the basin are vague, but the Rio Grande Rift near Albuquerque is a major bounding factor. The basement uplift of the northwest-trending Zuni Mountains, where the Abo Formation of Permian age rests on Precambrian granitic rocks, forms the southwestern limit of the basin. Finally, the northwest-trending Defiance Uplift and the adjacent Chuska Mountains define the western edge of the basin. Permian red

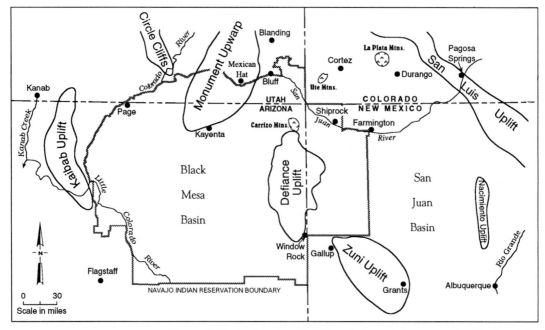

Figure 48. The San Juan and Black Mesa Basins and the structural uplifts that define and separate them.

beds of the lower Cutler (also called Abo by New Mexicans, and Supai by Arizonans) rest on Precambrian granite, and locally quartzite, along the Defiance Uplift. All these bounding basement uplifts were high and shedding sand into the basin in Pennsylvanian time, only to be buried by red beds of Early Permian age (Figure 48).

Pre-Cretaceous Rocks

Rocks of Pennsylvanian age occur in the subsurface of the San Juan Basin but are exposed at the surface only in the mountains near Albuquerque and Durango. Deep drilling, however, indicates that a marine inlet connected the seaway in southern New Mexico with the Paradox basin throughout much of Pennsylvanian time.

Rocks of Permian age are exposed only on the uplifts surrounding the basin, but deep drilling shows that the section is well developed beneath younger rocks throughout the basin. New Mexican geologists have long had their own terminology for these rocks of Permian age, and the tradition is still adhered to today. For example, red beds in every way similar to and continuous with the lower Cutler Formation are called the Abo Formation in New Mexico and are widespread across the basin. The DeChelly Sandstone of Monument Valley and Canyon de Chelly fame clearly extends into and across the San Juan Basin, but here the red windblown sandstone is

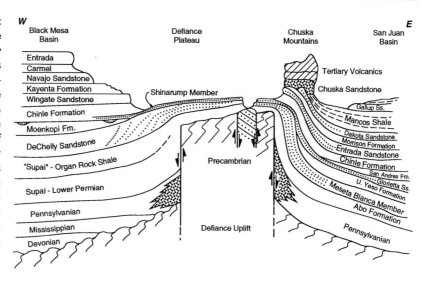

Figure 49. Diagrammatic cross section showing the relationships of the many rock formations across the ancient Defiance Uplift, which separates the San Juan and Black Mesa Basins. The uplift was high during most if not all of Paleozoic time, as many formations pinch out against the flank of the fault block and others received coarse-grained sediments derived from the highland. Most of the formations change names as they cross from Arizona into New Mexico, where local names dominate.

called the Meseta Blanca Sandstone Member of the Yeso Formation. The upper member of the Yeso Formation, the San Ysidro Member, is largely marine strata that form the northern limits of a seaway in southern New Mexico, extending northward only into the southern half of the San Juan Basin. It is approximately equivalent to the Toroweap Formation of the Grand Canyon and is not present, at least in this marine aspect, elsewhere to the north on the Colorado Plateau. Above the Yeso Formation a light-colored waterlaid (probably marine) sandstone unit, the Glorieta Sandstone, is the lateral equivalent of the Coconino Sandstone of Arizona. Like the upper Yeso Formation, the Glorieta extends only across the southern half of the San Juan Basin. The youngest rocks of Permian age are marine limestone deposits of the San Andres Formation. The San Andres is in every way the equivalent of the Kaibab Limestone of Arizona. Its northern shoreline is in the subsurface immediately north of the Zuni Mountains, where marine limestone grades abruptly into red beds, and is indistinguishable throughout most of the San Juan Basin. These relationships are shown in Figure 49.

The Lower Triassic Moenkopi Formation pinches out along the western flank of the Defiance Uplift, but a possible extension, called the "Moenkopi Formation(?)," is found only along the southern flank of the San Juan Basin south of Interstate 40. Little or none of the red beds extend northward into the basin. The overlying Chinle Formation, in contrast, is thick and well developed across the entire San Juan Basin. An erosional surface everywhere underlies the Upper Triassic Chinle Formation. The lower sandstone and conglomerate member of the Chinle, the Shinarump Member, is found only in sporadic exposures along the southernmost San Juan Basin. The probable stratigraphic equivalent, called the Agua Zarca Sandstone, occurs along the eastern margin of the basin. Upper members of the Chinle Formation are widespread in northwestern New Mexico and include, in ascending order, the Monitor Butte, Petrified Forest, and Owl Rock. Rocks equivalent to the Petrified Forest Member extend across the San Juan Basin to the east but are there given local member names. To the north along the northern boundary of the basin, red beds of similar age as the Chinle Formation are called the Dolores Formation. The Dolores is somewhat darker red in color and is more coarse-grained than the typical Chinle Formation, but it is approximately the same age as the Chinle to the south and west.

The Wingate Sandstone of Early Jurassic age was named for exposures near Fort Wingate, New Mexico, in cliffs just north of Interstate 40. Later studies indicated that only the lowermost sandstone cliffs at Fort Wingate were in the Wingate Sandstone and that the upper cliffs were in the Entrada Sandstone. As all things seem to go, still later work showed that there is no Wingate Sandstone at Fort Wingate; the entire red, cross-bedded sandstone that forms the prominent cliffs is the Entrada Sandstone. So now we have no type section for the Wingate Sandstone, although the name is still used for the massive, cliff-forming sandstone that so dominates the scenery of the western and northern Colorado Plateau. The early definition of the cliff-forming sandstone was the Lukachukai Member of the Wingate, for exposures in the Lukachukai Mountains that lie just northwest of the Chuska Mountains, but any rocks with affinities to this member pinch out in northwesternmost New Mexico.

The Entrada Sandstone of Middle Jurassic age, however, is widespread across the San Juan Basin of northwest New Mexico. The prominent red cliffs of highly cross-bedded sandstone are apparently of a windblown origin and directly overlie the Chinle Formation.

Above the Entrada Sandstone a peculiar unit of basal limestone and upper gypsum is called the Todilto Formation. The Todilto is largely restricted to the San Juan Basin, although a limestone equivalent in the San Juan Mountains of southwestern Colorado is called the Pony Express Limestone Member of the Wanakah Formation. The gypsum beds of the Todilto are best displayed along the eastern margin of the San Juan Basin from about San Ysidro north to near Cuba, New Mexico. The formation is peculiar with regard to the interpretation of its depositional environment. No definitive fossils have been found in the Todilto to indicate whether the formation is a lake deposit or marine in origin. Fossil fish have been recovered from the Todilto, but, like salmon, they are species that are believed to live in the sea but spawn in fresh water; so they don't help. Ostracodes, tiny bivalve arthropods, found in the Todilto, suggest a nonmarine (lake) environment of deposition. The Morrison Formation, primarily stream-deposited sandstone similar to the Salt Wash and Westwater Canyon Members of latest Jurassic age, is widespread throughout the San Juan Basin and elsewhere on the Colorado Plateau.

Rocks of Late Cretaceous Age

The Cretaceous System is, without a doubt, the most stratigraphically complex sequence of sedimentary rocks on the Colorado Plateau. The stack of strata is preserved from erosion only in the deeper basins (the San Juan, Black Mesa, and Uinta Basins) and in the down-faulted blocks of the High Plateaus to the west. By Late Cretaceous time the main intercontinental seaway had been pushed eastward by massive thrust-faulting along the Sevier Orogenic Belt in western Utah. Major thrust faults formed uplands in the Basin and Range Province that blocked off the sea and shed great volumes of gravel, sand, and mud by means of streams flowing eastward into the shoreline of the Late Cretaceous sea.

The main part of the seaway lay to the east of the present-day Rocky Mountains in the Midcontinent, and as the earliest shoreline advanced across the Colorado Plateau, nearshore sand deposits of the Dakota Formation accumulated along the westward-advancing shoreline. The sediments consist of complex assortments of nearshore sandstone, some conglomerate, interbeds of mudstone, and in some cases coal. The Dakota usually forms low, light-colored cliffs and ridges where it is exposed at the surface.

Black mudstones of the Mancos Shale gradationally overlie the Dakota Formation and extend across the entire Colorado Plateau and on eastward into the main seaway. Thinner in the western Colorado Plateau where it is locally called the Tropic Shale, the Mancos Shale thickens to the east and northeast, where it attains thicknesses of more than 5,000 feet. The black to dark gray color of the shale is due to a high organic content, but oxidation near the weathered surface often reduces the organic material to a tawny, yellowish color. But dig in—the rock is black! That the Mancos Shale has a marine origin is attested by the presence of a wide variety of marine fossils, including shark teeth, ammonitic cephalopods, clams and oysters, and ubiquitous millions of microscopic foraminifera. Highways, airports, and towns are commonly built on the thick shale, as it forms broad plains and gentle slopes, easily bulldozed. But beware! These structures sink into the mire and sometimes nearly disappear because the "gumbo," and its high swelling clay content, shrink and swell impressively. And don't get caught on a dirt road on the Mancos Shale during the rainy season; you might lose your car!

All was not peace and quiet in these Late Cretaceous seas, for sea level rose and fell wildly as shorelines fluctuated with time. Wherever and whenever a shoreline was present, nearshore beach and offshore bar sand

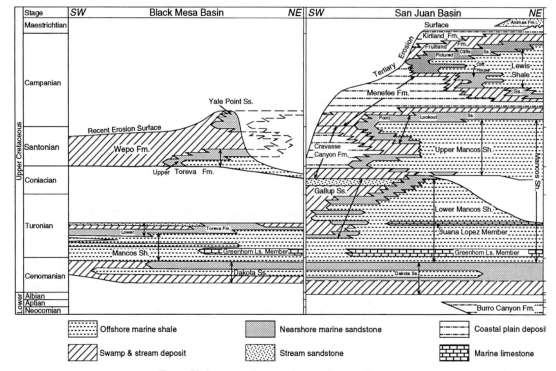

Figure 50. A time-rock cross section showing the complexity of formation names used for rocks of Cretaceous age in the San Juan Basin *(right)*, and the names and comparative ages of similar rocks in the Black Mesa Basin *(left)*. Time lines extend horizontally across the entire illustration. The strange names in the left-hand column are stages in the Upper Cretaceous Series that were derived from European terminology but are used pretty much worldwide. Modified from Peterson and Kirk 1977.

deposits accumulated. The resulting tongues of sandstone migrated to and fro across the countryside in a willy-nilly fashion, causing great complexities in the stratigraphic record—and causing myriad names to be applied to each variation that fluctuated in time and space. Nowhere is this more evident than in the San Juan Basin (Figure 50). Dozens of geologists have spent their lifetimes making sense of the various fluctuations, as the sandstone bodies often contain oil and natural gas. Indeed, the San Juan Basin contains the second largest gas field in the United States.

The history of local Cretaceous stratigraphic terminology began in and near Mesa Verde National Park along the northern margin of the San Juan Basin in southwestern Colorado. There thick sandstone and coal-bearing formations occur above the Mancos Shale and are collectively called the Mesaverde Group. The lowermost of the massive sandstone beds is the Point Lookout Sandstone, which consists of nearshore sandy shoreface and shallow marine deposits that formed as the shoreline at this locality withdrew toward the northeast. Behind the shoreline, marshes and coal swamps followed the migrating beach to form the Menefee Formation. The shoreline once again advanced toward the southwest, and another shoreface sandstone deposit, the Cliff House Sandstone, was deposited above the Menefee swamps. The term *Mesaverde Group* is used pretty much all across the Colorado Plateau, although the age varies considerably from basin to basin. Consequently, the sandstone units are given different formation names in each of the basins where Cretaceous rocks are found (see Figure 50). But this is only the simplest part of the story.

When the study of exposures of the Cretaceous formations was extended along the margins of the San Juan Basin, it was soon realized that there were complexities in the overall patterns. For example, it was discovered that shoreline fluctuations did not begin or end with the classic formations of the Mesaverde. Beach and bar sandstone beds began their complex fluctuations in the southwestern San Juan Basin, with the several beds of the Gallup Sandstone, and waxed and waned through time as indicated by numerous tongues of sandstone within the Mancos Shale. And the complexities worked their way up through the entire Late Cretaceous section (see Figure 50). I was never forced to struggle with this stratigraphic mess, and I certainly don't consider myself an expert on Cretaceous stratigraphy. This discussion, and the diagrams of others, will suffice to set the stage for travels through the San Juan and other basins by the nonspecialist.

Tertiary Deposits   As the San Juan Mountains of southwestern Colorado began to rise in latest Cretaceous to Early Tertiary time (the Laramide Orogeny), erosion attacked the uplands, and sediments began to flow with the streams away from the mountains and into adjoining basins. Coarse-grained sand and pebble deposits accumulated near the northern margin of the San Juan Basin, generally between Durango, Colorado, and Aztec, New Mexico. Sediments were transported farther into the basin by slowing streams, and fine-grained siltstone and mudstone dominates the Tertiary section. The central parts of the basin are now buried in thick lake and stream deposits, mostly of rather dull but interesting colors. The broad badlands thus formed are somewhat bleak with regard to vegetative cover but are dotted with oil and gas wellheads, pipeline facilities, and separator/refinery decorations. Formation names vary considerably from one geographic area to another, as will be sorted out in the accompanying roadlogs.

18. Gallup, New Mexico, to Monticello, Utah, via U.S. 666

Gallup, New Mexico, is in the southwest corner of the San Juan Basin. The town was named in 1882 for the paymaster of the Atlantic and Pacific Railroad, later known as the Santa Fe. Formerly a coal distribution center, Gallup is now known as the Indian Capital of the Southwest. Gallup is nestled between the northwesterly plunging nose of the Zuni Uplift and the southern termination of the Defiance Uplift. The structural depression between the two basement uplifts is known as the Gallup sag. Because the town lies in a structural depression, it is located high in the Cretaceous section, and the stratigraphy is rather complex, comprising interbeds of sandstone and dark-colored shale with numerous intervening coal beds. U.S. 666 follows the western margin of the San Juan Basin and is in the Upper Cretaceous section throughout the length of the highway.

Mile 0.0: Intersection of U.S. 666 and Jefferson Street in north Gallup, a short distance north of Interstate 40. Head north on 666 (Figure 51).

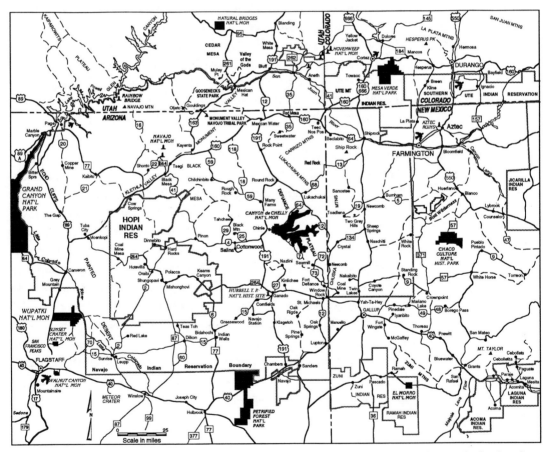

Figure 51. Road map of Navajo Country showing most roads tying the San Juan Basin to the Black Mesa Basin to the west.

Figure 52. Window Rock, eroded from the Cow Springs Member of the Entrada Sandstone, is near the Navajo tribal headquarters in Window Rock, Arizona.

Mile 1.0: Roadcuts in the Point Lookout Sandstone (Upper Cretaceous) are right and left for the next 4 miles. The view to the left at **Mile 4.5** is of the Defiance Uplift with the southern Chuska Mountains ahead. The highway is here in the Menefee Formation.

Mile 6.7: Intersection of U.S. 666 and New Mexico 264, which goes to Window Rock, Arizona, at Yah Ta Hey (a Navajo expression meaning "welcome" or "hello," literally "it is beautiful"). Turn left to Window Rock for a short detour. **Mile 15.0:** Back-filling from coal mining in the Menefee Formation and Crevasse Canyon Formation (Late Cretaceous) is on the right and left. Mining equipment is visible in the distance to the north. The Dakota Formation hogback is visible on the left and right at **Mile 21.8.** The steeply dipping Dakota marks the western edge of the San Juan Basin in this area.

Mile 22.5: Entering Window Rock, Arizona. The Arizona–New Mexico state line is at **Mile 22.9.**

Exposures of the nearly white Cow Springs Sandstone (Jurassic) are on the right at **Mile 23.1.** Window Rock, the capital of the Navajo Nation, is located on the southeastern margin of the Defiance Uplift. The natural arch in the Cow Springs Sandstone on the east edge of town is called Window Rock (Figure 52). Return to U.S. 666.

Restart your odometer at **0.0** at Yah Ta Hey and head north on U.S. 666.

Mile 0.7: Exposures of sandstone in the Menefee Formation are seen left and right. From here to **Mile 55** the highway is in the Menefee Formation (Mesaverde Group, Late Cretaceous), consisting of interbedded thin sandstone beds within dark-colored, organic-rich shale. The Menefee represents

Figure 53. Narbona Pass, recently renamed from Washington Pass, in the Chuska Mountains along the Arizona–New Mexico border. The pass is the notch to the left, the site of numerous skirmishes between the Navajos and Spaniards, Mexicans, and Americans in the late 1700s and early 1800s. The rocks above and to the right are volcanic rocks of Tertiary age.

nearshore stream and swamp deposits, with a few coal beds, that formed following the easterly retreat of the Late Cretaceous seaway from the Four Corners region. The formation overlies the Point Lookout Sandstone and underlies the Cliff House Sandstone, both nearshore marine sandstone deposits of the Mesaverde Group. The highland ahead at **Mile 2.5** is the south end of the Chuska Mountains, consisting largely of Tertiary volcanic rocks.

Mile 4.7: Milepost 13. The road to the right at **Mile 6.8** goes to Crownpoint, New Mexico. **Mile 7.7: Milepost 16.** The view ahead at **Mile 9.3** is of the south end of the Chuska Mountains; the rocks of the Menefee Formation, capped high on the left by cliffs of the Tertiary Chuska Sandstone, obviously dip into the San Juan Basin. The road to Tohatchi, meaning "water is dug out with one's hand" in Navajo, is to the left at **Mile 15.2.** Dark gray shale exposures with thin sandstone interbeds of the Menefee Formation make up the country. For the next 7 miles, Cretaceous beds are dipping strongly toward the east into the San Juan Basin with considerable landslide debris from the Chuska Mountains on hillsides to the left (west).

Mile 28.8: Entering San Juan County, New Mexico. Naschitti, a Navajo word meaning "badger," is on the left at **Mile 32.8**. The road to the left at **Mile 39.1** goes through Narbona Pass (formerly known as Washington Pass) in the Chuska Mountains to the villages of Crystal and Navajo (Figure 53). The high country to the left is the Chuska Mountains, which consist of Tertiary-age Chuska Sandstone capped by volcanic rocks.

Mile 41.9: Milepost 50: Prominent volcanic necks are visible ahead and to the left. The Chuska Mountains form the high skyline to the left. The Beautiful Mountain anticline is visible on the left at 10:00 at **Mile 44.5**. A road to Two Gray Hills is on the left at **Mile 48.7**. The Chuska Mountains are visible at 9:00, the Lukachukai Mountains at 10:00, and the Carrizo Mountains at 10:30 to 11:00 at **Mile 49.5**. The road to Toadlena is on the left at **Mile 52.3**; the road to the right goes to Burnham. Prominent volcanic necks are visible ahead on the left and right. The prominent volcanic neck on the left at **Mile 54.6** is Bennet Peak, and the smaller one on the right at 2:00 is Ford Butte. These are both diatremes, explosive volcanic vents.

Mile 55.0: Milepost 63.0: The highway crosses down through the Point Lookout Sandstone, the basal formation of the Mesaverde Group (Upper Cretaceous). The highway is on the Upper Mancos Shale at **Mile 56.0**; the Point Lookout Sandstone forms the prominent hogback to the right along the Hogback monocline. The highway is now on the Four Corners Platform, a relatively flat structural area between the Hogback monocline and San Juan Basin to the southeast and the Paradox basin to the north.

Mile 57.3: Ship Rock, the granddaddy of diatremes, is on the left at about 10:30 (Figure 54).

The Carrizo Mountains are visible on the left at **Mile 57.6**. The highway passes under power lines. The road to the left at **Mile 57.9** goes to Tocito. **Mile 58.5**: The Carrizo Mountains are at 9:00, the Lukachukai Mountains are at 8:30, and the Chuska Mountains are at 7:00 to 9:00. The road to the left at **Mile 59.0** goes to Sanostee. **Mile 60.2**: Little Water Trading Post. Table Mesa, at **Mile 63.0**, consisting of the Upper Mancos Shale, capped by the Point Lookout Sandstone, is visible ahead. Ship Rock is obvious at 10:30.

Mile 63.6: The Hogback monocline is visible at 2:00; the Mancos Shale is capped by the Point Lookout Sandstone. A small volcanic neck lies between Table Mesa and the Hogback. A view of Ship Rock and the southern dike is to the left at **Mile 67.7**. The Carrizo Mountains are visible in the distance to the left; Table Mesa is ahead with a small volcanic neck to the right, and the Hogback monocline is at about 2:00 to 3:00. Table Mesa is on the left at **Mile**

Figure 54. Aerial view of Ship Rock in the northwest corner of New Mexico, showing the three radiating dikes. This is the neck of an explosive, gaseous igneous eruption, called a *diatreme*, that occurred in about mid-Tertiary time. The surrounding plains are in the Mancos Shale of Late Cretaceous age, and the Chuska Mountains are in the distance.

69.9 with another small volcanic neck ahead at 11:00. A dike from the second small neck is on the left at **Mile 70.3.** There is a nice view at **Mile 74.5** of Ship Rock at 9:00. The road to the left at **Mile 77.1** goes to Red Rock Valley. The Carrizo Mountains are on the skyline to the left.

The road to the right at **Mile 80.0** goes to the Hogback Oilfield. The discovery well was completed in 1922, when oil was pumped from the Dakota Sandstone at a rate of 375 barrels a day from a depth of 776 to 696 feet. Deeper wells were drilled in the field in 1952, producing lesser amounts of oil and gas from Mississippian and Pennsylvanian rocks. Shiprock, New Mexico, is visible ahead and to the right at **Mile 81.4.**

Mile 83.3: The road to the left goes to the Rattlesnake Oilfield, the Four Corners Monument, and Mexican Water and Kayenta, Arizona. A bridge over the San Juan River is at **Mile 83.6. Mile 84.2**: Intersection with U.S. 64, which heads to the right (east) to Shiprock and Farmington, New Mexico;

continue left on U.S. 666. Leaving Shiprock at **Mile 84.6.** Exposures on the right are Mancos Shale, covered by terrace gravels of the San Juan River. **Mile 86.4: Milepost 94.** The Ute Mountains, a laccolithic range, are visible ahead on the skyline at **Mile 86.7.**

The Ute Mountains are ahead at **Mile 90.2**, and Mesa Verde is on the right at 2:00. The rock formations of Mesa Verde consist of lower gray slopes of the Mancos Shale, overlain by cliffs of the Point Lookout Sandstone and ledgy slopes of the Menefee Formation, capped by the Cliff House Sandstone; this is the type section of the Mesaverde Group of Late Cretaceous age. **Mile 92.4**: A small volcanic vent is on left at 9:00. **Mile 95.5: Milepost 103** is in broad exposures of the Mancos Shale. Buttes in the vicinity of **Mile 96.7** are the Mancos Shale capped by the Point Lookout Sandstone. **Mile 98.6**: Chimney Rock, visible at 2:00, is the Mancos Shale capped by the Point Lookout Sandstone.

Mile 99.8: Entering Colorado, Montezuma County. **Mile 100.8: Milepost 1.** The highway crosses the Mancos River at **Mile 101.7.** There is a good view at **Mile 102.7** to the right up the Mancos River valley of the entire Mancos Shale and Mesaverde Group; the Ute Mountains (Sleeping Ute) are ahead. Chimney Rock is on the right at **Mile 105.0.**

Mile 106.2: Junction with U.S. 160, which goes left to Four Corners Monument; Aneth and Bluff, Utah; and Kayenta, Arizona (see Roadlog 27). The road to the left at **Mile 113.0** goes to Towaoc, the Ute Mountain Ute tribal headquarters.

Another road to Towaoc is at **Mile 114.5.** Ute Mountain Casino is on the left at **Mile 114.7.** The entire Mesaverde Group is well exposed to the right at **Mile 115.5.** This is the northwest corner of Mesa Verde and the San Juan Basin. **Mile 120.0**: Lone Cone is visible ahead at 12:00 and the La Plata Mountains at 2:00. The road left at **Mile 123.3** goes to McElmo Canyon and Hovenweep National Monument. The highway crosses McElmo Creek at **Mile 123.9.**

Mile 124.0: Cortez, Colorado, city limits. Intersection at **Mile 126.0** with U.S. 160 going east (right) to downtown Cortez and Durango (see Roadlog 21). Continue north on U.S. 666 toward Dove Creek, Colorado. Leaving Cortez at **Mile 127.6.**

Dakota Sandstone exposures are ahead on the right. From Cortez to Monticello the highway is at or just above the Dakota Formation in the basal Mancos Shale, with occasional exposures of Dakota in roadcuts and stream gullies.

Mile 133.4: Entering Arriola. The road to the right at **Mile 136.5** goes to Dolores. Red soils in the vicinity of **Mile 140.0** are windblown dust from the Monument Valley region, providing excellent conditions for growing pinto beans in this area. The road to the left at **Mile 140.9** goes to Yellow Jacket. **Mile 145.6**: Entering Pleasant View. The Dolores anticline is visible to the right. The La Sal Mountains are visible in the distance ahead; the Abajo Mountains are to the left. Entering Dolores County at **Mile 149.4**. The Dolores anticline on the right is capped by the Dakota Sandstone. The highway crosses Cajon Creek at **Mile 150.6** with Dakota exposures in the gully.

Mile 151.2: Entering Cajon. The Abajo Mountains, known locally as the Blue Mountains, are on the left at **Mile 156.0. Mile 160.5**: Entering Dove Creek, the "pinto bean capital of the world," elevation 6,843 feet. **Mile 161.7**: Leaving Dove Creek. The Abajo Mountains form the backdrop. From Dove Creek to Monticello the highway is on or about the Dakota-Mancos contact.

Mile 163.0: The road to the right goes to Egnar. The highway crosses Coalbed Creek at **Mile 168.3**. Entering Utah, San Juan County, at **Mile 169.4**. The road to the right at **Mile 170.7** goes to Ucolo. **Mile 184.8**: Entering Monticello, Utah. The intersection of U.S. 666 and U.S. 191 is at **Mile 186.6**. Turn right to Moab, Utah (Roadlog 8) or left to Blanding, Utah (Roadlog 12).

19. Albuquerque to Bloomfield, New Mexico, via U.S. 550

The rocks in the valley floor north of Albuquerque are totally obscured by talus from the Sandia Mountains and river terrace gravels. The route crosses the "rift valley" from Bernalillo to about San Ysidro and then crosses north-south faults in the southern Nacimiento Range that expose rocks of Permian through Mesozoic age. Near Cuba, New Mexico, the road crosses up-section through Cretaceous marine rocks of the Mancos Shale and Mesaverde Group and continues up into rocks of Tertiary age that extend across the eastern San Juan Basin from Cuba to Bloomfield, New Mexico. **Mile 0.0**: Intersection of San Mateo Boulevard and Interstate 25 in Albuquerque. Head north on I-25 toward Santa Fe with the Sandia Mountains to the right along the eastern boundary of the Rio Grande Rift. The Sandia Mountains (*sandia* means "watermelon" in Spanish) consist of a thick core of Precambrian metamorphic rocks with a thin "rind" of Mississippian and Pennsylvanian rocks at the high rim. The west face of the mountain is a major fault zone at the edge of the Rio Grande Rift.

Mile 4.6: Milepost 235 on I-25. The highway is on a field of outwash gravels. **Mile 9.6: Milepost 240. Mile 10.0**: Bernalillo city limits. **Mile 11.6**: Exit 242, to Placitas via New Mexico 165 to the east, and to Bernalillo, Cuba, and Farmington via U.S. 550. Take exit and turn left onto U.S. 550. Entering Bernalillo. **Mile 12.8: Milepost 1**. Bridge across the Rio Grande at **Mile 13.3**. The road to Coronado State Monument is to the right at **Mile 13.6**.

Mile 16.8: Milepost 5: The countryside is covered with sand and dirt, sagebrush and weeds. Erosional features to the right at **Mile 18.4** exhibit Pleistocene gravels. **Mile 19.4**: Entering Santa Ana Indian Reservation. Lava-capped Tertiary sediments are to the right, and the Nacimiento Mountains are ahead at about 11:00. The valley floor is mainly sand. **Mile 25.2**: Entering the Zia Indian Reservation.

Mile 26.9: Milepost 15: More sand and gravel everywhere. The Nacimiento Mountains are very prominent ahead at **Mile 27.7.** Zia Pueblo is to the right at **Mile 29.8.** The white cliffs ahead are in the Jurassic Todilto Formation, here largely varved (laminated) gypsum.

Mile 31.3: The road crosses the north-south Jemez fault, marking the western boundary of the Rio Grande Rift. To the right at about 2:00 is a ridge of upturned Paleozoic rocks, consisting of the thin San Andres Formation, the cliff-forming Glorieta Sandstone, the red Yeso Formation below, and the dark red Abo Formation at the base of the cliff, all of Permian age. The ridge forms the east edge of a flat-bottomed syncline formed by drag between two north-south faults. The peaks in the distance are in the Jemez Mountains and the Valles Caldera, a Tertiary volcano. **Mile 31.9: Milepost 20.** The road is along the Jemez fault at **Mile 33.0.** The Todilto Formation forms the skyline to the left above the pink Entrada Sandstone, with the upturned ridge of Permian rocks ahead at about 2:00.

Mile 34.7: Crossing the Rio Salado and entering San Ysidro. The intersection of New Mexico 4 and U.S. 550 is at **Mile 35.5.** Highway 4 goes north to Jemez Pueblo and eventually to the Valles Caldera (recently designated a national preserve) and Los Alamos. **Mile 35.9**: Leaving San Ysidro. Exposures of the Moenkopi Formation, capped by the Agua Zarca Member of the Chinle Formation (Shinarump equivalent), are on the right, and cliffs of the Todilto Formation, largely gypsum deposits, are on the left. The road is in the middle of the flat-bottomed, fault-bound syncline at **Mile 36.7.**

Mile 37.0: Milepost 25: Red beds of Triassic age are exposed in the structural lows of the flat-bottomed syncline. Rocks of Permian age on the right at **Mile 38.0** are dipping strongly eastward toward the flat-bottomed syncline. Exposures of limestone in the San Andres Formation extending upward into the Moenkopi red beds are on the right at **Mile 38.6.**

Mile 39.6: The road crosses the Nacimiento fault, a high-angle reverse fault, with Permian rocks on the right and Triassic rocks on the left. Gypsum beds of the Todilto Formation are highly contorted along the fault. The Todilto exposures on the left at **Mile 41.2** are rolling over into the fault and then rising in the distance to about halfway up the cliff, where it is faulted against the red Abo and Yeso Formations. **Mile 41.7**: The road crosses exposures of collapsed gypsum of the Todilto Formation. **Mile 42.0: Milepost 30**: The road runs along a valley in red beds of Triassic age (Chinle Formation).

The cliffs on the left at **Mile 43.3** are capped by gypsum of the Todilto Formation, cliffs of the Entrada Sandstone, and lower slopes of the Chinle Formation and Agua Zarca Member, with Moenkopi red beds at the base. The Nacimiento fault block is to the right at **Mile 44.2**; dark gray rocks of Precambrian age are exposed below the Paleozoic section and rise to the north toward San Pedro Mountain, which is visible on the skyline at about 2:00. **Mile 47.0: Milepost 35**: The Todilto Formation forms the white escarpment ahead and to the right and left. San Pedro Mountain is visible in the background. Rocks of Pennsylvanian age thin northward above the Precambrian basement from about 800 feet thick near Jemez Pueblo to zero on the south flank of San Pedro Mountain. The regional northward pinchout of Pennsylvanian rocks is by onlap onto basement rocks, indicating that San Pedro Mountain was a topographic highland in Pennsylvanian time.

Mile 48.7: The site of the abandoned Warm Springs Resort is to the right. Rocks of Mississippian and Early Pennsylvanian age were found here to pinch out abruptly across the fault, indicating that the Nacimiento fault zone had movement in late Paleozoic time as well as later during Laramide (Late Cretaceous to Early Tertiary) time. ("Old faults never die . . .") The road now crosses upward over the ridge of Todilto gypsum.

The view ahead at **Mile 49.8** is across the eastern margin of the San Juan Basin. Exposures of the Morrison Formation and Dakota Sandstone are to the right at **Mile 50.1**.

The road at **Mile 50.5** crosses the Dakota Sandstone outcrops. **Mile 51.0: Milepost 39**: The road is now in the basal part of the Mancos Shale (Cretaceous). San Pedro Mountain is obvious to the right at **Mile 53.0**. Holy Ghost Springs Recreational Area is to the right at **Mile 54.5**. The gray rolling hills around the countryside at **Mile 55.3** are in the Mancos Shale (Cretaceous). **Mile 57.0: Milepost 45.** Sandstone beds of the Mesaverde Group cap the plateaus on the right and high on the left at **Mile 58.8**. The road to the left at **Mile 60.1** goes to Dragonfly Recreation Area. The intermediate-level sandstone ahead and to the right is the Gallup Sandstone (Cretaceous).

Mile 62.0: Milepost 50: The Gallup Sandstone is above the highway to the right and left. Roadcuts are in the Gallup Sandstone at **Mile 62.6**, and then Mesaverde sandstone beds cap the bluffs to the right and left above the Mulatto Tongue of the Mancos Shale. **Mile 64.0: Milepost 52**: Sandstone beds of the Point Lookout Sandstone of the Mesaverde Group are above road level in all directions. The road at **Mile 65.3** is the Point Lookout Sandstone in the lower Mesaverde. **Mile 66.6**: The road crosses through the Cliff House Sandstone of the Mesaverde Group. The middle Mesaverde Menefee Formation is pretty well obscured by cover in this area. **Mile 67.1: Milepost 55.**

Exposures of the nearly black Lewis Shale are to the right and left at **Mile 68.0.** The sandstone capping the mesa to the left is the Pictured Cliffs Sandstone, exposed low to the left and right at **Mile 70.3.** The high country ahead is in Tertiary-age sedimentary rocks. **Mile 72.0: Milepost 60**: The sandstone exposures to the right and left are in the Ojo Alamo Sandstone, the oldest rocks of Tertiary age (Paleocene) in the San Juan Basin. Sedimentary rocks of the Nacimiento Formation (Paleocene) are exposed on the left, and San Pedro Mountain is to the right.

Mile 75.6: Entering Cuba, New Mexico. A Santa Fe National Forest road is to the right. Crossing the Rio Puerco at **Mile 76.8.**

Mile 77.1: Milepost 65: The road to the right goes to the San Pedro Mountains Wilderness Area on New Mexico 126. The bluffs ahead are in the Cuba Mesa Member of the San Jose Formation (Eocene). **Mile 78.1: Milepost 66**: The road is in the Cuba Mesa Member sandstones. The road at **Mile 78.7** passes upward through the first sandstone of the Cuba Mesa Member into the second of three Cuba Mesa sandstone beds. The road crosses San Jose Creek at **Mile 79.7.** At **Mile 80.3** New Mexico 96 goes right to Gallina, New Mexico.

Mile 80.6: The road passes up-section through the upper Cuba Mesa Member into shales of the Regina Member of the San Jose Formation (Eocene). **Mile 82.1: Milepost 70**: The road is in varicolored mudstones of the Regina Member of the San Jose Formation. Soft yellowish-brown sandstones exposed at **Mile 83.8** are stream-channel deposits in the Regina Member. **Mile 85.8**: The road crosses into light gray, mottled pink mudstones of the Regina Member of the San Jose Formation. **Mile 87.1: Milepost 75.** A white channel sandstone in the Regina Member is visible to the right at **Mile 87.7.**

Mile 88.9: The Continental Divide, elevation 7,275 feet. The highway is in the light-colored channel sandstone of the Regina Member. Entering the Jicarilla Apache Reservation at **Mile 90.1.** The road is now crossing the "Largo Plains" region of the central San Juan Basin. **Mile 92.2: Milepost 80.** Banded mudstone at **Mile 95.0**, capped by the channel sandstone of the Regina Member, is to the left. **Mile 97.2: Milepost 85.** The road to the right at **Mile 97.8** goes to a rest area and Lindrith, New Mexico. **Mile 102.2: Milepost 90.**

Channel sandstones and mudstones of the Regina Member form buttes to the right at **Mile 106.7.** Leaving the Jicarilla Apache Reservation at **Mile 107.2: Milepost 95**. Buttes composed of banded mudstone and sandstone in the Regina Member are to the right and left at **Mile 108.6.**

Mile 109.5: Entering Counselors Trading Post and School, established in 1930 by Jim Counselor.

A view into the Canyon Largo drainage is ahead at **Mile 109.8.** The region around Canyon Largo was the traditional homeland of the Navajo Nation, called Dinetah. Many topographic features in the region are considered holy by Navajos.

Mile 111.3: Entering Rio Arriba County from Sandoval County. Roadcuts are in the Cuba Mesa Member of the San Jose Formation. Roadcuts to the right and left at **Mile 112.3** are in a channel sandstone in the Regina Member. The view to the left at **Mile 114.0,** toward the radio towers, is of the Cuba Mesa Member in the lower bluffs, overlain by mudstones of the Regina Member of the San Jose Formation.

Mile 115.1: Lybrook Plant gas refinery for the Lybrook Oilfield, discovered in 1957. The field has produced more than 3 million barrels of oil from offshore sandbar deposits in the Tocito Sandstone (Cretaceous). Sandstone exposures to the right are in the Cuba Mesa Member. Escrito Trading Post is on the left at **Mile 116.8.** The road is on the Cuba Mesa Member for the next 2 miles.

Mile 117.5: Milepost 105: Light gray to purplish banded mudstones of the Nacimiento Formation (Paleocene) are exposed in the vicinity. The road is in the Nacimiento Formation from here to Bloomfield. This area has been called the Bisti Badlands. **Mile 120.1:** Entering San Juan County. **Mile 122.5: Milepost 110.** The road to the left at **Mile 125.4** goes to Chaco Canyon National Historic Park. **Mile 127.5: Milepost 115.**

Mile 128.0: Entering Nageezi, meaning "squash" in Navajo. **Mile 132.5: Milepost 120.**

The Blanco Trading Post is to the left at **Mile 136.0.** Huerfano (meaning "orphan" in Spanish) is on the right at **Mile 139.9. Mile 142.6: Milepost 130.** At **Mile 147.5** the La Plata Mountains, west of Durango, Colorado, are visible at about 2:00. **Mile 147.7: Milepost 135.** The road to the right at **Mile 149.4** goes to Angels Peak Recreation Area. **Mile 152.4:** To the right is the Navajo Irrigation Project, bringing water from Navajo Dam on the San Juan River. **Mile 152.7: Milepost 140.** The road at **Mile 157.7: Milepost 145** descends through badlands weathered in the Nacimiento Formation (Paleocene), nearly white mudstone terrain. Bloomfield and the San Juan River are visible low and ahead, with the La Plata (meaning "silver" in Spanish) Mountains in the distance ahead. The road crosses a wash; oilfield facilities are to the right and left.

Mile 162.7: Entering Bloomfield. **Mile 162.8: Milepost 150.** The Giant and Mustang petroleum refineries are to the right at **Mile 163.4.** A bridge over the San Juan River is at **Mile 163.6.** At **Mile 164.5** is the intersection of U.S. 64, going to the left to Farmington, and New Mexico 544, going ahead to Aztec, New Mexico. (Continue north on Roadlog 20.)

20. Bloomfield, New
Mexico, to Aztec,
New Mexico, and
Durango, Colorado,
on New Mexico 544
and U.S. 550

The route from Bloomfield, New Mexico, to Durango, Colorado, crosses the northern flank of the San Juan structural basin. It begins in the structurally deepest part of the basin and climbs onto the southern flank of the San Juan Mountains uplift. Rocks of Tertiary age at Bloomfield are gradually removed from the rim of the basin by erosion and nondeposition as the route progresses northward until the hogbacks of the Cretaceous Mesaverde Group are crossed immediately south of Durango. A short trip north of the town of Durango takes the traveler down the stratigraphic section through the Mesozoic and Paleozoic sequence of sedimentary rocks to the Precambrian basement in only some 20 miles. This is a journey back through nearly all of geologic time, one rock layer at a time.

Mile 0.0: Intersection of U.S. 550, New Mexico 544, and U.S. 64. Head north from Bloomfield toward Aztec on 544.

Mile 1.0: River terrace gravels are above the road to the right and left above the San Juan River. The San Juan Gas Plant is on the right at **Mile 1.5**. Nearly white mudstone of the Nacimiento Formation is visible. **Mile 5.0: Milepost 5**: The La Plata Mountains (laccolithic mountains of mid-Tertiary age) are ahead. Rolling hills are nearly white mudstone of the Nacimiento (Animas) Formation.

Mile 6.7: Welcome to Aztec, New Mexico, the site of Aztec Ruins National Monument, restored ruins of an Anasazi village, including a giant kiva. **Mile 8.3**: Intersection of New Mexico 544 and U.S. 550, going left to Farmington, New Mexico, and right, or north, to Durango, Colorado. Turn right toward Durango. The road to the right at **Mile 9.4**, New Mexico 173, goes to Navajo Dam on the San Juan River.

Mile 13.6: Milepost 19: The highway is in the lower Animas River valley, with the river to the left (west). The Animas River joins the San Juan River near Farmington. The Farmington Sandstone is exposed to the right at **Mile 14.7**. The Animas River is visible on the left at **Mile 15.3**. The Spanish name *Río de las Animas Perdidas* means "river of lost souls." **Mile 15.6: Milepost 21**.

The high cliffs ahead at **Mile 17.5** are in the San Jose Formation of Early Tertiary age. **Mile 18.0**: Bridge over the Animas River. The village of Cedar Hill is at **Mile 18.3**. Terrace gravels of the Pleistocene Animas River are in roadcuts at **Mile 18.8**. Sandstone of the Tertiary San Jose Formation is exposed in cliffs to the right. **Mile 19.6: Milepost 25**: Tertiary-age stream deposits of the San Jose Formation are exposed to the left and form the canyon walls from here to Bondad Hill.

Mile 22.9: Entering Colorado. At **Mile 23.5** entering Southern Ute Indian lands. The "reservation" comprises scattered parcels of Indian land interspersed with homestead lands. **Mile 24.0: Colorado Milepost 1**.

The Animas River is on the right at **Mile 25.0.** The Bondad Store is on the right at **Mile 26.2.** A bridge over the Animas River is at **Mile 26.6.** The highway at **Mile 26.9** is on terrace gravels from the early Animas River drainage. The road to the right at **Mile 27.4** goes to Ignacio, Colorado, the tribal headquarters for the Southern Ute Indian Reservation. At **Mile 27.7** the highway ascends Bondad Hill.

Mile 28.0: Milepost 5: The highway climbs onto Florida Mesa, a high terrace gravel of the Animas River formed during the first Pleistocene glaciation and resting for the most part on the Animas Formation (Early Tertiary). **Mile 29.0:** The La Plata Mountains are obvious ahead and to the left. Mount Hesperus, the highest peak in the La Platas, is the Navajo holy mountain of the north. **Mile 30.6:** The San Juan Mountains are visible to the right at about 2:00; the La Platas are to the left. **Mile 32.9: Milepost 10.**

The road to the right at **Mile 35.0** goes to Ignacio, Colorado, and the Durango–La Plata County Airport. **Mile 35.8:** Hogbacks formed on sandstone beds of the Mesaverde Group are obvious ahead and to the left at the northern edge of the San Juan Basin. **Mile 37.9: Milepost 15.**

Mile 38.5: The highway descends from Florida Mesa to the intersection of U.S. 160 at **Mile 39.4** through exposures of the Animas Formation. U.S. 160 to the right goes to Bayfield and Pagosa Springs, Colorado; to the left to Durango. Turn left toward Durango.

Mile 39.7: Durango city limits, elevation 6,512 feet. The strata ahead and to the left, dipping strongly toward the south into the San Juan Basin, are the purple beds of the McDermott Formation, an undated unit at the top of the Cretaceous or base of the Tertiary section, with the brownish-colored Animas Formation exposed to the left (south) of the McDermott. To the right, or north, of the purple beds are rather poor exposures of the Fruitland Formation (latest Cretaceous), containing black coal-bearing and carbonaceous beds, with a middle sandstone ridge of the Farmington Sandstone Member. The overlying Kirtland Shale contains no black beds. The next prominent sandstone hogback is the Pictured Cliffs Sandstone (Late Cretaceous), overlying dark gray shale slopes of the thick, black Lewis Shale. These hogbacks form the northern edge of the San Juan Basin and the southern edge of the San Juan dome (San Juan Mountains).

The large landslide to the left at **Mile 40.9**, across the Animas River, is at the top of the Pictured Cliffs Sandstone. It was no doubt triggered in historic time by fires in a coal mine in a thick coal seam that occurs in the Fruitland Formation just above the Pictured Cliffs ledge.

An exit to east Durango is at **Mile 41.1.** A bridge over the Animas River is at **Mile 41.2**, and to the right is the steeply dipping hogback of the

Figure 55. Aerial view of Durango, Colorado, looking toward the northwest. The gray slopes beyond the town and in the left center of the photo are in the marine rocks of the Mancos Shale. The first cliffs above the Mancos slopes are in the Point Lookout Sandstone of the Mesaverde Group, all Cretaceous in age. The low cliffs in the right center of the photo are in the Dakota Sandstone, dipping toward the south from the San Juan Mountains into the San Juan Basin. The La Plata Mountains, a laccolithic intrusive igneous range, form the far skyline. Igneous intrusive bodies were emplaced in Late Cretaceous time.

Pictured Cliffs Sandstone, with an exposure of the coal bed just above and the Lewis Shale below.

The industrial park in south Durango is at **Mile 41.6.** The infamous Animas–La Plata Project would pump water from the Animas River up 1,000 feet and over the high ridge to the left. The steeply dipping sandstone beds ahead and to the left are formations in the Mesaverde Group (Late Cretaceous): in ascending order, the Point Lookout, Menefee, and Cliff House Formations. The highway is along a fault here, with the Point Lookout Sandstone high on the left and low on the right along the river. The black beds on the right are coal and carbonaceous mudstone beds in the Menefee Formation.

Recrossing the Animas River at **Mile 42.9.** The proposed damsite for the Animas–La Plata Project is to the left here. **Mile 43.3**: The high peak ahead (Perins Peak) is capped by the Point Lookout Sandstone above the unstable black slopes of the Mancos Shale (both Late Cretaceous in age) (Figure 55).

Mile 43.4: Intersection of U.S. 550 and 160 in south Durango. A drive toward the north on U.S. 550 for some 15 miles allows one to view about 16,000 feet of stratigraphic section from Pleistocene glacial deposits down to granite of Precambrian age. Such a complete section, with few or no structural complications, is rarely found. For detailed descriptions of the rocks, see *The American Alps* (Baars 1992). (To continue west on U.S. 160 to Cortez, Colorado, see Roadlog 21.)

21. Durango to Cortez, Colorado, via U.S. 160 (including Mesa Verde National Park)

The highway from Durango to Cortez follows the northern border of the San Juan Basin, exposing rocks of Cretaceous age for the entire route. The strata dip toward the south into the San Juan Basin. From Durango to Mancos the road is mostly in the Mesaverde Group, and coal mines are noticeable along the way. The La Plata Mountains, igneous rocks intruded into sedimentary rocks of late Paleozoic to Mesozoic age, are prominent to the north. These are the Navajos' sacred mountain of the north.

Mile 0.0: Intersection of U.S. 160 and 550 on the south side of downtown Durango, Colorado. Turn west on U.S. 160 toward Cortez. Bridge over the Animas River. Exposures of the nearly black slopes of the Mancos Shale (Upper Cretaceous) are in all quadrants.

Perins Peak, to the right at **Mile 0.9**, consists of dark gray slopes of the Mancos Shale, capped by the Point Lookout Sandstone of the Mesaverde Group (Upper Cretaceous) near the crest of the Durango anticline. The mountain ahead at **Mile 1.5** is the Mancos Shale, with the Point Lookout Sandstone forming the intermediate cliff, the Menefee Formation forming the upper slope, and the Cliff House Sandstone forming the skyline cliff. **Mile 2.0**: The Wildcat Canyon road is to the left. **Mile 2.2: Milepost 81.**

Twin Buttes, on the right at **Mile 2.4,** are capped by the Point Lookout Sandstone. **Mile 3.3: Milepost 80.** Lightner Creek Road is to the right at **Mile 3.4.** At **Mile 4.3** the highway is climbing up-section through the Point Lookout Sandstone. The highway is in the Menefee Formation at **Mile 4.6.** A road to a coal mine producing from the Menefee is to the right. The highway at **Mile 5.1** climbs up-section through the Cliff House Sandstone. The highway is in the Lewis Shale (Late Cretaceous) at **Mile 6.0.** The Durango West housing development is on the right at **Mile 7.0.**

The La Plata Mountains, a laccolithic mountain range of Late Cretaceous age, are visible to the right at **Mile 7.9. Mile 8.3: Milepost 75.** The highway crosses a saddle at **Mile 9.8** and begins a descent. Hesperus Junction is at **Mile 10.6.** A ranch of the late author Louis L'Amour is located in the valley to the left. The road, Colorado 140, goes to Farmington, New Mexico. **Mile 10.7**: Crossing La Plata Creek. The road to the right at **Mile 11.0** goes into La Plata Canyon, an abandoned mining district. The Hesperus Ski Area is on the left at **Mile 11.7.** The rocks along the highway are thoroughly covered with boulders, soil, and brush. **Mile 13.3: Milepost 70.** The highway at **Mile 17.2** is back into the Mancos Shale. The peak to the left is in the Point Lookout Sandstone. **Mile 18.3: Milepost 65.** The peaks and ridges to the left at **Mile 19.8** are capped by the Point Lookout Sandstone.

The road crests at **Mile 22.0** and begins the descent of Mancos Hill in the Mancos Shale. Exposures of the Point Lookout Sandstone cap the ridges on the left. **Mile 23.3: Milepost 60:** The bottom of Mancos Hill. **Mile 24.2:** The topographic features Point Lookout and Mesa Verde are visible in the distance; Sleeping Ute Mountain, a laccolithic range, is visible in the far distance beyond Cortez.

Mile 26.4: Welcome to historic Mancos. The highway crosses the Mancos River at **Mile 26.6.** A matchstick factory on the left at **Mile 28.1** uses local aspen wood. There is a good view of the Mancos River Valley on the left at **Mile 29.3** and a good view of Point Lookout at **Mile 31.6. Mile 33.6: Milepost 50. Mile 34.6: Milepost 49:** Take the exit to Mesa Verde National Park. Stop sign at **Mile 34.7:** Turn left to Mesa Verde National Park and reset odometer to 0.0.

Mesa Verde National Park, Colorado

Mile 0.0: Intersection of U.S. 160 and Mesa Verde National Park. **Mile 0.1:** Entering Mesa Verde National Park. Point Lookout is very obvious ahead at **Mile 0.3**; the type section of the Point Lookout Sandstone of the Upper Cretaceous Mesaverde Group forms the cliffs above the Mancos Shale. The Point Lookout Sandstone, mostly shoreface deposits, was deposited in shoreline settings as the Late Cretaceous sea withdrew toward the northeast. **Mile 0.8:** Entry pay station; fee area.

Roadcuts are in the Mancos Shale at **Mile 2.1** with beautiful exposures of the very unstable Mancos Shale ahead. The road crosses a major landslide area in the Mancos Shale at **Mile 3.0.** The road is often closed by mud and rock slides here. There is a good view at **Mile 3.3** of the La Plata Mountains to the east. **Mile 3.5:** Viewpoint on the left. The road passes through a saddle in the Point Lookout Sandstone. This region of the park was heavily damaged by wildfires in 2000.

Morefield Campground is on the right at **Mile 4.0.** The valleys are in the Mancos Shale with ridges all around of Point Lookout Sandstone. **Mile 5.0:** Tunnel. A viewpoint to the right at **Mile 5.7** overlooks the Mancos Shale

Figure 56. Cliff Palace, an Anasazi village in Mesa Verde National Park, southwestern Colorado. The ruins occur in an alcove eroded in the Cliff House Sandstone of the Mesaverde Group, Late Cretaceous in age.

slopes, the valley below, and the town of Cortez and the Sleeping Ute Mountains to the left. The road at **Mile 6.5** is going stratigraphically upward through the Point Lookout Sandstone, as revealed in roadcuts on the right. The road crosses upward into the Menefee Formation at **Mile 7.4**. The Menefee consists of coastal carbonaceous swamp deposits, including coal, deposited when the Late Cretaceous shoreline lay some distance to the northeast. The road crosses upward at **Mile 8.4** into the Cliff House Sandstone, shoreline sand deposits of the next advancing Late Cretaceous sea, in roadcuts on the right.

Mile 9.2: The road is now on the mostly covered Cliff House Sandstone. The beds on the left at **Mile 9.6** are in the top of the Cliff House Sandstone. Views to the right are into the Mancos-Cortez valley. Roadcuts in the Cliff House Sandstone are on the right at **Mile 9.9**. Above the Cliff House Sandstone everything is covered with soil, rocks, and lots of brush. The crest of the hill is at **Mile 10.7** with a view to the south across Mesa Verde. The roadcuts on the left and exposures on the right at **Mile 12.8** look like Menefee Formation, with Point Lookout Sandstone exposed down the cliff. **Mile 14.3: Milepost 14.**

Mile 15.1: The parking lot to the right is for the Visitor Center and Far View Lodge. Bus transportation to Wetherill Mesa, open during the summer only, is based here. **Mile 17.4: Milepost 17.**

Mile 20.3: Stop sign. Go straight ahead to Spruce Tree House and the museum; parking is at **Mile 20.7,** elevation 7,000 feet. Spruce Tree House is behind the museum in a large alcove in the Cliff House Sandstone at the top of the Menefee Formation. **Mile 21.5**: Back at the stop sign at the top of the parking loop. Go straight ahead to Cliff Palace.

The road to the left at **Mile 21.9** goes to Cliff Palace (Figure 56). The Cliff Palace parking area is at **Mile 23.5.** Park Service guided tours are by reservation. Cliff Canyon is to the right at **Mile 24.0.** The Balcony House parking area is at **Mile 25.3**; visitation is by reservation. The Soda Canyon overlook is at **Mile 25.8.** The mesa-top loop is to the left at **Mile 27.5.** A pit house dated at A.D. 600 is at **Mile 28.7. Mile 29.0**: Square Tower House parking area. Oak Tree House, at **Mile 30.2,** is dated at A.D. 1250. **Mile 31.1**: Sun Temple parking area. Back to the "mousetrap" intersection at **Mile 32.7.**

Head back to the park entrance. Restart the roadlog for Durango to Cortez.

Mile 0.0: Intersection of Mesa Verde National Park entrance and U.S. 160. Turn westbound toward Cortez. The highway from here to Cortez is near the base of the Mancos Shale, with Mesaverde Group cliffs to the left; very little is to the right except occasional exposures of the Dakota Sandstone in stream gullies.

The Sleeping Ute Mountains, another laccolithic mountain range, are obvious ahead at **Mile 1.8. Mile 4.5: Milepost 44**: The rounded hill ahead and to the right (north) of the Sleeping Ute is McElmo Dome, the site of decades of production of carbon dioxide from the Mississippian Leadville Limestone for the production of dry ice, making it "the dry ice capital of the world." The highway crosses McElmo Creek at **Mile 5.2. Mile 5.6: Milepost 43.**

Mile 7.8: Entering Cortez, Colorado, elevation 6,200 feet. The road to the right at **Mile 8.3** is Colorado 144 to Dolores, Colorado. Turn right at **Mile 10.4** to U.S. 666.

Mile 10.7: Intersection of U.S. 160 and 666. (Join Roadlog 18, Gallup, New Mexico, to Monticello, Utah, on U.S. 666 at **Mile 124.**)

22. Flagstaff, Arizona, to Albuquerque, New Mexico, via Interstate 40 (including Petrified Forest National Park)

Interstate 40 through Arizona crosses a broad, rather bleak plain floored by the resistant Kaibab Limestone (Permian) with scattered exposures of red beds of the Triassic Moenkopi and Chinle Formations. Highlights along the way are Meteor Crater (Coon Butte), the Painted Desert, and Petrified Forest National Park. Near the New Mexico border the highway crosses the south-plunging nose of the Defiance Uplift in colorful exposures of Jurassic formations. At Gallup, New Mexico, the highway crosses the northwesterly plunging nose of the Zuni Uplift and follows along the north flank for many miles in rocks that are at or near the top of the San Andres (Kaibab

Figure 57. Numerous volcanic cinder cones such as this one are scattered about the landscape near Flagstaff, Arizona. They formed by volcanic eruptions that occurred only a few hundred years ago.

equivalent) Formation. Recent volcanic rocks floor the valley at the Malpais, below Mount Taylor, near El Malpais National Monument. From about Grants, New Mexico, the highway crosses several obscure faults and badly exposed terrain and drops down into Albuquerque across the fault zone of the Rio Grande Rift. I-40 crosses the southern reaches of the Black Mesa (Arizona) and San Juan (New Mexico) Basins and thus constitutes a tie between the geologic and geographic structural and physiographic features.

Mile 0.0: Easternmost on-ramp to I-40 in Flagstaff, heading east. A cinder cone to the left (north) at **Mile 0.5** (Figure 57) is being torn down for road metal. **Mile 0.8: Milepost 202.**

  Mile 1.1: Roadcuts are in the Kaibab Limestone (Permian) with more partially destroyed cinder cones to the left. **Mile 3.4**: Exit 204, to Walnut Canyon National Monument, 5 miles. Numerous small Indian ruins are under ledges in the Kaibab Limestone in Walnut Canyon. There are good exposures of the Kaibab and "Coconino Sandstone," which is here actually a light-colored, highly cross-bedded, windblown sandstone unit equivalent to the Toroweap Formation and indistinguishable from the real Coconino Sandstone of Grand Canyon fame. Roadcuts are in the Kaibab Limestone at **Mile 3.6. Mile 4.0: Milepost 205.**

  Roadcuts at **Mile 5.7** are in brown mudstone beds of the Moenkopi Formation (Lower Triassic) and then back into the more resistant Kaibab Limestone. **Mile 8.9: Milepost 210**: Deep roadcuts into the Kaibab Limestone with more cinder cones to the left that are under destruction for construction purposes. **Mile 13.9: Milepost 215**: The highway is on the Kaibab Limestone with still more cinder cones to the left.

Mile 17.3: Leaving Coconino National Forest. Crossing Padre Canyon at **Mile 17.7.**

Mile 19.0: Milepost 220. At **Mile 21.0** the Hopi Buttes are visible to the left at about 10:00 to 11:00. The buttes consist of a swarm of diatremes (explosive gaseous volcanic vents) of Middle Tertiary age with no apparent structural control on their geographic distribution. Volcanic peaks along the Mogollon Rim are visible in the distance to the right, forming the southern boundary of the Colorado Plateau Province. **Mile 24.0: Milepost 225**.

Mile 29.0: Milepost 230: The highway crosses Canyon Diablo, cut into the Kaibab Limestone. The exit goes to Two Guns. The highway is going up-section at **Mile 31.0** into the low red shale hills of the Moenkopi Formation.

Mile 32.0: Milepost 233: Meteor Crater is visible to the right in the distance. **Mile 32.6**: Exit 233, to Meteor Crater, 5.9 miles. The crater is a bona-fide, documented Recent meteorite impact feature. A *meteor* is a chunk of material sailing through space that enters the earth's atmosphere; a *meteorite* is such a chunk that strikes the earth. Properly named, this feature would be "Meteorite Crater." Low exposures of Moenkopi Formation occur along the highway. The Hopi Buttes are still visible off to the left. **Mile 34.0: Milepost 235** is just beyond a rest area.

Mile 34.8: There is a good view of the Hopi Buttes to the left (north). A railroad overpass is at **Mile 36.3.** The highway is still in the Moenkopi Formation. **Mile 39.0: Milepost 240**: The highway is still in the basal Moenkopi Formation. **Mile 44.1: Milepost 245**: To the left the Moenkopi hills are capped by the Shinarump Conglomerate Member of the Chinle Formation, stream-deposited sandstone and conglomerate deposited on an unconformity at the base of the Chinle Formation. Exit 245 goes to Leupp, 16.4 miles to the north. **Mile 44.8**: Elevation 5,000 feet. The Hopi Buttes are still visible to the left. **Mile 49.1: Milepost 250**: Small volcanic features are visible to the right and left.

Mile 50.2: Entering Winslow, Arizona. Exit 252 is the highway to Payson, Arizona. **Mile 52.4**: Exit 253 leads to the heart of "beautiful downtown Winslow." The Hopi Buttes are still obvious to the north. **Mile 54.1: Milepost 255**: East end of Winslow.

Mile 56.0: Milepost 257: The highway crosses the Little Colorado River, here very little.

Mile 59.1: Milepost 260: The Shinarump Member is exposed in the hills on the left. **Mile 62.5**: Exposures of Moenkopi Formation, capped by the Shinarump Member, are on the left. **Mile 64.1: Milepost 265**: The highway is in the Moenkopi Formation. **Mile 65.4**: The Shinarump is exposed in the low buttes to the left. **Mile 69.1: Milepost 270.**

Mile 73.4: Exit 270, to Joseph City, Arizona. **Mile 74.1: Milepost 275.** The

Cholla Power Plant is on the right at **Mile 76.5. Mile 77.4:** The Shinarump Member is exposed to the right and left. **Mile 79.2: Milepost 280:** Elevation 5,000 feet. The Moenkopi Formation is capped by the Shinarump Member in the low hills.

Mile 82.7: Holbrook, Arizona, city limits. The Shinarump Member is exposed in low hills. **Mile 84.1:** Exit for Arizona 77 to Snowflake and Show Low, Arizona. The highway crosses through the top of the Moenkopi Formation into the Shinarump Member in the town of Holbrook at **Mile 85.0. Mile 89.3: Milepost 290:** East edge of Holbrook.

Mile 91.9: Exit 292, Arizona 77 to Keams Canyon to the north. **Mile 94.3: Milepost 295.** The low hills to the left at **Mile 97.5** are in the mudstone members (Petrified Forest) of the Chinle Formation. **Mile 99.3: Milepost 300.** Exposures of the Shinarump Member are to the left at **Mile 100.0. Mile 104.3: Milepost 305.** The highway at **Mile 105.5** is back down-section to the top of the Shinarump Member. **Mile 106.7:** Entering Apache County. **Mile 109.4: Milepost 310.**

Mile 110.6: Take Exit 311 to Petrified Forest National Park. Reset odometer to 0.0.

Petrified Forest National Park

Mile 0.0: Entering Petrified Forest National Park. The road to the right at **Mile 0.1** goes to the Visitor Center.

Mile 0.8: Tiponi Point, an overview of the badlands of the Chinle Formation exposures, generally known as the Painted Desert. The flat-topped mesa to the left is capped by a lava flow.

Milepost 1. The road goes up onto the lava-capped mesa. **Mile 1.6:** Tawa Point, which offers a better view of the Painted Desert badlands from the

Figure 58. Broken fragments of petrified logs lie scattered about the countryside at Crystal Forest. The silicified wood, colored red by iron minerals in the silica, is weathering out of the much softer Petrified Forest Member of the Chinle Formation of Late Triassic age.

top of the lava-capped mesa. The Tawa viewpoint and trailhead are at **Mile 1.7. Mile 2.1**: Tewa Desert Inn and viewpoint. **Mile 2.3**: Chinde Point. **Mile 2.7**: A lava-capped butte is visible to the right in the middle distance. Pintado Point is at **Mile 2.9.**

Milepost 3: The road is still on the lava-capped mesa. The road descends the lava-capped mesa at **Milepost 4** to Nizhoni Point, another Painted Desert overlook. Whipple Point (a good Navajo name) is at **Mile 4.3. Mile 4.7**: Lacey Point.

Mile 6.0: Overpass across I-40. **Mile 10.4**: Overpass across the Burlington Northern Santa Fe Railroad line.

Mile 10.9: Bridge over the Rio Puerco. There is a turnout for Puerco Ruin at **Mile 11.0**. The road is now on top of the Shinarump Member of the Chinle Formation. The road ahead is climbing up on the dipslope of the Shinarump Member. The road to the right at **Mile 12.0** goes to Newspaper Rock.

Milepost 12: The road is on top of the Shinarump Member. **Mile 13.0**: Exposures of the lower shale member (Monitor Butte?) of the Chinle Formation are dusted by volcanic cinders. Chinle Formation badlands are well developed around **Mile 13.8**. The Tepees are at **Mile 14.0.**

Blue Mesa road, 3.5 miles, is to the left at **Mile 15.7.** At **Mile 17.7** the road crosses through a fluvial sandstone in the Chinle Formation. **Mile 17.9**: Agate Bridge road is to the left.

Mile 18.6: Jasper Forest is to the right. Crystal Forest is on the left at **Mile 20.4.** Petrified wood fragments and logs are scattered about profusely in the Petrified Forest Member of the Chinle Formation (Figure 58). **Mile 23.3**: Entering the Flattops.

Mile 26.0: The road to the right goes to the Visitor Center and Rainbow Forest Museum; to the left are the Long Logs (Figure 59). Return to I-40 and restart the roadlog from Flagstaff to Albuquerque.

Mile 0.0: Milepost 311: Intersection of Petrified Forest turnoff and I-40. Head east toward Gallup, New Mexico. **Mile 1.4: Milepost 313**: Exposures of the varicolored mudstone of the Chinle Formation are to the left.

Mile 8.5: Milepost 320: Miles and miles of sagebrush desert. **Mile 12.2**: Roadcuts are in the brown Moenkopi Formation. **Mile 14.1**: Exit 325, to Navajo, Arizona. **Mile 18.6: Milepost 330**: Low hills to the right and left are Moenkopi Formation capped by the Shinarump Member.

Mile 21.8: Exit 333, to Chambers, Arizona. Intersection with U.S. 191 north to Ganado, Arizona (see Roadlog 25).

Mile 28.0: Exit 339, to Sanders, Arizona. The road to the right, U.S. 191, goes to St. Johns, Arizona (see Roadlog 26). The rocks behind the village are the Moenkopi Formation, capped by the Shinarump Member. **Mile 28.7: Milepost 340.**

Mile 29.5: Entering the Navajo Indian Reservation. Roadcuts are in the Moenkopi Formation at **Mile 32.0**. At **Mile 32.5** roadcuts are in the Shinarump Member for some distance ahead. There are good exposures of the Shinarump Member in Querino Wash at **Mile 33.1**.

Mile 33.7: Milepost 345: Crossing Box Canyon; good exposures of the Shinarump Member. **Mile 38.7: Milepost 350**: Exposures of Moenkopi Formation, capped by the Shinarump Member, are to the left.

Mile 43.5: The highway is beginning to cross the southern plunging nose of the Defiance Uplift; the strata are dipping gently toward the south. The escarpment ahead from **Mile 43.6** is capped by the Dakota Sandstone (Cretaceous), overlying slopes of the Morrison Formation, cliffs of the Cow Springs/Entrada Sandstone, and lower red slopes of the "beds at Lupton," a local unit of uncertain affinities, all of Jurassic age.

Mile 46.1: Exit 357, to Lupton, Arizona, and Navajo Route 12 to Window Rock, Arizona. A steeply dipping hogback of the Shinarump Member(?) crosses the highway at **Mile 46.7.**

At **Mile 47.0** the massive red-colored cliffs on the left are probably the Entrada Sandstone; the light-colored cliffs above are in the Cow Springs Sandstone. The regional relationships of these massive sandstone units have been in question for years, as their correlations with other known formations are obscure. Some geologists now believe the Cow Springs is a light-colored upper part of the Entrada Sandstone, and here they are both present. The Glen Canyon Group (the Navajo, Kayenta, and Wingate Formations) is not present here, having pinched out farther to the west and northwest.

Mile 48.0: Leaving the Navajo Indian Reservation. Cliffs of Cow Springs Sandstone are to the left and right, overlain by the Morrison Formation. Entering New Mexico at **Mile 48.4**: the Arizona–New Mexico state line.

Mile 0.0: Reset the trip odometer at the New Mexico state line to match the mileposts.

The highway is crossing a small anticline at **Mile 2.5**, exposing the Cow Springs Sandstone and Morrison Formation on both sides of the highway. **Mile 3.2**: Rest area. Nice alcoves eroded from the Cow Springs Sandstone are to the right. A bridge crosses Manuelito Wash at **Mile 4.9**. Manuelito was a highly respected Navajo headman. **Mile 5.1**: The highway climbs up through the Dakota Sandstone at the crest of a small anticline.

Mile 8.0: Milepost 8: Bridge across the Rio Puerco ("Dirty River" in Spanish). The valley walls are set in the Gallup Sandstone (Upper Cretaceous) above gray slopes of the lower Mancos Shale. The roadcut is in the Gallup Sandstone at **Mile 10.0**. A truck weigh station is on the right at **Mile 12.0**. The Torrivio anticline is visible to the left. **Mile 12.8**: The very steeply dipping beds northeastward toward the San Juan Basin are the Gallup Sandstone on the lower Mancos Shale on a sharp little anticline. Exposures of the Gallup Sandstone are to the right and left at **Mile 14.3** as the highway crosses the Torrivio anticline. On the right at **Mile 15.2** are the Twin Buttes volcanics, now much depleted for road aggregate.

Exit 16, the west exit to Gallup, New Mexico. The roadcuts on the right are in the Crevasse Canyon Formation (Cretaceous) with coal seams exposed. **Mile 17.8**: Crossing the Rio Puerco. The rock exposures around the town of Gallup are in the Menefee Formation of the Mesaverde Group (Upper Cretaceous). Exit 20 goes to downtown Gallup and is the intersection of I-40 and U.S. 666 going north to Shiprock, New Mexico (see Roadlog 18). New Mexico 602 goes south to Zuni, New Mexico.

The steeply dipping rocks to the right and left at **Mile 24.0** constitute the northwesterly plunging nose of the Zuni Uplift in the Nutria monocline. The dominant cliffs are formed of the Gallup Sandstone, which is nearly vertical at **Mile 24.5**, dipping westward along the Nutria monocline. The hogback at **Mile 24.9** is the Dakota Sandstone. Exit 26 is the east Gallup exit.

Mile 27.2: The cliffs on the left have been puzzling to geologists for decades. The original type section of the Wingate Sandstone was along these cliffs at Fort Wingate. Later the upper cliffs were considered the Entrada Sandstone, and only the lower cliffs were thought to be the Wingate. It is now realized that the entire section is Entrada Sandstone and that the rocks known elsewhere as the Wingate Sandstone are not present here. Thus, the Wingate Sandstone, as known throughout much of the Colorado Plateau (the former Lukachukai Member of the Wingate), is not present at

Fort Wingate and therefore has no type section. Nevertheless, the sandstone long known as the "Wingate," which is widespread across the country, is still officially called the Wingate Sandstone. The Entrada Sandstone here dips toward the north into the San Juan Basin along the northeastern flank of the Zuni Uplift. The Zuni Uplift is to the right as the highway runs along exposures in the Chinle Formation at **Mile 29.7.**

Mile 30.0: The crest of the Zuni Uplift is the tree-covered skyline to the right. The Zuni Uplift is a large, northwest-trending anticline: the Permian Abo Formation rests directly on the Precambrian basement rocks, and no Paleozoic rocks older than Permian are present along the structure. The northward-dipping beds just below the tree-covered skyline at **Mile 31.5** are the San Andres Formation of Permian age, equivalent to the Kaibab Limestone of Arizona.

Exit 33 goes to Fort Wingate and McGaffey 13 miles to the right on New Mexico 400. McGaffey, and the beautiful associated campground, is on the crest of the Zuni Uplift. The route to McGaffey passes down-section through the entire Permian System of New Mexico: limestone of the San Andres Formation, the Glorieta Sandstone, the Yeso Formation, and red beds of the Abo Formation that rest directly on Precambrian metamorphic rocks near McGaffey.

Roadcuts are in the Chinle Formation at **Mile 34.2.** At **Milepost** 37 Exit 36 leads to Iyanbito ("Buffalo Spring" in Navajo) to the north. The highway here skirts the northeast flank of the Zuni Uplift at or about the top of the San Andres Formation. **Mile 41.0**: Exposures of the San Andres Formation are to the right and left as the highway climbs to the top of the formation.

Mile 48.2: The Continental Divide, elevation 7,268. **Mile 49.0**: Mount Taylor, the Navajo sacred mountain of the south, is visible ahead at about 11:00 (Figure 60). It is a composite volcano of Late Tertiary age, having erupted some 4.5 to 2.3 million years ago. Exit 53 goes to Thoreau, New Mexico. New Mexico 371 goes north to Crownpoint and Farmington, New Mexico.

The highway is still skirting the Zuni Uplift at **Mile 59.0** with roadcuts in the San Andres Formation. The cliffs to the left are the Entrada Sandstone above the Chinle Formation; Mount Taylor is ahead. Exit 63 goes to Prewitt; New Mexico 412 goes south to Bluewater Lake State Park on the Zuni Uplift. **Mile 69.2**: Quaternary lava flows are exposed in roadcuts.

Mile 71.0: The high country to the south is the Zuni Mountains; the single mountain to the left is Mount Taylor. The highway is still at or about the top of the San Andres Formation. Exit 72 goes to Bluewater Village. Mount Taylor in the San Mateo Mountains is on the left at **Mile 76.0**, with a Tertiary-age lava flow forming the long ridge toward the south. Roadcuts here

Figure 60. Mount Taylor, a composite volcano of Tertiary age, is the Navajos' sacred mountain of the south.

are in the San Andres Formation. Exit 79 goes to Milan; New Mexico 605 goes north to San Mateo and the inactive Ambrosia Lake uranium mining district.

Mile 81.2: Grants, New Mexico, city limits. The high ridge behind town is a lava flow of Tertiary age. Very recent aa-type lava flows of the Malpais country are to the right and left at **Mile 81.6**. *Aa* is a Hawaiian term for very rough, broken surfaces on lava flows. Exit 81A goes to El Malpais National Monument to the south on New Mexico 117. Exit 81B goes to Grants. The highway crosses the Malpais for the next several miles; the valley is covered by the very recent and well-preserved lava flows. Exit 85 goes to Grants and Mount Taylor via New Mexico 547.

Mile 87.0: A section of sedimentary rocks of Cretaceous age, capped by a lava flow of Tertiary age, is exposed to the left. The northward-dipping rocks to the right are the Morrison and Entrada Formations, covered in the valley by recent lavas. Exit 89 goes to Quemado and Pietown via New Mexico 117.

Mile 91.0: Entering the Acoma Indian Reservation. The highway is in the Dakota Sandstone, and the valley floor is still filled with recent lavas. Sedimentary rocks of Cretaceous age are exposed all around at **Mile 95.0** as the highway leaves the flank of the Zuni Uplift and crosses the southern San Juan Basin at about the stratigraphic level of the Dakota Sandstone. The Cretaceous sedimentary rocks at **Mile 99.5** are capped by Tertiary lava flows emanating from Mount Taylor.

Mile 102.4: Leaving the Acoma Indian Reservation. Entering the Laguna Indian Reservation.

The highway at **Mile 103.8** is on the northward-dipping Dakota Sandstone, capped by volcanic rocks. **Mile 104.5**: The highway descends through the Dakota Sandstone into the Morrison Formation. At **Mile 106.0** the highway crosses the railroad overpass; the Morrison Formation, capped by the Dakota Sandstone, is exposed across the countryside.

The highway is in the Entrada Sandstone at **Mile 109.0**. At **Mile 111.0** the highway is back on top of the Entrada Sandstone and in the lower Morrison Formation. Exit 114 goes to Laguna Pueblo.

The highway is back into the Entrada Sandstone at **Mile 115.2**, and Chinle-equivalent red beds are exposed ahead. **Mile 117.0**: The Todilto Formation is exposed in roadcuts to the left. The light-colored gypsum and dolomite of the Todilto are believed to have been deposited as lake beds on top of the Entrada Sandstone.

Mile 124.0: The roadcuts on the left are Moenkopi-equivalent, capped by Agua Zarca Sandstone(?), a probable equivalent to the Shinarump Member of the Chinle Formation.

Exit 126 goes to Los Lunas via New Mexico 6. The highway here is in the lower red beds (Moenkopi?).

The highway crosses a fault at **Mile 127.0** and is back into Cretaceous rocks again. Exit 131 goes to To'hajiilee (formerly Cañoncito) to the north. **Mile 132.4**: Bernalillo County line.

Mile 133.0: The highway begins its long descent into the Albuquerque basin and the Rio Grande Rift over a series of faulted steps. The Sandia Mountains are ahead, and the Manzano Mountains are to the right in the distance across the rift. Thank heavens the geology is all covered here with grass, sagebrush, and a few cactus.

The highway crosses the Rio Puerco at **Mile 141.0**, volcanic rocks are to the left in a half-mile, and a volcanic blowout, called Cerro Colorado, is on the right at **Mile 142.7.**

Mile 149.0: The highway makes its final descent into Albuquerque. Spanish place-names are common: the Sandia ("watermelon") Mountains are obvious ahead, Tijeras ("scissors") Canyon is ahead at about 1:00, and the Manzano ("apple") Mountains are to the right. The city of Albuquerque is obvious low, ahead, and to the right at **Mile 151.0.**

Mile 155.0: Entering Albuquerque. A complex interchange of I-40 and I-25 is at **Mile 160.0.**

III

The Black Mesa Basin

The Defiance Uplift and the adjoining Chuska Mountains separate the Black Mesa Basin from the San Juan Basin to the east. The Defiance Uplift is a northwest-trending basement highland that was certainly present as a minor source of sediments in Pennsylvanian time but may have been a structurally high feature since Precambrian time. There red beds of the Cutler Group (Supai of Arizona or Abo Formation of New Mexico) rest directly on Precambrian rocks. Just south of Hunters Point, between Lupton and Window Rock, the Cutler/Abo/Supai rests on pink granitic rocks, and west of Fort Defiance the red rocks rest directly above upturned quartzite of apparently Precambrian age. Overlying the red mudstone and siltstone of Permian age, the DeChelly Sandstone, also Permian, forms massive orange sandstone cliffs, capping most of the Defiance Plateau. The DeChelly Sandstone, named for 825-foot high cliffs in Canyon de Chelly, thins across the Defiance arch and includes the northern pinchouts of the upper Yeso and Glorieta Formations as it is mapped near Window Rock and Fort Defiance. Red beds of the Triassic Moenkopi Formation thin and pinch out against the western flank of the uplift, and the Shinarump Member of the Chinle Formation fills deeply eroded stream channels carved into the top of the DeChelly Sandstone, especially apparent in Canyon de Chelly. The adjacent Chuska Mountains, a northerly-trending range just east of the Defiance Plateau, consist mainly of volcanic rocks of Tertiary age.

The western flank of the geographically broad Black Mesa Basin is along the East Kaibab and Echo Cliffs monoclines, which form the eastern limits of Marble and Grand Canyons. Rocks of Late Cretaceous age, the Mancos Shale and Mesaverde Group, are preserved from erosion in the central, deepest part of the basin on Black Mesa, south of Kayenta, Arizona. Picturesque exposures of the Triassic Chinle Formation are the most noticeable geologic formation surrounding Black Mesa, forming the so-called Painted Desert. North of Black Mesa the Monument Upwarp is the most visible structural feature bounding the basin. A gentle structural bench, the Four Corners Platform, forms the southern limits of the Paradox basin. To the south is the Mogollon Rim, which also serves as the southern border of the Colorado Plateau Province. Unlike the San Juan Basin to the east, the Black Mesa Basin is decorated by several northwest-trending anticlines and synclines that are most apparent in the Hopi Indian Reservation on Black Mesa.

That the Black Mesa Basin has a Paleozoic ancestry is made clear by the presence of a salt basin of Permian age buried beneath its southern region. Although not apparent at the surface, the Permian section changes to evaporites, including salt, beneath the approximate location of Holbrook, Arizona, thus the name Holbrook basin. Salt and gypsum occur in a red bed sequence that is confined largely to the DeChelly Sandstone interval. To the north and east the DeChelly Sandstone is an orange-colored, windblown sandstone that pinches out against the East Kaibab monocline to the west. To the south, where it emerges from the salt basin, it is an orange, very fine-grained sandstone that is much more massive and less cross bedded and perhaps waterlaid. Because of the differences in bedding and its remote occurrences, the sandstone along the Mogollon Rim has been named the Schnebley Hill Formation. The presence of the salt basin is known entirely from deep drilling for oil.

The Late Cretaceous Mesaverde Group caps Black Mesa proper. Above the steep gray slopes of Mancos Shale, the Toreva Sandstone forms the lower prominent cliff seen south of Kayenta, Arizona, overlain by the Wepo Formation. The high cliff consists of the Yale Point Sandstone. The nearshore sandstone formations pinch out toward the south, and the coastal swamp deposits of the Wepo cap much of the high plateau. The Wepo Formation's origin is demonstrated by its vast deposits of coal, which occur mostly along the northeastern margin of Black Mesa. Peabody Coal Company has strip-mined the deposits for many years. The coal is transported by slurry pipeline to the Mojave Power Plant in Nevada and by electric train to the Navajo Generating Station at Page, Arizona.

23. Bluff, Utah, to Chinle, Arizona, via U.S. 191

Shortly after leaving Bluff, Utah, U.S. 191 turns south. Heading south from the intersection of U.S. 191 and U.S. 163, the highway first crosses the San Juan River and the northern boundary of the Navajo Indian Reservation. Before reaching Mexican Water, it is mostly in strata of the San Rafael Group (Middle Jurassic), passing the Tohonadla and Akah Oilfields and the western plunging nose of the Boundary Butte anticline. South of the Mexican Water Trading Post, the highway crosses extensive exposures of the Navajo, Kayenta, and Wingate Sandstones on the northwesterly plunging nose of the Defiance Uplift and continues down into the Triassic Chinle Formation in Chinle Valley. The village of Chinle lies at the gate to Canyon de Chelly National Monument.

Mile 0.0: Leaving Bluff, Utah, heading west on U.S. 191. The road is on Pleistocene terrace gravel benches cut onto brown, crinkly beds of the Carmel Formation (Middle Jurassic). The road left at **Mile 2.4** goes to the Sand Island Recreation Area campground and boat ramp. River trips from Bluff to Mexican Hat through the upper canyon of the San Juan River depart from here.

Mile 3.2: U.S. 191 turns left (south) toward Mexican Water. The highway ahead becomes U.S. 163 to Mexican Hat and on south through Monument Valley to Kayenta, Arizona (see Roadlog 36). U.S. 191 crosses down into the top of the Navajo Sandstone at **Mile 3.6**. At **Mile 3.9** a bridge crosses the San Juan River, which marks the northern boundary of the Navajo Indian Reservation, established by the Treaty of 1868. At **Mile 4.2** the highway crosses onto the top of the massive, cross-bedded Navajo Sandstone. The road ahead is on the Carmel Formation, and the brown cliffs ahead are in the Bluff Sandstone Member of the Morrison Formation (Late Jurassic). There is a Navajo ranch on the right at **Mile 7.0.**

Mile 9.7: Utah-Arizona border. The Tohonadla Oilfield is on the left. Discovered by Shell Oil Company in 1957, Tohonadla is expected to produce some 2 million barrels of high-grade oil from algal bank reservoirs in the Paradox Formation of Middle Pennsylvanian age (mostly from the Lower Ismay cycle) from depths of around 5,000 feet.

At **Mile 13.3** the Boundary Butte volcanic neck, a small diatreme, is visible to the right at 2:00. The Boundary Butte anticline is visible ahead. The road to the left at **Mile 14.9** goes to the Akah Oilfield. Originally called the North Boundary Butte Field, the small field was discovered by Shell Oil Company in 1955, when the discovery well flowed about 500 barrels of high-grade oil per day from a depth of about 5,300 feet in the Paradox Formation (Middle Pennsylvanian). When the name *North Boundary Butte* was denied by the state of Utah, a Shell employee asked a local resident Navajo what the place was called. The Navajo replied "*Akah*," pronounced "aw-kye," the Navajo word for "oil"—so now we have the "Oil Oilfield." The Boundary Butte volcanic neck, a diatreme of Tertiary age, is dead ahead.

The westerly plunging nose of the Boundary Butte anticline is visible on the left at **Mile 16.3**. White rocks that cap the structure are in the Navajo Sandstone. The top of the Navajo Sandstone is at road level at **Mile 17.9.** The Boundary Butte volcanic neck is obvious to the left. The hills to the left are composed of the Morrison Formation, capped with small remnants of

Figure 61. Round Rock, a prominent landmark of the Wingate Sandstone, Early Jurassic, rises majestically above a "painted desert" of the Chinle Formation, Late Triassic in age.

the Dakota Sandstone. The plateau at 9:00 is Toh Atin Mesa ("No Water Trail" in Navajo), capped by the Dakota Formation, with the Carrizo Mountains beyond. (Dakota-capped buttes across the Navajo country were commonly called "lava-capped buttes" in the best-selling mystery book series by Tony Hillerman.)

The red badlands ahead at **Mile 24.7** are in the San Rafael Group (Jurassic). The butte to the left is capped by cliffs of the Bluff Sandstone Member of the Morrison Formation. The highway crosses the Arizona border once again at **Mile 25.7.** The odd-looking buttes to the right at **Mile 26.5** are in the red Entrada Sandstone. At **Mile 27.7** the road crosses through low, dome-shaped hills of the Morrison Formation, known unofficially as the "Maidenform Hills," named for their appearance by Shell Oil Company field geologists in 1952, including myself—it was a long, hot summer.

Mile 30.8: Junction with U.S. 160. Go to the right (west). The highway is at or near the top of the Navajo Sandstone. The Mexican Water Trading Post is on the left at **Mile 32.3.** U.S. 191 turns left (south) toward Chinle, Arizona, at **Mile 33.1.** U.S. 160 goes ahead to Kayenta and Tuba City, Arizona. U.S. 191 is in the Navajo Sandstone. (See Roadlog 27, **Mile 53.3.**)

Entering the Navajo village of Rock Point at **Mile 48.1;** leaving Rock Point at **Mile 48.6.** The townsite is in the Wingate Sandstone; south of here the road is on the varicolored mudstone of the Chinle Formation (Triassic). The highway crosses Lukachukai Creek at **Mile 55.3.** The Lukachukai Mountains are visible to the left at **Mile 58.0.** The Navajo word *Lukachukai* means "the place of slender reeds." The road to the left at **Mile 65.8** goes to

Round Rock village, near the landmark of Round Rock (Figure 61). At **Mile 66.3** U.S. 191 turns right; the road ahead goes to the village of Lukachukai.

Boulders of the Shinarump Member of the Chinle Formation are on the right at **Mile 66.5**; the highway crosses down into brown mudstone of the upper Moenkopi Formation (Lower Triassic). The highway crosses the Shinarump Member and ascends onto the Chinle again at **Mile 67.9**. The road is crossing the northwesterly plunging nose of the Defiance Uplift. At **Mile 71.9** the highway heads down into Chinle Valley in badlands in the Chinle Formation. The highway crosses Chinle Wash at **Mile 73.7**.

Entering the Navajo village of Many Farms at **Mile 81.3**. Leaving Many Farms at **Mile 82.7**.

Entering the town of Chinle at **Mile 95.3**. The name is Navajo, meaning "place where the water flows out of the mountain." The first trading post was established here in 1882 by Naakaii Yazzie. A stoplight is at **Mile 95.9**. The road to the left is Navajo Route 7, which goes to Canyon de Chelly National Monument and beyond. (To continue south on U.S. 191, see Roadlog 25.)

24. Canyon de Chelly National Monument

Although inhabited by the Navajo people for the past few hundred years, this haven has domiciled various groups of Native Americans for many centuries. The very nature of the spectacular canyon system—its precipitous walls, its fine microenvironment, and its isolation—provides a comfortable, well-protected homeland. The land at Canyon de Chelly is vastly more ancient, more complex, and more interesting than we can imagine. A look into the story behind the rocks of this fabulous national monument with its sheer, massive, red canyon walls provides insight into the origin of this fascinating land and its people.

Geology. Canyon de Chelly (pronounced d'*shay*), and its several tributary canyons, located just east of Chinle, Arizona, has been eroded over the millennia from the western flank of the high country known as the Defiance Plateau. The plateau is the present-day expression of an ancient geologic arch called the Defiance Uplift. Although this feature appears to be a simple, broad fold, it is a composite structure consisting of individual anticlines (upfolds) and synclines (downfolds) that trend generally in a northwesterly direction. This structural fabric is inherited from a ubiquitous system of faults that originated nearly two billion years ago in the Precambrian Era. Many geologic features of the Colorado Plateau were localized by the basement fault system.

It seems likely that the Defiance Uplift has been a highland ever since Paleozoic time began, as all the regionally distributed sedimentary layers of pre-Permian age thin and pinch out along the flanks of the highland. Deep drilling has revealed that rock layers of Cambrian, Devonian, Mississippian, and Pennsylvanian age are present surrounding the uplift but are missing over its higher parts. Atop the high structure, near Fort Defiance and Hunter's Point, rocks of Permian age rest directly on metamorphic rocks of Precambrian age (see Figure 49). Thus, the Defiance Uplift has a similar history to that of the ancient Ancestral Rocky Mountains of Colorado and neighboring areas.

Sedimentary Rocks. The oldest known sedimentary rocks on the Defiance Uplift and in Canyon de Chelly are red mudstone and siltstone of Early Permian age. These "red beds" may be seen in the floor of Canyon de Chelly only at the base of Spider Rock at the confluence with Monument Canyon. The reddish-brown, fine-grained sediments that form this distinctive layer were deposited on continental lowlands and intertidal mudflats as streams and tides shaped the coastal lowlands in Permian time.

The red beds at the base of Spider Rock in Canyon de Chelly are, without doubt, the same strata as those that form the base of the buttes, mesas, and spires of Monument Valley. That layer in Monument Valley is called the Organ Rock Shale and is a formation in the Cutler Group. Thus, the sedimentary rocks immediately beneath the massive cliff-forming sandstone in Canyon de Chelly should rightly be called the Organ Rock Shale. Older Permian red beds of lower Cutler affinities are believed to be present along the crest of the Defiance Uplift near Fort Defiance and Hunter's Point.

The massive cliffs that comprise the pinkish-colored, high canyon walls in Canyon de Chelly have been named the DeChelly Sandstone for these exposures. The formation is included in the Cutler Group by most geologists. It is the same layer of sandstone that forms the high cliffs in the buttes and mesas of Monument Valley.

The DeChelly Sandstone reflects the shape of the Defiance Uplift in Permian time. The impressive cliffs of sandstone at Spider Rock, the only place in Canyon de Chelly where both the top and the bottom of the DeChelly Sandstone are exposed, reveal that the sandstone is 825 feet thick (Figure 62). A few miles to the east, along the crest of the Defiance Uplift near Fort Defiance and at Hunter's Point, the DeChelly Sandstone has thinned to a mere 200 feet. That is a very significant change in thickness indeed. A lighter-colored sandstone, believed to be the Glorieta Sandstone of New Mexico, occurs above the DeChelly near Fort Defiance and Hunter's Point. The thickest deposits of DeChelly Sandstone occur in the Black Mesa Basin to the west, where more than 1,000 feet have been drilled, and in the San Juan Basin on the eastern side of the Defiance Uplift. This relationship implies that the Defiance Uplift was a high topographic feature during the

Figure 62. Spider Rock, the home of Spider Woman in Navajo mythology, stands guard over the confluence of Canyon de Chelly and Monument Canyon in Canyon de Chelly National Monument, Arizona. The prominent spire consists of the entire section of the DeChelly Sandstone of Permian age with a thin cap of the Shinarump Member of the Chinle Formation of Late Triassic age.

time the DeChelly Sandstone was being deposited, and was only barely buried by the sand.

But why all the very obvious cross bedding in the DeChelly Sandstone? It is because the sand was deposited, in Permian time, by wind as sand dunes on a vast desert. Individual sets of cross-beds represent bedding surfaces on the lee slopes of sand dunes that wandered aimlessly across broad, dry plains. In the walls of Canyon de Chelly, cross-bed sets vary between 10 and 50 feet in thickness, indicating that the dunes were mature and rather large. Prevailing winds that formed the sand dunes in the DeChelly Sandstone were from the north and slightly northeast, blowing southward.

The sedimentary rocks seen in Canyon de Chelly are of the early half of the Permian Period. No rocks of Late Permian age are present on the Defiance Uplift. There is no way to know what transpired here during the Late Permian, but if any sedimentary rocks were deposited, they were removed by erosion before Triassic time.

The basal unit of the Late Triassic Chinle Formation, the Shinarump Conglomerate Member, rests directly atop the DeChelly Sandstone throughout the Defiance Uplift. These irregular beds, seen at the rims of the canyons as dark brown conglomeratic sandstone, form the harder caprock protecting the less resistant DeChelly Sandstone from erosion. The roads and viewpoints in Canyon de Chelly National Monument are all built on top of the Shinarump Conglomerate Member of the Chinle Formation. The

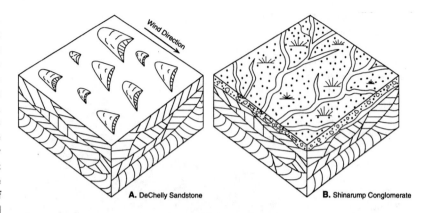

Figure 63. A. Sand dunes drifting across the desert floor in Permian time formed high-angle cross beds seen in the canyon walls carved from the DeChelly Sandstone. B. Meandering streams eroded the upper surface of the DeChelly Sandstone in Late Triassic time and deposited the Shinarump Member of lenticular beds of sand and pebbles that form the rimrock of Canyon de Chelly. From Baars 1998d.

Shinarump (properly pronounced "shin-*air*-rump") beds were deposited by streams that coursed the southern Colorado Plateau at the beginning of Late Triassic time. The scoured contact between the DeChelly and Shinarump beds is an erosional surface on which the stream sand and gravel was deposited. The irregular contact is known as an *unconformity*, a surface that represents erosional activity and a missing record of lost time.

A close examination of the Shinarump Member reveals that the cross bedding is on a much smaller scale than that of the DeChelly and that the angles of cross-bedding inclinations are much flatter. Cross-bed sets are planar in the DeChelly but lens-like in the Shinarump, representing sand-filled channels formed by streams. The sand grains in the DeChelly are nearly identical and very fine in size whereas the Shinarump consists of more coarse sand grains with scattered pebbles throughout. Occasional thin lenses of pebbles may be seen scattered through the exposures. Thickness of the Shinarump Member varies greatly throughout its area of deposition, ranging from zero to tens of feet thick, depending on the shape of the surface over which the streams flowed. These features are characteristic of sediments deposited by streams (Figure 63).

The remainder of Triassic time is represented by the upper several members of the Chinle Formation. The variously colored shale beds of the Chinle are seen only in isolated erosional remnants within Canyon de Chelly National Monument but are very prominent west and south of the Defiance Uplift, constituting the rocks of the "Painted Desert." Sedimentary rocks of Jurassic age are not seen in Canyon de Chelly National Monument but are prominent in the scenic landscape west and north of the Defiance Uplift.

The Defiance Uplift and the surrounding areas were the site of numerous volcanic eruptions during Tertiary time. These were not quiet, lava-like eruptions but rather highly explosive, gaseous blowouts that were probably very powerful and noisy. The vents, or necks, of these blowout centers are

Figure 64. Mummy Cave, Anasazi ruins tucked into an alcove in the DeChelly Sandstone in Canyon del Muerto, Canyon de Chelly National Monument. The tower complex in the upper center of the photo was built in the 1280s, and the entire village was evacuated in about A.D. 1300.

scattered about the countryside in apparent random distribution. Volcanic rocks, mostly ash deposits, comprise the bulk of the Chuska Mountains to the east.

Canyon-cutting. It is a well-known fact that canyons are formed by the erosional powers of rivers, and Chinle Creek is the obvious culprit in Canyon de Chelly. It is equally clear in this case that the canyon is not now being deepened, but instead the canyon floor is being choked with sand washed in by seasonal floodwaters. The canyon must have been carved by flooding streams during the previous "ice ages" when the climate was much wetter in this region. Meltwaters during inter-glacier episodes of the past million years or so were much more powerful than the trickles of Chinle Wash today.

Even so, the magnificence of Canyon de Chelly is in perfect harmony with the Anasazi ruins (Figure 64) and the Navajo culture so beautifully displayed here today: "Beauty all around me, with it I wander."

To continue south on U.S. 191, see Roadlog 25. To go north on U.S. 191, see Roadlog 23.

25. Chinle to Chambers, Arizona, via U.S. 191

The highway south from Chinle, Arizona, travels through the wide-open Chinle Valley, the area from which the Chinle Formation of Late Triassic age was named. It is here a varicolored shale and mudstone with considerable amounts of petrified wood, although most of the fossil wood has been removed over the past decades. The petrified wood so well exposed and well preserved in Petrified Forest National Park occurs in the Chinle Formation.

Nearly white mudstone and sandstone beds of the Bidahochi Formation, Tertiary (Pliocene) in age, appear above the Chinle Formation about 25 miles south of Chinle. These beds are lake and stream deposits, forming soft-weathering rounded terrains most of the way to Chambers.

Mile 0.0: Stoplight at the intersection of Navajo Route 7 and U.S. 191 in Chinle, Arizona. Proceed south on U.S. 191. **Mile 2.0:** Exposures of the Chinle Formation are to the right and left as the highway heads south in Chinle Valley.

The road to the right at **Mile 6.1** goes to Pinyon. The view to the left is of the Defiance Plateau with the Chuska Mountains in the distance. A nice view of Beautiful Valley is on the left at **Mile 10.0.** A junction with a road to Nazlini is at **Mile 18.8.**

At **Mile 26.0** Ganado Mesa is to the left, capped by the light-colored Bidahochi Formation of Tertiary age. The hills to the right are in the Wingate Sandstone, capped by the Bidahochi Formation. Beds of the Bidahochi Formation are light in color, nearly white, and consist of mudstone and sandstone deposited in stream and lake environments. It has been postulated that the Bidahochi represents sediment deposits from an ancestral course of the Colorado River before the opening of the Grand Canyon.

Mile 30.5: Intersection with Arizona 226; turn left for continuation of U.S. 191. The hills are in the Chinle Formation at **Mile 31.3**, capped by thin remnants of Bidahochi Formation.

Mile 34.7: Entering Ganado (meaning "livestock" or "cattle" in Spanish). Turn right at **Mile 35.8** to Hubbell Trading Post National Historic Site. The Hubbell Trading Post was established by John Lorenzo Hubbell in 1876; in 1967 it was designated a national historic site to preserve a realistic picture of the old Navajo trading days. Return to the highway and turn right. Turn right again onto U.S. 191 at **Mile 37.1.**

Exposures of the Chinle Formation with thin remnants of Bidahochi Formation are on the right at **Mile 37.5.** The top of the plateau, in the Bidahochi Formation, is reached at **Mile 37.8.** The road to the right at **Mile 38.7** goes to Cornfields. Roadcuts at **Mile 43.8** are in the Bidahochi Formation. The highway is on the Bidahochi Formation for the next 15 miles, with numerous roadcuts showing the nature of the formation. There are views of the Defiance Uplift to the left at **Mile 48.0.**

Mile 51.3: Entering Klagetoh, the Navajo word for "water in the ground" or "spring." The road to Wide Ruins is at **Mile 57.0.** After crossing a wash at **Mile 58.9** which exposes the Chinle Formation, the highway climbs back into the Bidahochi Formation. A liquor store on the left at **Mile 73.5** indicates we have left the Navajo Reservation.

Mile 75.1: Chambers, Arizona, local end of U.S. 191. Turn left on Interstate 40. (To continue south on U.S. 191, see Roadlog 26. For I-40, join Roadlog 22 at **Mile 21.8** after the Petrified Forest National Park exit.)

26. Chambers to Springerville, Arizona, via U.S. 191

South of Chambers, Arizona, U.S. 191 crosses poorly exposed rocks of the Moenkopi Formation (Early Triassic in age) for about 35 miles and then passes up-section into the Chinle Formation (Late Triassic). The Tertiary-age Bidahochi Formation caps the Triassic red beds in places. Young lava flows become more numerous on exposures of the Chinle Formation as the highway progresses southward, until volcanic cinder cones become numerous across the open countryside. The White Mountains south of Springerville are volcanoes of Late Tertiary and Pleistocene age and mark the southern boundary of the Colorado Plateau Province in this region.

Mile 0.0: Intersection of U.S. 191 and Interstate 40 at Chambers, Arizona. Proceed eastward on I-40 to rejoin U.S. 191 in 6.1 miles.

The knob on the left at **Mile 0.9** is brown mudstone and siltstone of the Moenkopi Formation, capped by a thin erosional remnant of the Shinarump Member of the Chinle Formation (both Triassic in age).

Mile 6.1: Leave I-40 at Exit 349; turn right to continue south on U.S. 191. Exposures of the Moenkopi Formation are on the left. At **Mile 6.5** the highway crosses the Puerco River; Burnham Trading Post is just south of the river. An Indian village is at **Mile 6.9.** The countryside is badly covered, but the road is probably on the Bidahochi Formation. There are roadcuts in the Bidahochi Formation at **Mile 13.2.** The Navajo Springs road is on the right at **Mile 17.9**; the white bluffs ahead and to the left are in the Bidahochi Formation. **Mile 30.3:** Junction with Arizona 61, which goes to Zuni, New Mexico.

Exposures of the varicolored Chinle Formation are to the left at **Mile 37.9**; the highway is still on the white Bidahochi Formation. The mesa to the left has a thin lava cap. There is a lava flow low on the left at **Mile 39.6.** Exposures of the Chinle Formation are to the right at **Mile 39.8** in "painted desert" hues.

Exposures of the Chinle Formation are on the left at **Mile 40.3,** capped by channel-filling lava. Lava flows are common for the next 3 miles. The highway crosses Zuni Wash at **Mile 43.3.** Lava-capped mesas are all around for the next 3 miles. Outcrops of the varicolored gray and purple Chinle Formation are capped by erosional remnants of a channel-filling, light-colored sandstone for the next 5 miles.

A large power plant, the Coronado Power Station, is visible to the left at **Mile 51.9.** The road on the left goes to the power plant. **Mile 55.8:** Chinle badlands dominate the scene in all quadrants.

The St. Johns, Arizona, city limits are at **Mile 59.8** in the Chinle Formation. The highway crosses the Little Colorado River in a tenth of a mile. **Mile 60.0**: Junction of U.S. 180 and U.S. 191 in St. Johns; turn left (south) on 191. **Mile 60.9**: Leaving St. Johns.

The highway is on the Chinle Formation at **Mile 61.4**, capped by volcanics on the left. Volcanic peaks are occasionally visible on the skyline ahead. There are exposures of the Moenkopi Formation at **Mile 64.8**, often capped by remnants of the Shinarump Member of the Chinle Formation for the next 5 miles. The mesa on the right at **Mile 70.2** is capped by lava. The highway is near the Moenkopi-Shinarump contact for the next 4 miles, with scattered remnants of lava flows.

Apache State Prison Unit is on the right at **Mile 74.4.** Lava flows and scattered cinder cones, Quaternary in age, are common from here to the White Mountains. The Springerville Generating Station road is to left at **Mile 77.4. Mile 83.1: Milepost 392**: The road is on a lava-capped plateau.

Mile 84.4: Springerville, Arizona, city limits. The intersection of U.S. 191 and U.S. 60 is at **Mile 85.7.** The White Mountain volcanic field ahead to the south marks the southern boundary of the Colorado Plateau.

27. Chimney Rock, Colorado, to Kayenta, Arizona, via U.S. 160

The route basically follows what is called the Four Corners Platform, a structural hinge that separates the Paradox basin to the north, the San Juan Basin to the southeast, and the Black Mesa Basin to the southwest. The road is in the base of the Mancos Shale of Late Cretaceous age from the Chimney Rock Junction to the San Juan River in the southwestern corner of Colorado and then follows exposures of the Morrison and Navajo Formations (Jurassic) across northern Arizona. The East Boundary Butte and Boundary Butte oilfields and anticlines are passed en route, and several diatremes are visible near the southern plunging nose of the Monument Upwarp at Kayenta.

Mile 0.0: Intersection of U.S. 666 and U.S. 160 in southwestern Colorado. Head west on U.S. 160 toward Teec Nos Pos and Kayenta, Arizona. Chimney Rock, consisting of the dark gray Mancos Shale capped by cliffs of the Point Lookout Sandstone of the Mesaverde Group (Late Cretaceous), is just east of the intersection. Ship Rock, a volcanic blowout pipe (diatreme) in New Mexico, is visible to the south in the distance. Ahead are the Carrizo Mountains, a laccolithic range; the Sleeping Ute Mountains, another laccolithic range, are to the northwest. The road is near the base of the Mancos Shale in the Ute Mountain Indian Reservation.

Mile 1.2: The road crosses Navajo Springs Wash. The Sleeping Ute Mountains are to the right. The Carrizo Mountains are ahead at about 11:00

at **Mile 4.0**, the Lukachukai and Chuska Mountains are visible at about 10:00, and Ship Rock is at about 9:00. Mesa Verde is behind to the east. There is a deeply entrenched wash to the right at **Mile 7.0**, cut into recently deposited sediments. **Mile 7.5**: The Juana Lopez (Sanostee) Sandstone Member of the Mancos Shale, with abundant fossil mollusks, forms a low bench to the right. The road at **Mile 10.8** crosses Aztec Creek.

The road to the right at **Mile 13.4**, Utah 41, goes to Aneth, Blanding, and Bluff, Utah. **Mile 15.4: Milepost 3**: The road is still in the lower Mancos Shale. **Mile 15.9**: The top of the cliff-forming Dakota Sandstone is exposed in the gully to the left. The road at **Mile 17.2** crosses down-section through the Dakota Sandstone (Late Cretaceous) and Burro Canyon Formation (Early Cretaceous) into the poorly exposed Brushy Basin Member of the Morrison Formation (Late Jurassic).

Mile 18.1: The bridge across the San Juan River is in the Brushy Basin Member, a varicolored mudstone. Entering the Navajo Indian Reservation. The road to the right at **Mile 18.9** goes to the Four Corners Monument, 0.4 miles. This is the only place in the United States where four states come together at a common point.

Mile 19.6: Entering Arizona. **Mile 20.1: Milepost 470**: The road is still in the upper Morrison Formation. **Mile 21.3**: Elevation 5,000 feet. **Mile 22.1: Milepost 468.** The buttes to the right and left at **Mile 22.6** are capped by the Dakota and Burro Canyon Sandstones. The Carrizo Mountains are the prominent laccolithic mountain range ahead.

Mile 24.4: Entering Teec Nos Pos, Arizona. The name is Navajo, meaning "circle of cottonwood trees." The original trading post was south of the present highway. **Mile 24.7**: Intersection: U.S. 64 to the left goes to Shiprock and Farmington, New Mexico, and U.S. 160 to the right goes to Kayenta, Arizona. Turn right on 160 toward Kayenta. The village is in the Brushy Basin Member of the Morrison Formation, and the Dakota Sandstone caps the small buttes and mesas. Teec Nos Pos now hosts a Navajo boarding school.

Roadcuts are in the Brushy Basin Member at **Mile 26.2**, capped by river terrace gravel deposits. The gravel beds are Pleistocene river deposits along the ancient course of the San Juan River. **Mile 26.8**: The road is about at the top of the Westwater Canyon Sandstone Member of the middle Morrison Formation for the next 3 miles. **Mile 30.1: Milepost 460.**

At **Mile 32.0** the countryside is covered with Recent sand. Red Mesa is on the right at about 2:00, White Mesa is in the right distance at about 3:00, and Toh Atin Mesa is to the left at about 10:00.

Mile 36.1: Milepost 454: Red Mesa is to the right, capped by the Bluff Sandstone Member of the Morrison Formation above red beds of the

Wanakah and Entrada Formations. Toh Atin Mesa, capped by the Dakota Sandstone above the Morrison Formation, is to the left.

Mile 39.1: Milepost 451: Red Mesa, on the right, is on the eastern nose of the Boundary Butte anticline and is known as the East Boundary Butte Oil-field, discovered by Shell Oil Company in the 1950s. It produces oil and gas from rocks of Middle Pennsylvanian age. The saddle to the left between Toh Atin Mesa and the Carrizo Mountains is the Toh Atin anticline, trending roughly east-west; it produced small amounts of oil from the Mississippian Redwall (Leadville) Limestone.

The road to the right at **Mile 40.2**, Navajo Route 35, goes to Montezuma Creek, Utah, in the Aneth area. The Aneth Oilfield, to the north of here, was discovered in 1956 by Texaco and spread in a couple of years to become one of the nation's giant oilfields. It is still producing from rocks of Middle Pennsylvanian age (Desert Creek zone) and is expected ultimately to produce about 500 million barrels of high-paraffin oil. A Shell Oil Company pipeline delivers oil to southern California, and a Texaco pipeline goes to Texas, where the oil is processed for motor oil.

Mile 40.3: Red Mesa Trading Post is on the right. **Mile 41.2: Milepost 449.** The road to the right at **Mile 41.9** goes to the Red Mesa Boarding School.

Mile 42.7: The light-colored sandstone ridge in the middle distance to the right is the Navajo Sandstone on top of the Boundary Butte anticline. The Boundary Butte Oilfield atop the anticline was officially discovered in 1948 in the Western Natural Gas Company Number 2 English well. Continental Oil Company (Conoco) had found flowing oil and the deeper gas in 1930 but lost the hole because of mechanical failure. The field produces oil from the DeChelly Sandstone in 29 holes drilled to a depth of about 1,500 feet. Prolific amounts of natural gas were found in the Middle Pennsylvanian Paradox Formation (lower Ismay zone) at a depth of about 5,500 feet; the gas was never produced for lack of a market.

Mile 44.2: Milepost 446: The west end of Toh Atin Mesa, capped by the Dakota Sandstone, is to the left. From here to the west the Dakota caprock has been removed by erosion, and the remaining rounded hills of the Brushy Basin Member of the Morrison Formation are unofficially known collectively as the "Maidenform Hills."

The countryside around **Mile 45.4** is generally covered with sand, but exposures of the Salt Wash Member of the Morrison Formation, stream-deposited sandstone, peek out here and there. **Mile 45.7**: Boundary Butte, the neck of an old volcanic blowout (diatreme), is visible in the middle distance at about 2:00. The name is derived from the location of the neck on the Utah-Arizona boundary. It is also known as "Rabbit Ears" from its appearance. Beyond the butte in the distance is Cedar Mesa. The Bears Ears are to the right at about 3:00 on top of the Monument Upwarp. The Maidenform Hills are to the left and ahead. **Mile 47.3: Milepost 443.** The road at **Mile 48.4** crosses between two of the Maidenform Hills.

Mile 49.2: Milepost 441: The road is passing through a field of "boon-docks." A *boondock* is a nontechnical term for a sand dune that is partially stabilized by plant growth to enable one to load livestock into the rear of a wagon without building a wooden ramp. **Mile 49.6**: Black Mesa, capped by sandstone cliffs of the Mesaverde Group, is visible on the distant skyline ahead and to the left. The western Maidenform Hills are to the right. At **Mile 52.0** the road is on top of the Navajo Sandstone. The road to the right at **Mile 53.1** is U.S. 191, going north to Bluff, Utah. **Mile 53.3: Milepost 437.** (See Roadlog 23, **Mile 33.1**.)

The mesa to the left at **Mile 53.8** is on the Walker Creek anticline, the northwestern nose of the Defiance Uplift. **Mile 54.7**: Mexican Water Trading Post (Nokaitoh) is on the left. The highway crosses Walker Wash at **Mile 54.9**. **Mile 55.3: Milepost 435.** The road to the left at **Mile 55.4** is U.S. 191, heading south to Chinle, Arizona, and Canyon de Chelly National Monument. (See Roadlog 23, **Mile 33.1**.) The road ahead on U.S. 160 is near the top of the Navajo Sandstone. Well-developed "boondocks" are to the right and left at **Mile 56.7**, and exposures of the nearly white Navajo Sandstone are ahead. **Mile 61.0**: At Tes Nez Iah ("tall cottonwood stands") the road crosses Chinle Wash, here cut deeply into the Navajo Sandstone and displaying beautiful eolian (windblown) cross bedding. **Mile 61.9**: Beautiful exposures of the Navajo Sandstone in Chinle Wash are to the left.

Mile 65.3: Milepost 425: The many dirt roads leading away from the highway go to Navajo hogans and ranches—or perhaps nowhere. Anyway, they are off limits to travel without a permit from the Navajo Nation. The red exposures on the right at **Mile 66.4** are in the Carmel Formation; the top of the white Navajo Sandstone is visible to the left.

The highway is back on top of the Navajo Sandstone at **Mile 67.5**. Black Mesa is ahead and to the left. **Mile 69.5**: Elevation 5,000 feet. Crossing Laguna Creek at **Mile 70.2**. Dinnehotso is on the right at **Mile 72.0**. Another road to the right to Dinnehotso is at **Mile 72.7**. The name is Navajo, meaning "upper ending of the meadow." Exposures of the Navajo Sandstone are ubiquitous.

The flat-topped mesa to the immediate right at **Mile 74.0** is capped by a thin limestone in the Navajo Sandstone. Such thin limestone beds occur often in the Navajo, representing playa lake deposits in the Navajo desert.

The red middle cliff to the left at **Mile 78.5** is the siltstone facies (aspect) of the Entrada Sandstone, known here as the "Beds at Baby Rocks." Above

Figure 65. Church Rock, the neck of a gaseous, highly explosive volcano, or diatreme, of Middle Tertiary age, seen east of Kayenta, Arizona. Comb Ridge, the eastern flank of the Monument Upwarp, is visible in the left distance.

the silty Entrada Sandstone are red beds of the Wanakah Formation and, above an unconformity, more red beds of the Tidwell Member of the Morrison Formation. The upper cliffs are the Salt Wash Member of the Morrison Formation. **Mile 80.4: Milepost 410:** Comb Ridge is visible to the right. Agathla Peak, a prominent volcanic blowout vent (diatreme), is at about 2:00. The Organ Rock anticline is ahead in the distance. Baby Rocks is on the left at **Mile 82.7.** The peculiar carved "baby rocks" are in the siltstone facies of the Entrada Sandstone, topped by the light-colored Salt Wash Sandstone Member of the Morrison Formation.

Mile 85.4: Milepost 405: Black Mesa is ahead, capped by sandstone beds in the Mesaverde Group (Late Cretaceous); the lower cliff is the Yale Point Sandstone. There is a good view of Agathla Peak at about 2:30; the small craggy buttes at about 2:00 are also diatremes. The road to the left at **Mile 88.6** is Navajo Route 59 to Many Farms, Arizona.

Church Rock, a large diatreme, is on the right at **Mile 89.1** (Figure 65). The roadcuts ahead are in the Salt Wash Member of the Morrison Formation. Crossing Church Rock Wash at **Mile 89.9.**

Mile 90.4: Milepost 400. The southern plunging nose of the Monument Upwarp is obvious on the right at **Mile 93.5: Milepost 397.** The Organ Rock anticline is dead ahead, and Black Mesa is to the left.

Mile 96.5: Entering Kayenta, Arizona; the name means "fingers of water" in the Navajo language. In English it means "the farthest place from anywhere." **Mile 97.0:** Stoplight; intersection of U.S. 160 and U.S. 163, which heads north toward Monument Valley and Mexican Hat, Utah. (To go north on U.S. 163, see Roadlog 36.)

28. Kayenta to Cameron, Arizona, via U.S. 160

Heading west from Kayenta, U.S. 160 skirts the north side of Black Mesa and then crosses the magnificent Organ Rock monocline with its many crags and deep rocky canyons in the Navajo Sandstone. At Marsh Pass the highway enters Long House Valley and continues through desolate plains filled with blow-sand and rare exposures of the Cow Springs Member of the Entrada Sandstone of Middle Jurassic age. At Tuba City the highway begins its step-like descent through the Moenave Formation across "painted desert" exposures of the very thick Chinle Formation to the junction of U.S. 89 near Cameron, Arizona.

Mile 0.0: Intersection of U.S. 163 and U.S. 160 at Kayenta, Arizona. Proceed west on U.S. 160.

Mile 0.5: Leaving Kayenta. **Milepost 393:** Black Mesa is high to the left with slopes of Mancos Shale and capped by sandstone beds of the Mesaverde Group (Cretaceous). Mudstone of the Morrison Formation (Late Jurassic) is at the base of the cliffs. Ahead and to the right is the Organ Rock monocline (Figure 66); the light-colored cliffs are steeply dipping Navajo Sandstone. The low hills to the left beneath Black Mesa at **Mile 2.2** are in the Morrison Formation with Dakota Sandstone caps. **Mile 3.5: Milepost 390.** The Organ Rock monocline is very obvious ahead and to the right at **Mile 4.8.** The light-colored hogback is Navajo Sandstone. The Morrison Formation, capped by Dakota Sandstone, forms the low hills on the left.

Mile 6.5: Milepost 387. Elevation 6,000 feet. Roadcuts are in ancient stream deposits of the Westwater Canyon Sandstone Member of the Morrison Formation at **Mile 7.7.** The Organ Rock monocline is really impressive to the right. Steep dipslopes and the caprock at the skyline are the nearly white Navajo Sandstone; erosional gullies expose the Kayenta and Wingate Sandstones. The highway is beginning to skirt the southern plunging

Figure 66. The Organ Rock monocline, west of Kayenta, Arizona. The light-colored rocks rolling over sharply to the right are in the Navajo Sandstone (Jurassic), where they have been draped across a deep-seated basement fault.

nose of the Organ Rock monocline at **Mile 10.0.** Beautiful exposures of the steeply dipping Navajo Sandstone are to the right. **Mile 10.5: Milepost 383.**

Tsegi (meaning "rocky canyon" in Navajo) is on the right at **Mile 11.3.** It is a rocky canyon indeed, cutting down through the Navajo and Kayenta Sandstones into the Wingate Sandstone. Marsh Pass, elevation 6,750 feet, is at **Mile 12.1.** Entering Long House Valley ahead.

Rounded slopes of the Navajo Sandstone are to the right at **Mile 12.2** on the southern plunging nose of the Organ Rock anticline. The low cliffs to the left are in the Westwater Canyon Sandstone Member of the Morrison Formation, with occasional exposures of thin Salt Wash Member, followed by the underlying Cow Springs Member of the Entrada Sandstone. **Mile 17.3:** Exposures of the light-colored Cow Springs Member are in the low cliffs to the left. **Mile 17.6: Milepost 376:** The Navajo Sandstone to the right is plunging toward the south.

Mile 19.3: Intersection of U.S. 160 and Arizona 564 to the right. Route 564 goes 9.5 miles to Navajo National Monument, site of the well-preserved Betatakin Anasazi ruins.

The road to the left goes to coal strip mines on top of Black Mesa; the coal comes from the middle section (Wepo Formation) of the Mesaverde Group. The coal is processed into a slurry and carried by an 18-inch pipeline some 275 miles to the Mohave Power Plant in southern Nevada. Water for the slurry comes from deep water wells drilled on Black Mesa. Other coal is transported by conveyer to the railhead a mile ahead at U.S. 160 and then is carried by electric rail some 80 miles to the Navajo Generating Station at Page, Arizona.

The stratigraphic section to the left consists of the basal white cliffs of the Cow Springs Member beneath ledges of the Morrison Formation (both Jurassic). The Morrison is here subdivided into the lower Recapture Member and the upper Westwater Canyon Member, the Salt Wash Member having pinched out a short distance northeast of here. The Morrison is overlain by the Cretaceous Dakota Sandstone and the dark gray Mancos Shale, topped by cliffs of the Cretaceous Mesaverde Group, consisting here of the Toreva and Wepo Formations. **Mile 19.6: Milepost 374.**

Mile 20.1: The coal conveyor system crosses the highway from Black Mesa to the electric train loading dock. The electric railroad follows the highway on the right for several miles. Low ledges on the left at **Mile 24.6** are in the Morrison Formation with the Dakota Sandstone caprock at the skyline. The Navajo Sandstone forms the light-colored knobs on the right across Long House Valley. **Mile 24.7: Milepost 368:** Scattered exposures of the Cow Springs Member of the Entrada Sandstone show through the valley cover in the vicinity. **Mile 28.7: Milepost 365.** The road to the right at **Mile 32.1** is Arizona 98, leading to Page, Arizona (see Roadlog 40).

Mile 33.7: Milepost 360: The highway is still in Long House Valley, which is heavily covered with windblown, partly vegetated sand. The mesa to the right at **Mile 35.5** is capped by the white Cow Springs Member, which is also exposed on the left at the base of the cliff. Exposures to the left at **Mile 36.7** are in the Cow Springs Member at the base of the cliff. The old abandoned Cow Springs Trading Post is on the left at **Mile 40.1.** The type section for the Cow Springs Member is near here. **Mile 43.4: Milepost 350**: The electric railroad turns north toward Page. At **Mile 44.8** the San Francisco Peaks are visible in the far distance ahead. The low cliffs on the left at **Mile 45.5: Milepost 348** and the low hills in the valley are the Cow Springs Member, now known to be a bleached upper part of the Entrada Sandstone.

Mile 48.5: Milepost 345: The Elephant's Feet, on the right, are remnants of the Cow Springs Member. **Mile 49.6: Milepost 344**: Red Lake Trading Post is on the left. The highway climbs through sandstone of the Cow Springs Member. The road to the right at **Mile 50.0** goes to the town of Tonalea, in Navajo "water flows and collects." The Tonalea General Store is on the left. The wide valley floor is thickly covered with dune sand with a few scattered "boondocks." Exposures of the Cow Springs Member are on the left at **Mile 51.4. Mile 55.7: Milepost 338**: The San Francisco Peaks are visible way to the west. **Mile 59.1**: A vast field of well-developed "boondocks," especially to the left.

Mile 62.9: Milepost 331: The San Francisco Peaks are visible ahead at 11:00, and the Kaibab Plateau is visible at 1:00 to 2:00 in the distance. The site of a former uranium processing mill, now covered with black basaltic boulders, is near **Mile 65.0** on the east side of the highway. **Mile 66.0: Milepost 328**: In the valley bottom to the left are exposures of Navajo Sandstone with the Cow Springs Member holding up the surface. Everything else is covered with windblown sand. **Mile 68.1: Milepost 326**: The Kaibab Uplift forms the distant skyline ahead. Tuba City is visible in the middle distance. Many "boondocks" are to the right and left at **Mile 69.0.** The highway is at about the top of the Navajo Sandstone.

Mile 71.1: Entering Tuba City, Arizona. The village was named for Tueva, a Hopi Indian chief, and the spelling was corrupted by later settlers. The village is apparently in the Navajo Sandstone but is heavily covered by windblown sand. A stoplight is at **Mile 72.3.** The road to the right goes to the city center; the road to the left, Arizona 264/Navajo Route 3, goes to Moenkopi, Keams Canyon, and several Hopi pueblos on Black Mesa.

Mile 72.5: The road descends through the lower part of the Navajo Sandstone, goes into the red silty Kayenta Formation, and then goes into the

Figure 67. The
stratigraphic relationships
between the Wingate-
Kayenta-Navajo Forma-
tions of the Monument
Valley region to the east
and the Moenave Forma-
tion of the western
Colorado Plateau near
Tuba City, Arizona. Drawn
from written communica-
tions with Fred "Pete"
Peterson, 2001.

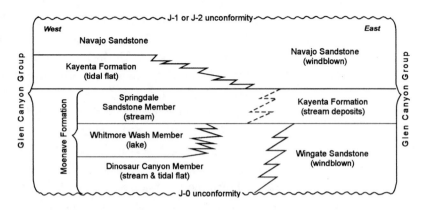

Figure 67. The stratigraphic relationships between the Wingate-Kayenta-Navajo Formations of the Monument Valley region to the east and the Moenave Formation of the western Colorado Plateau near Tuba City, Arizona. Drawn from written communications with Fred "Pete" Peterson, 2001.

Moenave Formation. The Moenave Formation is a series of generally red stream deposits of Early Jurassic age that are the lateral equivalent of the Wingate Sandstone and the Kayenta Formation (Figure 67). The term *Moenave Formation* is used for this section from here to the west across the Colorado Plateau.

Mile 73.1: Milepost 321: The highway is in the Chinle Formation (Upper Triassic) with limestone beds of the Owl Rock Member at the top of the grade. The road to the right at **Mile 77.6** goes to Moenave. The dinosaur tracks advertised on the sign are in the Moenave Formation. Hamblin Wash is at **Mile 81.9.** The highway is in the lower Chinle Formation. Exposures of the Chinle here are considered an extension of the Painted Desert.

Mile 82.6: Intersection with U.S. 89. The road to the right goes to Marble Canyon; the road to the left goes to Cameron (14 miles) and eventually to Flagstaff, Arizona. (To continue on Arizona 64 to Grand Canyon National Park, see Roadlog 29. For Flagstaff or Page on U.S. 89, join Roadlog 31 at **Mile 62.5.**)

IV

Grand Canyon Country

I do not describe the geology of Grand Canyon here, as I have covered the topic in some detail in *The Colorado Plateau: A Geologic History* (Baars 2000) and *Navajo Country* (Baars 1995). That is not to say that the geology of the grandest canyon is unimportant. Rather, the scope of this book is the geology visible along the roads leading to the canyons. But it is undeniable that the exposures of rocks of Paleozoic and Precambrian age in Grand Canyon are unequaled anywhere else on earth. They tell a fascinating but complex story that is basic to understanding the geology of the Colorado Plateau (Figure 68).

It is certainly awe-inspiring to view the canyon from the rim or to add a third dimension by hiking one of the trails into the magical gorge. To fully appreciate Grand Canyon, however, one must see it by riverboat. Most of the Precambrian and Paleozoic section is seen at close hand by river, from the Precambrian basement rocks up through the Permian limestone. Of course, there are no known rocks of Ordovician or Silurian age present in the canyon or elsewhere on the Colorado Plateau (Figure 69). In Grand Canyon country the tales of advances and retreats of the seas through millions of years of earth history, from the great oceans that covered western Utah and Nevada onto the higher, shallower shelves of what is now the Plateau Country, are beautifully demonstrated. But more time has been lost to the various and sundry unconformities than is represented by the rocks. Here is a geologic fantasyland.

The Colorado River has carved these canyons across high, up-arched plateaus as if the structures did not exist. First there is the Echo Cliffs monocline, which brings the Paleozoic rocks abruptly up into view where Marble Canyon dissects the Marble Plateau. Where the river then encounters the East Kaibab monocline, the strata are again sharply uplifted, exposing the stratigraphic section down into rocks of Precambrian age at the head of

Figure 68. The Grand Canyon in all its glory. Grand exposures of rocks from Precambrian through Paleozoic age are seen here, for an unprecedented 188 river miles of pages in an "open book," as John Wesley Powell described it.

Grand Canyon proper. Exactly how and when the river decided to ignore the regional geologic structures and carve these magnificent canyons is still open to debate among geologists. Then, in western Grand Canyon, several major north-south faults are exposed that drop the strata, step by step, back down to the level of the Great Basin of western Utah and Nevada.

The roads that lead to canyon viewpoints generally skirt the major up-lifts, crossing the monoclines and faults where the barriers have the least topographic significance and then crossing the very broad plateaus that form the crests of the huge anticlinal folds. In the case of Grand Canyon country, the roads are largely constructed at or near the top of the Kaibab Limestone, of Permian age. The younger overlying rocks of Mesozoic age, especially the Triassic Moenkopi and Chinle Formations, being less resistant to erosion, have been stripped back from the uplifted areas to form the beautifully colored escarpments of the Echo Cliffs and Vermilion Cliffs along the way on the east and north. Because the Kaibab Limestone is far more resistant to the weathering processes, it has been left in place to cap the plateaus and rim the canyons.

Figure 69. The
stratigraphic section and
names applied to the
rocks in Grand Canyon,
from the Precambrian up
through rocks of Paleo-
zoic age. From Potochnik
and Reynolds 1990.

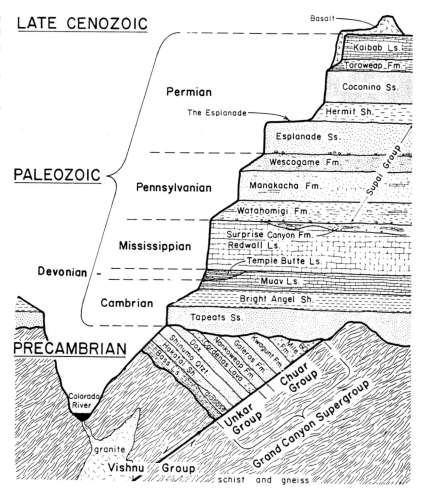

Figure 69. The stratigraphic section and names applied to the rocks in Grand Canyon, from the Precambrian up through rocks of Paleozoic age. From Potochnik and Reynolds 1990.

This situation leads to rather dull geology along the highways, as nearly every exposure and roadcut is in the Kaibab Limestone. Yet the breathtaking viewpoints sprinkled around the Grand Canyon justify the drive there.

29. Cameron, Arizona, to East Entrance to Grand Canyon National Park via Arizona 64

Arizona 64 west from Cameron, Arizona, ascends the East Kaibab and Coconino Point monoclines to the east rim of Grand Canyon. Most of the road is along the top of the Kaibab Limestone (Permian in age), the rimrock of much of Grand Canyon, with occasional roadcuts in the overlying Moenkopi Formation of Triassic age. Views to the right into the Little Colorado River Canyon and other tributaries expose rocks of Paleozoic age that are found in Grand Canyon proper at Desert Viewpoint.

Mile 0.0: Intersection of Arizona 64 and U.S. 89 at Cameron, Arizona. Turn west toward Grand Canyon, East Entrance. The highway is in the lower Chinle Formation.

Mile 0.9: Exposures of the Shinarump Member of the Chinle Formation (Triassic) are on the right and left. **Mile 1.2: Milepost 294.** Shinarump exposures are to the right and left at **Mile 1.8.**

There are exposures of the Moenkopi Formation (Triassic) at **Mile 3.7** below Shinarump ledges on the left. **Mile 4.2: Milepost 291**: The prominent monocline ahead is a southeastern extension of the East Kaibab monocline. At **Mile 4.6** the highway is going down-section into the Moenkopi Formation. **Mile 5.2: Milepost 290.** The highway at **Mile 5.8** parallels the monocline.

The top of the Kaibab Limestone (Permian) is on the right at **Mile 6.1.** The highway is near the top of the Kaibab Limestone for the next few miles. The Kaibab is here seen to roll over the crest of the East Kaibab monocline, the steep fold facing the northeast, and then flatten at road level. **Mile 8.2: Milepost 287**: The Moenkopi is exposed on the right, above the top of the Kaibab Limestone. The road to the right at **Mile 9.5** goes to the Little Colorado River overlook, where the Kaibab is at the surface.

Mile 10.2: Milepost 285: The highway is at the top of the Kaibab Limestone. The DeChelly Sandstone, as seen in Monument Valley and Canyon de Chelly, pinches out against the East Kaibab monocline near here and is not seen in Marble or Grand Canyons. If it were present in the canyons, it would lie between the Coconino Sandstone and the Hermit Shale. A tributary canyon at **Mile 12.3** is cut downward into the Kaibab Limestone on the right.

Mile 13.2: Milepost 282: Roadcuts are in the Moenkopi Formation for the next mile, close against the monoclinal fold on the left. **Mile 14.2: Milepost 281**: The highway is running along the base of the monoclinal flexure. Moenkopi red beds are exposed in the roadcuts, and the Kaibab Limestone swoops upward onto the monocline to the left. At **Mile 14.9** a tributary canyon can be seen to the right, cut down into the Kaibab Limestone. The highway begins a steep climb at **Mile 15.9** up onto the top of the monocline.

Mile 16.5: Milepost 279: The highway is in roadcuts of Kaibab Limestone on the crest of the East Kaibab monocline for the next 2 miles. **Mile 18.4**: Entering Kaibab National Forest. There is another small monoclinal flexure, the Coconino Point monocline; the Kaibab Limestone rises to the top of the monocline to the left. **Mile 25.8: Milepost 270:** Roadcuts are in the Kaibab Limestone for the next 4 miles.

Mile 27.9: Milepost 268: Entering Grand Canyon National Park. **Mile 29.8: Milepost 265**: The vegetation has changed from juniper-pinyon forest to heavily wooded ponderosa pine. **Mile 31.1**: East entry pay station for Grand Canyon National Park, a fee area. Desert View Campground is to the right at **Mile 31.3.**

Figure 70. Desert Watchtower, a landmark at the east end of Grand Canyon proper on the South Rim in Grand Canyon National Park. The structure is visible from the Colorado River.

Figure 71. The eastern Grand Canyon from Desert View overlook. The Colorado River is visible in the lower right of the photograph. The layered rocks above the river and in the lower half of the view are in the Grand Canyon Supergroup of Late Precambrian age, seen dipping rather strongly to the right (east). The flat-lying cliff that truncates the Precambrian section at the "Great Unconformity" is the Tapeats Sandstone of Cambrian age. Above the angular unconformity the entire section of layered rocks of Paleozoic age is beautifully exposed.

Mile 31.4: Desert View overlook and Desert Watch Tower are to the right (Figure 70). The view is of the eastern end of Grand Canyon proper, more than 4,500 feet deep here, with the East Kaibab monocline visible to the east. The Colorado River may be seen in the deepest part of the canyon, where it crosses exposures of the Late Precambrian Nankoweap, Cardenas, and Dox Formations, seen from right to left. The upturned ancient sedimentary rocks are truncated by the "Great Unconformity," with the Tapeats Sandstone of Cambrian age resting above the angular erosional surface. The section above the Tapeats cliffs includes the Upper Cambrian Bright Angel Shale and Muav Limestone, very thin beds of the Upper Devonian Temple Butte Limestone, the Redwall Limestone of Mississippian age, the Supai Group of Pennsylvanian and Permian age, and the Coconino Sandstone, the Hermit Shale, the Toroweap Formation, and the Kaibab Limestone caprock, all of Permian age (Figure 71). Cedar Mountain, an erosional remnant of Moenkopi Formation, capped by the Shinarump Member of the Chinle Formation, is about 3 miles to the east.   The Grand Canyon National Park brochure, handed out at the entry stations, contains a good map of the many viewpoints along the South Rim. For those interested in details of mileage between overlook points, the roadlog from Cameron to Grand Canyon Village, South Rim, continues as follows:

Mileage

31.6: Leaving the Desert View parking area.

32.4: The road to the right goes to Navajo Point.

33.0: Milepost 262.

33.5: The road to the right goes to Lipan Point.

33.7: Turnout for viewpoint.

33.9: Milepost 261.

34.8: The road to the left goes to Tusayan Ruins and Museum.

36.0: Milepost 259.

38.6: The road to the right goes to Moran Point.

40.0: The countryside is forested with ponderosa pine.

40.6: Roadcuts to the right are in the Kaibab Limestone.

41.0: Milepost 254.

41.3: Picnic area on the right.

42.6: Arizona Trail on the left. The road crosses the Grand View monocline with a structural relief on the Kaibab Limestone of about 400 feet.

44.0: Milepost 251.

44.6: The road to the right goes to Grand View Point.

45.0: Milepost 250.

46.8: Viewpoint parking on the right.

Figure 72. The layered rocks of Paleozoic age as seen from midway down the Kaibab Trail. The massive cliff just beyond the hiker is the Redwall Limestone, of Mississippian age. Below that cliff are beds of Late Cambrian age. The ledgy cliffs above the Redwall Limestone are in the Supai Group of Pennsylvanian and Permian age with a prominent slope-former above in the Hermit Shale. The next massive cliff above is in the Coconino Sandstone, capped by lesser cliffs of the Toroweap and Kaibab Formations, all of Permian age.

47.0: Milepost 248.

47.3: Picnic area on the right.

48.0: Milepost 247.

48.5: Viewpoint parking area on the right.

51.2: Picnic area on the left.

52.1: Milepost 243.

52.2: Yaki Point overlook and the top of the Kaibab Trail are to the right (Figure 72).

52.5: Overlook parking on the right.

53.1: Milepost 242.

53.3: T-intersection: south entrance road and Flagstaff to the left, Grand Canyon Village to the right. Turn right.

54.0: Mather Point overlook to the right.

54.6: Yavapai Point to the right.

55.2: Visitor Center and Yavapai Museum to the right. The parking area is usually full.

56.1: Stop sign at the head of the Grand Canyon Railroad. El Tovar Hotel is to the right.

56.2: Train depot.

56.3: Bright Angel Lodge on the right.

56.5: Entrance to Bright Angel Lodge and parking area, always full.

56.6: Bright Angel Trailhead to the right.

56.7: Loading station for West Rim Drive bus trains is on the right (no parking facilities). The West Rim Road, going west to Hermit's Rest, is closed to private vehicles most of the year. Through traffic must turn left and follow the loop back toward the South Rim entrance.

58.2: Road to the right goes to the Babbitt shopping center and lodge.

59.2: Yavapai Point parking to the left.

60.1: Leaving Yavapai parking area (full).

60.9: Intersection with East Rim Drive. Go straight (south) toward the south entrance gate.

62.2: The road to the right goes to the South Rim Clinic.

65.0: South Gate pay station.

65.5: Leaving Grand Canyon National Park, entering Kaibab National Forest. Flagstaff is 74 miles on U.S. 180 and Arizona 64.

66.6: Entering Tusayan, with tourist accommodations of all sorts.

(To continue south on U.S. 180 to Flagstaff, see Roadlog 30.)

30. South Rim of Grand Canyon National Park to Flagstaff, Arizona, via U.S. 180

The highway from the South Rim of Grand Canyon to Valle Junction is on open plains at the top of the Kaibab Limestone down the south flank of the Kaibab Uplift. From Valle Junction to Flagstaff, Arizona, the highway crosses the San Francisco volcanic field, mostly of Quaternary age.

Mile 0.0: South edge of Tusayan (Best Western Grand Canyon Squire Inn) on U.S. 180, heading south toward Flagstaff, Arizona. Tusayan Airport is on the right at **Mile 0.3**.

Mile 1.5: Milepost 234: The highway descends the south flank of the Kaibab Uplift, with roadcuts and occasional exposures of the Kaibab Limestone (Permian), for the next 20 miles.

At **Mile 5.2** roadcuts are in the Kaibab Limestone with a local cap of Moenkopi Formation.

Mile 5.5: Milepost 230. Small volcanoes are visible ahead in the distance at **Mile 8.0.** Numerous roadcuts along the way are in the Kaibab Limestone as the highway descends the southern nose of the Kaibab Uplift on the resistant Kaibab surface. Red Butte is visible occasionally to the left at **Mile 10.0.** Volcanoes are obvious ahead in the distance at **Mile 10.5: Milepost 225.**

Mile 11.5: Leaving Kaibab National Forest. The prominent butte to the left (east) is Red Butte, an erosional remnant of the Moenkopi Formation overlain by the Shinarump Member of the Chinle Formation. The butte is capped by an 8-million-year-old Tertiary basalt flow.

Mile 15.6: Milepost 220: The San Francisco volcanic field is ahead and to the left in the distance. The composite volcanoes consist of the San Francisco Peaks (Mount Humphreys, elevation 12,655 feet, the Navajo sacred mountain of the west), Kendrick Mountain (elevation 10,418 feet), and Sitgreaves Mountain (elevation 9,388 feet). Roadcuts at **Mile 18.7** are in the Moenkopi Formation. Bill Williams Mountain (elevation 9,255 feet) is straight ahead in the distance.

Mile 22.2: Valle Junction: straight ahead is Arizona 64 to Williams, Arizona; Flagstaff is to the left (50 miles) on U.S. 180. Turn left. Bedrock City is on the right. The highway and countryside are on top of the Kaibab Limestone.

Mile 22.9: Milepost 265: The San Francisco Peaks are ahead, with Mount Humphreys, the highest peak in Arizona, at about 11:00. The peak was named for Andrew A. Humphreys of the army's Topographical Engineers Office of Western Explorations and Surveys. A small cinder cone is on the skyline to the left at **Mile 25.1**. Red soil colors at **Mile 30.6** suggest that the highway is now in the Moenkopi Formation. **Mile 31.9: Milepost 256**: Small cinder cones are to the right, and Mount Humphreys is ahead at about 1:00.

Mile 33.0: Entering Kaibab National Forest. Volcanic rocks and cinder cones are exposed on the right for the next 7 miles. **Mile 37.8: Milepost 250**: The highway is on a volcanic surface. **Mile 40.5**: Mount Humphreys is straight ahead, and volcanic features are all around the area. **Mile 40.9: Milepost 247**: There is a good view to the right into the dissected cinder cone of Red Mountain; the structure is well exposed because of weathering. **Mile 43.1**: Elevation 7,000 feet and climbing between two cinder cones to a saddle at **Mile 43.7.**

Mount Humphreys is dead ahead at **Mile 45.0** with cinder cones to the right. At **Mile 48.1** the highway is on a basaltic lava flow. **Mile 50.8**: Kendrick Recreation Area in Coconino National Forest is on the right (Figure 73). **Mile 51.0: Milepost 237**: Entering Kendrick Park. This area was farmed by homesteaders around the turn of the century. The Kendrick General Store is to the left at **Mile 51.7.** The road to the east circles the north flank of the San Francisco Peaks to U.S. 89. The White Horse Hills at about 10:00 have rocks of the Redwall through Kaibab Formations domed by a 0.85 million-year-old rhyolite intrusive body. **Mile 52.5**: Picnic area to the right. Hart Prairie Road is on the left at Mile 52.9

Mile 53.0: Milepost 235. The highway crosses a saddle at **Mile 54.4**, elevation 8,046 feet. **Mile 63.0: Milepost 225**: Leaving Coconino National Forest. Entering Fort Valley at **Mile 64.3**, named for old Fort Moroni. Mount Humphreys is to the left. **Mile 65.0: Milepost 223**: The road to the left goes to the Arizona Snow Bowl, a popular ski area. Lava flows are to the left and right for the next 3 miles.

Figure 73. Mount Humphreys, a composite volcano of Tertiary age, is the highest peak in the San Francisco Mountains. This view is from the Kendrick Recreation Area in the Coconino National Forest.

Mile 68.5: Entering Flagstaff, elevation 6,906 feet. Mile 69.5: The Museum of Northern Arizona is to the right. Mile 70.8: Milepost 217: Entering Flagstaff proper. Stoplight at Humphrey Boulevard at Mile 71.8. Turn right. Mile 72.4: Intersection of U.S. 180 and U.S. 66.

31. Flagstaff to Marble Canyon, Arizona, via U.S. 89

U.S. 89 heads north from Flagstaff, crosses the San Francisco volcanic field, and then progresses north across plains formed near the top of the Kaibab Limestone. "Painted desert" landscapes are extensive to the east for much of the trip. The route follows the west flank of the Echo Cliffs monocline for the last several miles and then crosses down-section to the top of the Kaibab Limestone at Navajo Bridge across the Colorado River. Beautiful exposures of the Kaibab and Toroweap Formations are seen below the bridge at Marble Gorge.

Mile 0.0: Intersection of Interstate 40 and U.S. 89 at the east end of Flagstaff. Turn left on U.S. 89 toward Marble Canyon to the north. Mile 1.0: Stoplight. The cinder cone on the right is under destruction for road metal. The view to the left at Mile 3.1 is of the San Francisco Peaks, composite volcanoes of Tertiary age, and several Quaternary cinder cones can be seen to the right and ahead. Leaving Flagstaff at Mile 8.3. Mile 8.4: Milepost 427.

Mile 9.8: Elevation 7,000 feet. Volcanic rocks and debris are in the roadcuts and hillsides to the right and ahead. The San Francisco Peaks are to the left. Schultz Pass Road is to the left at Mile 10.6. The road is in a ponderosa pine forest in volcanic terrain. Mile 11.4: Milepost 430.

The road to the right at Mile 11.9 goes to Sunset Crater National Monument and Wupatki National Monument (see Roadlog 33). Mile 12.4: Milepost 431: Elevation 7,282 feet. The road crosses a saddle between two cinder cones; the one to the left has been nearly torn down for gravel. There is a good view of Sunset Crater, a Pleistocene cinder cone, on the right at **Mile 12.7.**

Mile 13.5: Milepost 432: Elevation 7,000 feet. The highway heads down from the volcanic mountains toward a broad plateau. **Mile 16.5: Milepost 435**: The highway is on the plateau surface of the Kaibab Limestone, and a long string of cinder cones is to the left. The highway at **Mile 18.9** crosses a broad hill of volcanic cinders. **Mile 21.5: Milepost 440**: The highway descends from the cinder-covered hill. **Mile 21.9**: Elevation 6,000 feet. There are lava flows ahead on the left and right. The road to the right at **Mile 26.4** goes to Wupatki National Monument. **Mile 26.5: Milepost 445.**

Mile 31.6: Milepost 450: Lava knobs are to the right, and the Kaibab Plateau is visible to the left. **Mile 32.1**: The lava-capped Moenkopi Formation is in the roadcut and exposures to the left and right. A thin lava flow to the left at **Mile 32.8** thins and pinches out onto hills of the Moenkopi Formation. **Mile 35.4**: More lava-capped hills on the Moenkopi Formation. The lava flows are very thin. The northern limits of the lava flows appear to be at **Mile 36.2**; the highway is on Moenkopi red beds. **Mile 36.6: Milepost 455.**

Mile 38.4: Entering Gray Mountain, Arizona. Leaving Gray Mountain at **Mile 39.1** and entering the Navajo Indian Reservation. The view ahead is of dark red Moenkopi hills, capped by the Shinarump Conglomerate Member of the Chinle Formation; the Kaibab Plateau and monocline are to the left.

Mile 40.3: The Kaibab Limestone is exposed in a gully to the left, overlain by the Moenkopi Formation with an obvious contact around the countryside. The Kaibab-Moenkopi contact is obvious to the left and right at **Mile 40.9** in the numerous Moenkopi hills. **Mile 41.6: Milepost 460**: The highway is in the Moenkopi Formation.

The highway is passing up-section through the Shinarump Member of the Chinle Formation at **Mile 42.7**. There is a nice channel-filled surface at the base of the Shinarump. **Mile 45.2**: The Shinarump Member is exposed in roadcuts and natural outcrops to the left and right. **Mile 46.7: Milepost 465.**

The road left at **Mile 46.9**, Arizona 64, goes to Grand Canyon National Park, South Rim, east entrance (see Roadlog 29). The highway is in the basal Chinle, just above the Shinarump Member. Cameron, Arizona, is at **Mile 48.4.** There are exposures of the Moenkopi Formation on the left. **Mile 48.5**: Bridge over the Little Colorado River.

At **Mile 49.7** the highway climbs back on top of the Shinarump Member with broad exposures of the upper members of the Chinle Formation. There is a cinder cone, Shadow Mountain, to the left at about 10:00. **Mile 51.7: Milepost 470**: Painted Desert–like topography and colors in the upper members of the Chinle Formation are ahead and to the right with Shadow Mountain off to the left. **Mile 56.7: Milepost 475:** The road is still in the badlands of the Chinle Formation. **Mile 61.7: Milepost 480:** Still in the Chinle Formation.

The road to the right at **Mile 62.5** is U.S. 160, going to Tuba City and Kayenta, Arizona, and the Four Corners (see Roadlog 28). **Mile 66.8: Milepost 485:** Exposures to the right are varicolored shale of the Chinle Formation, capped by red sandstone beds of the Moenave Formation, a relatively thin-bedded equivalent of the Wingate Sandstone, and the Kayenta Formation of Early Jurassic age. The light-colored Navajo Sandstone caps the high peak to the right. This is the southern end of the Echo Cliffs monocline.

Mile 68.0: The top of the Shinarump Member of the Chinle Formation is exposed to the left. The strata here dip noticeably to the right (east) across the Echo Cliffs monocline. The relatively flat plateau to the left (west), between the Echo Cliffs monocline and the East Kaibab monocline, is called the Marble Platform. The Shinarump Member is exposed on the left at **Mile 69.4,** and the lower shale member of the Chinle is to the right. The beds are here dipping more strongly to the east. **Mile 71.4: Milepost 490:** The highway is about midway up into the Chinle Formation, near the sharpest fold of the Echo Cliffs monocline, capped on the right by beds of the Moenave Formation. **Mile 73.0:** Highly fractured exposures of the Navajo Sandstone form the skyline to the right.

The crest of the high cliff to the left at **Mile 74.6** is the Shinarump Member. The red Moenave Formation, overlain by the slope-forming Kayenta Formation, immediately below the highly fractured Navajo Sandstone, is exposed on the right. **Mile 76.4: Milepost 495.** The Gap, Arizona, is at **Mile 78.9,** a village and low saddle in the Echo Cliffs. Arizona 20 to the right goes to a copper mine and Page, Arizona, via an unpaved road. **Mile 81.4: Milepost 500.**

Mile 84.1: The Kaibab Limestone is exposed on the left and the Moenkopi Formation and Shinarump and shale members of the Chinle on the right. The highway is along the Moenkopi-Kaibab contact at **Mile 85.0.** The Echo Cliffs are to the right. **Mile 85.4:** Entering Cedar Ridge, Arizona, a Navajo village.

Mile 86.5: Milepost 505: The characteristic fold of the Echo Cliffs monocline is obvious here, with the Kaibab Limestone on the left and the Moenave and Kayenta Formations on the right and occasional glimpses of the Navajo Sandstone high on the right. The highway is back into the lower Chinle Formation. **Mile 91.5: Milepost 510:** Exposures of the Kaibab Limestone are to the left, and the Moenkopi Formation lies ahead. **Mile 95.8:** The hogback to the left is the Shinarump Member. The highway ahead crosses

Figure 74. The new Navajo Bridge across the Colorado River at Marble Canyon. The original bridge, preserved as a footbridge, is just beyond the structure and largely hidden in the photo. The bridge abutments are in the Kaibab Limestone of Permian age, and the entire formation is exposed in the cliff beneath the bridge. The lowest ledgy cliff is in the Toroweap Formation. The Echo Cliffs are visible in the distance.

down through the Shinarump and back into the Moenkopi Formation. The Chinle, Moenave, and Navajo Formations form the Echo Cliffs to the right. The highway crosses down through the Shinarump into the Moenkopi at **Mile 96.2** with exposures of the Kaibab Limestone to the left. **Mile 96.5: Milepost 515.**

Mile 97.0: The highway is on the top of the Kaibab Limestone here, and exposures up through the Navajo Sandstone are to the right. The Vermilion Cliffs form the skyline to the north, or ahead. **Mile 101.6: Milepost 520**: The highway is still on top of the Kaibab Limestone, the Echo Cliffs are to the right, and the Vermilion Cliffs are ahead. **Mile 104.7**: Entering Bitter Springs, Arizona. **Mile 105.5**: Intersection of U.S. 89 and 89A. Highway 89 goes to Page and crosses the Echo Cliffs (see Roadlog 41); Highway 89A goes left to Marble Canyon. Turn onto Highway 89A.

Mile 106.6: Milepost 525: The highway is still on top of the Kaibab Limestone. At **Mile 107.7** the highway begins the descent to the lip of Marble Canyon. The highway is on the Kaibab Limestone at **Mile 108.8**, and the Moenkopi Formation forms the broad valley to the right. The Shinarump Member forms the low bench below slopes of Chinle mudstone and Moenave, Kayenta, and Navajo high cliffs. The Vermilion Cliffs are ahead and to the left. The Kaibab-Moenkopi contact is at road level at **Mile 109.8**. The highway is here on the Moenkopi Formation. **Mile 111.1**: Exposures of the Kaibab Limestone are in the gully to the right. **Mile 111.6: Milepost 530.**

Mile 113.7: Elevation 4,000 feet. Exposures of the Kaibab Limestone are

Figure 75. A *hoodoo*, or pedestal rock, along the road from Marble Canyon to Lees Ferry. A large block of talus has fallen onto a shale slope and protected the underlying soft rock of the Chinle Formation from completely eroding away, supporting the pedestal.

in the gully to the left; the highway is in the basal Moenkopi Formation. **Mile 116.7: Milepost 535**: The Shinarump Member forms a strong bench to the right. **Mile 117.5**: The stratigraphic section in the Vermilion Cliffs to the left is the same as in the Echo Cliffs to the right, consisting of the Kaibab up through the Navajo Formation.

Mile 118.7: Entering Glen Canyon National Recreation Area and leaving the Navajo Indian Reservation. **Mile 119.2**: The highway descends through roadcuts in the Kaibab Limestone toward Navajo Bridge, where it crosses the Colorado River. Navajo Bridge over the Colorado River in Marble Canyon is at **Mile 119.6**. A new, wider bridge was recently constructed here to accommodate motorized traffic (Figure 74); the old Navajo Bridge was preserved for foot traffic and sight-seeing. This is the only road crossing of the Colorado River between Glen Canyon Dam and Hoover Dam. The road to the right goes to Lees Ferry, Arizona, the site of the old ferry crossing and launch site for Grand Canyon river trips (see Roadlog 32).

32. Marble Canyon to Lees Ferry, Arizona

Mile 0.0: Turnoff from U.S. 89A between Navajo Bridge and Marble Canyon Lodge. Turn right toward Lees Ferry. The road is on the Kaibab Limestone near the base of the Moenkopi Formation. The Shinarump Member of the Chinle Formation makes the obvious lower ledge. **Mile 0.6**: Pay station for entrance to Glen Canyon National Recreation Area. Beautiful exposures of the Moenkopi are all around.

The high cliffs ahead at **Mile 0.8** are the Navajo Sandstone atop the Vermilion Cliffs. The gully to the right at **Mile 1.6** exposes the Kaibab Limestone in the Paria drainage. Turn out on the left at **Mile 2.1** to view some well-developed "hoodoos," Shinarump boulders sitting on and protecting Moenkopi pedestals (Figure 75). The hill ahead at **Mile 2.7** is of the Moenkopi Formation, capped by Cenozoic river terrace gravels. The top of the Kaibab Limestone is at road level at **Mile 4.1.** Hills of Moenkopi at **Mile 4.6** are capped by terrace gravels. **Mile 4.8**: A ranger station is on the left.

Mile 5.2: Bridge over the Paria River; the river road is to the right. A fish cleaning station is on the left at **Mile 5.5**. The boat launching ramp is on the right at **Mile 5.8** for river trips through Marble and Grand Canyons. **Mile 5.9**: The old Lees Ferry cabins are just ahead.

33. Wupatki and Sunset Crater National Monuments

Mile 0.0: Intersection of U.S. 89 and Arizona 22 at the entrance to Wupatki National Monument in northern Arizona. Citadel Ruin is 4 miles, Wupatki is 13 miles, the Visitor Center at Sunset Crater is 31 miles, and back to U.S. 89 is 34 miles. Quaternary basaltic lava flows are to the right and left.

Milepost 1: The countryside is covered in grassy plains and volcanic soil. **Milepost 3**: A good view of the San Francisco Peaks, Tertiary composite volcanoes, is to the right. Exposures of lava flows are all around at **Mile 3.6.** The road to Lomaki Ruin is to the left at **Mile 3.8.** The road to Citadel Ruin is to the right at **Mile 4.1.** Surface ruins are scattered through exposures of volcanic rocks. **Milepost 5**: Sunset Crater is visible at about 1:00 ahead.

Mile 6.3: Entering Coconino National Forest, consisting of scattered juniper trees. **Milepost 9**: Four small cinder cones are to the right. The road to the right at **Mile 9.4** goes to a viewpoint and picnic area. **Milepost 10**: A field of black cinders is to the right and left. A hill of Kaibab Limestone is on the left at **Mile 10.4.** A small monocline is on the left (north) at **Mile 10.6**; the Kaibab Limestone is at the surface of the structure, and Moenkopi red beds fill in the low, down-folded side. The road is on the Kaibab. Rolling hills of Moenkopi red beds are ahead at **Mile 11.3**, with black cinders filling in the hollows. The road crosses a gully at **Mile 12.6**; the Kaibab Limestone is exposed in the wash, surrounded by the Moenkopi Formation; cinders fill in the low places.

Mile 13.6: Wupatki Ruins and the Visitor Center are to the right. Wupatki is a relatively large, multistoried surface ruin. The volcanic eruptions occurred in the year A.D. 1065. The road to the left at **Mile 13.9** goes to Wukoki Ruin, 2.5 miles. **Mile 15.9**: Leaving Wupatki National Monument.

Mile 16.4: Lava-capped mesas are to the right and left with cinders and other volcanic debris filling the low places in the Moenkopi Formation. The road is climbing onto a lava-capped ridge at **Mile 17.2**. More cinder cones are ahead and to the right at **Mile 18.3**, and the San Francisco Peaks are in

Figure 76. Sunset Crater, in Sunset Crater National Monument near Flagstaff, Arizona, a very recently formed volcanic cinder cone, as seen across Bonita Park.

the right distance. **Mile 20.7**: Cattle guard at the boundary of the Coconino National Forest.

Mile 23.8: A lava flow is exposed on the left with cinders covering the hillsides. Cinder cones are ahead and to the right. **Mile 25.3**: Painted Desert Vista and Forest Service Recreation Site to the right. There are good stands of ponderosa pines here. Recent lava flows and cinders are to the right and left at **Mile 26.0**. Sunset Crater is obvious ahead at **Mile 27.9.** Black cinders cover the landscape ahead.

Mile 29.2: Entering Sunset Crater National Monument. Cinder Hills overlooks are to the right and left. **Mile 29.5**: Elevation 7,000 feet. The road is skirting the base of the Sunset Crater cinder cone. Small satellite cinder cones are ubiquitous. The road to the left at **Mile 31.2** goes to Lava Trail. A recent lava flow (aa-type lava) is on the right at **Mile 31.5.**

Mile 32.8: Visitor Center on the left; campgrounds to the right. Leaving Sunset Crater National Monument and pay station at **Mile 33.0.** Entering Coconino National Forest. The road soon enters Bonita Park (Figure 76). **Mile 34.6**: Intersection with U.S. 89. Turn left to Flagstaff; elevation is 6,906 feet.

34. Flagstaff to Sedona, Arizona, via U.S. 89A

The floor of the plateau at Flagstaff, Arizona, is on volcanic rocks of Tertiary age, but U.S. 89A soon descends into Oak Creek Canyon along the Oak Creek fault. The stratigraphic section is crossed from the Coconino Sandstone down through the Schnebley Hill, Hermit, and Esplanade Formations (all of Permian age) at Sedona, Arizona. The region has been the setting for numerous western movies and is the southern margin of the Colorado Plateau.

Mile 0.0: Intersection of Interstate 40 and Interstate 17 in Flagstaff, Arizona. Go south on I-17 toward Sedona and Phoenix, Arizona.

At **Mile 0.6** the highway enters a series of roadcuts in the gray Kaibab Limestone, of Middle Permian age. **Mile 2.1**: Take exit 337 to Oak Creek Canyon and Sedona, heading south on U.S. 89A. **Mile 2.4**: Stop sign, turn left; stop sign. Roadcuts are in volcanic rocks for the next 3.5 miles. At **Mile 5.8** the highway begins its descent into Oak Creek Canyon. **Mile 6.4: Milepost 395**: Exposures of volcanic rocks to the right and left. The volcanic rocks are a mixture of light brown andesite and vesicular basalt. **Mile 8.4**: Entering Coconino National Forest, a dense ponderosa pine forest. The road to the left at **Mile 10.8** goes to Oak Creek Rim Campground.

Mile 11.2: A viewpoint is on the left. An andesite flow holds up the point, filling a paleo-valley and capping the rim on the left. High to the right (west), across the Oak Creek fault, is a section of nearly white Coconino Sandstone, capped by remnants of marine rocks of the Toroweap and Kaibab Formations in the highest hills. The highway descends through the lava cap into sedimentary rocks of Permian age. Elevation is 6,400 feet.

Mile 11.8: Milepost 390: As the highway descends through the andesitic lava flow and turns the corner here, there are views of Coconino Sandstone exposures. The lava abuts the Coconino Sandstone at the Oak Creek fault. **Mile 12.4**: The highway is back into a basalt flow that is resting on a red soil zone in the roadcut to the right with an andesitic flow below. Massive white Coconino Sandstone cliffs are ahead.

Mile 12.8: Milepost 389: A white and pink chalky zone (fault gouge) marks the fault as the highway descends along the fault zone. The highway crosses into bedded Coconino Sandstone at **Mile 13.2** and is back in the fault gouge at **Mile 13.6**, winding down along the fault zone. **Mile 13.8: Milepost 388**: The road is back into bedded Coconino Sandstone. Elevation is 6,000 feet at **Mile 14.0**. Exposures of Coconino Sandstone are on the left. Thick valley fill of volcanic boulders is seen here in a heavily forested valley along Oak Creek. **Mile 15.0**: Pine Flat Campground.

Mile 16.0: The road to Cave Spring Forest Camp is to the right. The Coconino Sandstone forms the light-colored cliffs high to the right. A wooded notch marks the contact of the light-colored Coconino Sandstone and the underlying red Schnebley Hill Formation (Permian), which is largely equivalent to the DeChelly Sandstone to the north in Canyon de Chelly and Monument Valley. The Schnebley Hill Formation is a finer-grained sandstone and siltstone than the DeChelly and is less highly cross bedded and thinner bedded. It may have been largely deposited in coastal environments. These rocks are *not* the Supai Formation (or Group), as originally mapped.

Exposures of the Schnebley Hill Formation are on the left at **Mile 16.4** with the Coconino Sandstone high above and ahead. **Mile 16.8: Milepost 385.**

Call of the Canyon Day Use Area, a fee area, is to the right at **Mile 17.1.** Exposures of the red Schnebley Hill Formation are on the left and across the canyon to the right. **Mile 18.0**: Canyon Market and cabins are to the left. **Mile 18.6**: Extensive exposures of the Schnebley Hill Formation glorify the canyon walls for the next 7 miles, topped up high by the white Coconino Sandstone.

Junipine Resort is on the right at **Mile 19.1.** The road to the right at **Mile 20.6** goes to Slide Rock State Park. Manzanita Campground is on the left at **Mile 21.4.** Roadcuts and exposures on the right are in the Schnebley Hill Formation. **Mile 21.7: Milepost 380**: Elevation 5,000 feet in the Schnebley Hill Formation. Ensinoso Campground is to the right at **Mile 22.4.**

Mile 23.6: Indian Gardens Historical Site, the homestead of the first settler in Oak Creek Canyon. Jim Thompson built a cabin here in 1876 or 1877 and stayed until his death in 1917.

Mile 24.7: The road is at the top of the Hermit Shale (Permian) with good exposures to the right and left at the base of the Schnebley Hill Formation. The ridges and pinnacles in the vicinity of **Mile 24.9** are capped by the light-colored Coconino Sandstone; the smooth, rounded red cliff below is in the Schnebley Hill Formation; and the dark red slopes of shale and siltstone at the base are the Hermit Shale. The entire section below the Coconino cliffs was formerly thought by Arizona state and U.S.G.S. geologists to be the Supai Group of Grand Canyon fame, but the true Supai lies stratigraphically beneath these exposures.

Mile 25.0: Milepost 377: The top of the true Supai Group is exposed beneath the Hermit Shale in the creek bottom downstream from this point. The contact of the red Hermit Shale and the underlying light-colored Esplanade Sandstone of the Supai Group is exposed in the roadcut to the right. Exposures of the Hermit Shale are to the right and left at **Mile 25.2.** Grasshopper Viewpoint is at **Mile 25.3.** Good exposures of the Esplanade Sandstone of the Supai Group are in the canyon below at **Mile 25.4.**

Mile 26.0: Milepost 376. Entering Sedona, Arizona. At **Mile 26.5** there are exposures in the roadcuts right and left of the top of the Supai Group. "Welcome to Sedona" sign. No place to park!

35. Sedona to Flagstaff, Arizona, via Interstate 17

Mile 0.0: Intersection of Arizona 179 and U.S. 89A in Sedona, Arizona. Turn left on 179 toward I-17. **Mile 2.4: Milepost 311:** Leaving the outskirts of Sedona. The Red Rock Scenic Road heads east in exposures of the Hermit Shale with the Schnebley Hill and Coconino Formations beautifully exposed in all directions. **Mile 4.4: Milepost 309.** In contrast to the symbols

Figure 77. A butte eroded from the red Schnebley Hill Formation, the approximate equivalent of the DeChelly Sandstone of Permian age, east of Sedona, Arizona, along Arizona 179. The rocks are largely water-deposited siltstone and fine-grained sandstone beds of a beautiful shade of red.

on some maps, the road is paved. **Mile 5.2**: Entering Yavapai County. The village of Oak Creek is at **Mile 6.0**.

Mile 8.2: Milepost 305: Roadcuts are in the red Hermit Shale. The classical red rock country of gorgeous rocks of Permian age, known from dozens of western movies and television commercials, is to the left and ahead (Figure 77), capped by a lava flow. **Mile 9.5: Milepost 304**: Elevation 4,000 feet. The road here skirts the very southern edge of the Colorado Plateau and the Mogollon Rim. Lava flows cover the mountainside to the left. Ahead are typical fault-block mountains of southern Arizona. The hillside to the left at **Mile 10.0** is covered with volcanic rocks, filling an old paleo-valley. The road crosses onto a hill of lava flows at **Mile 12.2**. Roadcuts at **Mile 12.9** are in outwash gravels of volcanic boulders. **Mile 13.5: Milepost 300**.

Mile 13.9: Intersection of Arizona 179 and I-17. Reset trip odometer.

Mile 0.0: Intersection of Arizona 179 and I-17. Turn left (north) on I-17 to Flagstaff. **Mile 0.6**: Elevation 4,000 feet. The highway climbs steeply through thick lava flows toward the Mogollon Rim and Colorado Plateau country. There are volcanic rocks in roadcuts and exposures to the right and left for the next 12 miles. **Mile 2.1: Milepost 301**.

Mile 4.9: Elevation 5,000 feet. Cliffs visible to the left (west) at **Mile 5.3** are near Sedona and in the Coconino Sandstone. **Mile 7.1: Milepost 306**: The road to the right goes to Stoneman Lake. **Mile 10.9**: Elevation 6,000 feet. **Mile 11.1: Milepost 310**.

Mile 12.7: Entering Coconino County. Roadcuts and hills ahead are in lava flows on red soil zones. The highway has leveled off onto the southern margin of the plateau at **Mile 15.8**, still on lava flows. **Mile 16.2: Milepost 315.** The exit right at **Mile 16.5** is to Rocky Park Road. Woods Canyon Road is at **Mile 18.2**. The roadcuts are still in volcanic rocks. **Mile 21.2: Milepost 320.** The Schnebley Hill Road, the type section of the Schnebley Hill Formation, is to the right at **Mile 21.5.** The exit to Munds Park is at **Mile 23.5.** A rest area is on the right at **Mile 24.9.**

Mile 26.3: Milepost 325. Willard Springs Road is to the right at **Mile 27.2.** The San Francisco Peaks are visible ahead. **Mile 28.4**: Elevation 7,000 feet. The Newman Park Road is at **Mile 29.8. Mile 31.3: Milepost 330.** Kelly Canyon Road is at **Mile 32.0.** The San Francisco Peaks are obvious ahead at **Mile 34.9.**

Roadcuts at **Mile 37.0** are in the Kaibab Limestone (Permian in age). **Mile 37.5**: Flagstaff, Arizona, city limits; roadcuts are in the Kaibab Limestone. **Mile 41.3**: Intersection of I-17 and I-40. The town of Flagstaff is built on the Kaibab Limestone bench, and the brown hills north of "motel row" are in reddish brown mudstone and siltstone of the Moenkopi Formation.

V

The Plateau Interior

There is a section in the central Colorado Plateau Province that does not lend itself to classification by geologic basin. Some of the most spectacular scenery and most interesting geologic features are found in this region. The Monument Upwarp borders the region to the east, and the Henry Basin occurs as a large structural depression to the west of the upwarp. The uplift that borders the Henry Basin on the west is Capitol Reef and the Waterpocket Fold, known generally as the Circle Cliffs Uplift. The strata dip gently westward from the Circle Cliffs Uplift into the Kaiparowits Basin. The High Plateaus subprovince marks the western limit to this region.

The northern half of the Monument Upwarp was described previously as being a north-plunging, gentle, broad anticlinal nose in Canyonlands National Park. In contrast, the southern half of the north-south-trending structure is a complex uplift consisting of an east-facing monocline and several individual anticlines and synclines riding piggyback on the large upwarp. The geological dividing line that distinguishes the northern from the southern Monument Upwarp is a northwest-trending basement fracture zone known as the Four Corners lineament. The surface expression of the Four Corners lineament is a highly fractured zone that extends northwestward from Comb Ridge along White Canyon to Hite Marina on Lake Powell and on up North Wash. The eastern margin of the southern Monument Upwarp is the Comb Ridge monocline, an abrupt upfold that lies between Bluff and Mexican Hat, Utah. The Navajo Sandstone is eroded to form a snaggle-tooth ridge that extends from Kayenta, Arizona, generally northward to the Abajo Mountains near Monticello, Utah. Comb Ridge also serves as the eastern flank of the Lime Ridge anticline (Figure 78). The southern plunging nose of the Organ Rock anticline, bordering the west flank of the Monument Upwarp, is just west of Kayenta, Arizona (Figure 79).

The structural geology of the southern Monument Upwarp is beautifully exposed by erosion, and barren Pennsylvanian and Permian rocks comprise

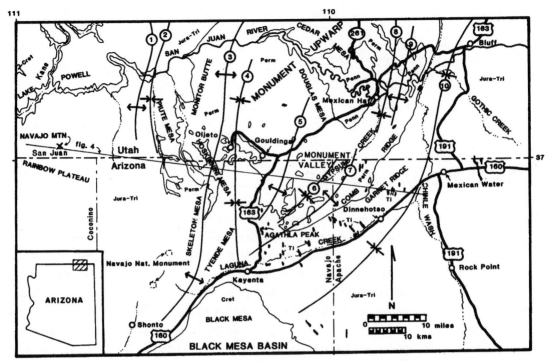

Figure 78. Geologic structures on the southern Monument Upwarp and surrounding areas:
(1) Balanced Rock anticline, (2) Nokai syncline, (3) Organ Rock anticline, (4) Oljeto syncline, (5) Douglas Mesa arch, (6) Mitten Butte syncline, (7) Cedar Mesa anticline, (8) Mexican Hat syncline, (9) Raplee anticline, (10) Tyende syncline. From Blakey and Baars 1987.

the surface. U.S. 163 between Bluff and Mexican Hat first crosses the Comb
Ridge monocline and climbs the western flank of the Lime Ridge anticline,
capped by the top of the Honaker Trail Formation (Pennsylvanian). The
uppermost Honaker Trail limestone bed is sufficiently resistant to erosion
to reveal the true, nearly textbook shape of the anticline. The highway then
crosses the equally well exposed northern plunging nose of the Raplee anticline, which lies immediately to the west of Lime Ridge, and descends into
the classic Mexican Hat syncline, with its preserved red beds of latest Pennsylvanian and Early Permian age. The San Juan River has carved a beautiful
canyon through the structures, providing a remarkable cross section of the
folds. The western flank of the Raplee anticline is just as steep and breathtaking as the Comb Ridge monocline. By now, anyone should be thoroughly
convinced that textbook anticlines and synclines surely exist.

West of the Mexican Hat syncline the rocks again rise onto the Cedar
Mesa anticline, known as the Mitten Butte anticline south of the San Juan
River, into which the river has carved a second deep canyon that exposes
Pennsylvanian rocks down through the Paradox Formation. The complexly

Figure 79. The southern plunging nose of the Organ Rock monocline, an anticlinal structure along the western margin of the Monument Upwarp, west of Kayenta, Arizona. The light-colored rocks dipping strongly toward the camera are in the Navajo Sandstone of Early Jurassic age.

Figure 80. Aerial view of the Goosenecks of the San Juan River, looking toward the northeast. The upper ledgy cliffs are in the Honaker Trail Formation, and the deeper, more massive cliffs are in the Paradox Formation, both of Pennsylvanian age. The overlook and Goosenecks State Park are in the upper left.

interbedded marine strata are well exposed to view from Goosenecks State Park (Figure 80), but a trip through the canyons by boat gives a much better perspective. The Cedar Mesa anticline forms the high, broad backbone of the southern Monument Upwarp.

South of the San Juan River, Monument Valley lies astride the Utah-Arizona state line in gentle folds that form the crest of the Monument Upwarp. The spectacular mesas, buttes, and pinnacles of the redrock fairyland consist of formations of Early Permian age. The red lower slopes in the buttes are the Organ Rock Shale, rising above a valley floor of Cedar Mesa

Sandstone, overlain by vertical cliffs of the orange-colored DeChelly Sandstone. In turn, the buttes are capped by thin remnants of the Triassic
Moenkopi Shale and Shinarump Conglomerate Member of the Chinle Formation. This gorgeous stratigraphic sequence has attracted movie producers and the makers of television commercials for decades (Figure 81).

West of the Monument Upwarp is the Henry Basin, named for the
prominent Henry Mountains, laccolithic mountains within the basin and
south of the San Rafael Swell (Figure 82). Hanksville, Utah, lies immediately
to the northeast of the basin, and Lake Powell (Bullfrog Marina) is near its
southern termination. The Waterpocket Fold forms the western flank of the
basin. Jurassic red beds, capped by a thick section of drab-colored Mancos-
and Mesaverde-equivalent strata, have been preserved in the deepest part of
the basin west and northwest of Hanksville.

The Waterpocket Fold, along the western flank of the Henry Basin, is yet
another spectacular monocline, with sharply eroded Navajo Sandstone,
capped by the Page Sandstone, forming the rugged western barrier. Capitol
Reef National Park lies along the monocline, providing outstanding vistas as
well as access to study this typical Colorado Plateau structural feature.
Rocks as old as the White Rim Sandstone and the overlying Kaibab Limestone, both of Permian age, are exposed in deep canyons near the crest of
the structure. The Triassic Moenkopi Formation is thicker here than to the
east and contains oil residues in the Torrey Sandstone Member, as seen in
the southern parts of the park. Rocks of the Moenkopi, Chinle, Wingate,

Figure 82. The Henry Mountains, seen from the Burr Trail. These are laccolithic ranges in the heart of the Henry Basin, visible across the sharply upturned beds of Jurassic age along the southern Waterpocket Fold in Capitol Reef National Park.

and Kayenta Formations form colorful cliffs along the highway near the Park Headquarters and Visitor Center. The strata dip gently westward from the crest of the huge north-trending fold into the Kaiparowits Basin, containing coal-bearing rocks of Cretaceous age preserved in the depths of the large syncline. The Burr Trail connects Boulder, Utah, with Bullfrog Marina on Lake Powell, passing through spectacular scenery in the Circle Cliffs and the Waterpocket Fold. This Scenic Backway is now a good paved road except for the switchbacks that cross the monoclinal fold down into the southwestern Henry Basin near Bullfrog. The Burr Trail and the canyon country near Escalante, Utah, provide insights into the wild country of the Grand Staircase–Escalante National Monument.

36. Bluff, Utah, to Kayenta, Arizona, via U.S. 163

The route from Bluff, Utah, to Kayenta, Arizona, is exciting because of the beautiful exposures of major geological structures and the geologically induced scenery. The town of Bluff is in the Blanding Basin, but Highway 163 soon encounters the spectacular Comb Ridge monocline, the eastern bounding structure on the Monument Upwarp. The Lime Ridge anticline is crossed on the clearly defined surface at the top of the Pennsylvanian Honaker Trail Formation, with colorful rocks of Permian age providing the backdrop. The road descends into the Mexican Hat syncline and oilfield before crossing the San Juan River into the Navajo Indian Reservation. There the road climbs again onto the Cedar Mesa anticline, and Monument Valley is soon dominating the view. Monument Valley is flanked on the west by the highway, which offers magnificent views of the spires, buttes, and mesas, made up of cliffs of the DeChelly Sandstone supported by sloping red beds of the Organ Rock Shale, both of Permian age. To the south the route crosses the southern plunging nose of the Monument Upwarp into the village of Kayenta, Arizona, once called "the farthest place from anywhere."

Mile 0.0. Intersection of U.S. 191 (south to Mexican Water) and U.S. 163 (west to Mexican Hat), just west of Bluff, Utah. Stay straight to the west on U.S. 163 toward Mexican Hat, Utah. U.S. 163, a Scenic Byway, is here on a high terrace gravel of Pleistocene age deposited along a former level of the San Juan River; a large gravel pit operation is on the left.

Roadcuts at **Mile 1.2** expose brown mudstone and siltstone of the Carmel Formation (Jurassic). Contact of the Carmel Formation on the nearly white Navajo Sandstone is at **Mile 1.3. Mile 1.4: Milepost 40:** The highway crosses Butler Wash and begins the ascent of the Navajo Sandstone dipslope onto Comb Ridge. The crest of Comb Ridge is obvious ahead and to the left and right at **Mile 1.9.** The highway crosses Comb Ridge at **Mile 2.5.**

The Navajo-Kayenta contact in roadcuts at **Mile 2.6** dips very steeply toward the east along the Comb Ridge monocline. The Kayenta-Wingate contact is in roadcuts at **Mile 2.7.** Drive with care: rockfalls are common. The pale reddish brown Wingate Sandstone (Early Jurassic) forms the massive cliffs to the left and right along Comb Ridge above varicolored mudstone of the Triassic Chinle Formation. The Wingate-Chinle contact is exposed to the right at **Mile 2.9.**

The highway crosses Comb Wash at **Mile 3.2.** The Moenkopi Formation (Early Triassic) is covered in the wash, and the DeChelly Sandstone (Permian) pinches out a short distance down the wash to the left (south). The highway is on the top of the reddish brown Organ Rock Shale (Permian) at **Mile 3.4** as the road begins its climb across Lime Ridge anticline, which forms the eastern margin of the Monument Upwarp. **Mile 3.5: Milepost 38.** At road level at **Mile 3.7** is the top of the Cedar Mesa Sandstone, here a pink-colored mudstone with interbeds and large nodules of gypsum in the lagoonal facies. The top of the Halgaito Shale, a dark brown mudstone and siltstone of latest Pennsylvanian age (latest Virgilian, or "Bursum"), is seen at **Mile 4.2.**

At **Mile 4.3** the top of the Honaker Trail Formation (Pennsylvanian) is visible on the left. The topmost thin, gray limestone forms the caprock across the flanks and crest of the Lime Ridge anticline, which is here the eastern flank of the Monument Upwarp. **Mile 4.5: Milepost 37.** The highway is on top of the Honaker Trail Formation at **Mile 4.8**, climbing the eastern flank of the Lime Ridge anticline.

Figure 83. The south end of Cedar Mesa, seen from north of Mexican Hat, Utah. The upper cliffs are in the type section of the Cedar Mesa Sandstone, of Early Permian age, overlying talus-covered slopes of the Halgaito Shale of latest Pennsylvanian age.

Mile 5.6: For the next quarter of a mile the facies change (change in sedimentary composition) in the Cedar Mesa Sandstone is visible in the skyline cliffs to the right: the sandstone facies to the west changes to the lagoonal, red bed facies to the east. **Mile 6.5: Milepost 35.** The highway crosses the crest of the Lime Ridge anticline at **Mile 7.2** and begins the descent of the west flank of the structure. Monument Valley is visible ahead in the distance. **Mile 7.5: Milepost 34.**

Roadcuts to the right and left at **Mile 7.8** expose basal brown beds of the Halgaito Shale. The high plateau to the right at **Mile 8.0** is Cedar Mesa, the type section of the Cedar Mesa Sandstone, which is exposed in the cliffs holding up the plateau (Figure 83). Monument Valley is visible ahead and to the left in the distance. **Mile 8.5: Milepost 33:** The shallow syncline visible to the left lies between the Lime Ridge anticline and the Raplee anticline. The butte nestled in the syncline to the left is capped by Cedar Mesa Sandstone on slopes of Halgaito Shale. The topmost limestone in the Honaker Trail Formation is exposed in roadcuts at **Mile 9.9** to the right and left. At **Mile 10.3** the Halgaito Formation is exposed in roadcuts and exposures to the right.

Mile 10.5: Milepost 31: The highway is circling the northern plunging nose of the Raplee anticline, heading into the Mexican Hat syncline. The Mexican Hat syncline, ahead at **Mile 11.4,** exposes red beds of Halgaito Shale, capped by thin remnants of basal Cedar Mesa Sandstone. Cedar Mesa is obvious high to the right.

The Valley of the Gods is visible on the right at **Mile 12.0.** The spires and buttes are composed of lower slopes of Halgaito Shale, capped by beds of basal Cedar Mesa Sandstone. **Mile 12.5: Milepost 29:** The dirt road to the right goes through the Valley of the Gods. The highway is now in the bottom of the Mexican Hat syncline at **Mile 13.5: Milepost 28.** At **Mile 14.5: Milepost 27** the strata dip eastward into the Mexican Hat syncline from the Cedar Mesa anticline, which is ahead and to the right. At **Mile 15.4** Mexican Hat Rock is visible in the distance to the left.

Mile 16.7: Turn right on Utah 261, which goes to Natural Bridges National Monument and Goosenecks of the San Juan State Park. **Milepost 0:** The road is climbing the east flank of the Cedar Mesa anticline. The road to the left at **Mile 17.5** goes to Goosenecks State Park on Utah 316. Turn left. **Milepost 4:** The road ahead goes to Natural Bridges National Monument up a steep grade of sharp switchbacks through the Halgaito Shale and Cedar Mesa Sandstone. The road to Goosenecks State Park is on the topmost limestone of the Honaker Trail Formation.

Mile 18.1: The unimproved road to the right goes to the top of Honaker Trail and on to Slickhorn Gulch. Four-wheel-drive vehicles are necessary. The road goes a few feet down below the top of the Honaker Trail Formation at **Mile 18.8.** *Slow down*: sharp curves are ahead at **Mile 19.4.** The road is going up and down through the top of the Honaker Trail Formation. Alhambra Rock, a diatreme-related dike, is visible to the left at **Mile 19.8.** There is a good distant view of the Raplee anticline to the left (east) at **Mile 20.0: Milepost 1.**

Mile 21.1: Milepost 0. Overlook and turnaround for Goosenecks of the San Juan State Park. Visible from the viewpoint are magnificent entrenched meanders of the San Juan River, here more than 1,000 feet below the rim (see Figure 80). The upper gray cliffs and slopes in the canyon walls are of the Honaker Trail Formation, and the lower, more massive cliffs near river level are in the Paradox Formation (Middle Pennsylvanian) (Figure 84). Vast amounts of oil and gas are produced by age-equivalent beds in the Four Corners area, such as the Aneth Oilfield. The Raplee anticline is visible to the east, and the Carrizo Mountains are to the southeast. Retrace the route back to U.S. 163.

The mountains visible ahead at **Mile 21.7** are the Abajos. **Mile 24.7:** Intersection with Utah 261. Turn right. Intersection with U.S. 163 at **Mile 25.6.** Turn right toward Mexican Hat.

The San Juan River exits the upper canyons and the beautiful Raplee anticline in the sharp canyon visible to the east (left) at **Mile 25.8.** The rocks exposed in the canyon and along the flank of the anticline are in the Honaker Trail Formation of Pennsylvanian age. **Mile 26.1:** Mexican Hat Rock and the Mexican Hat syncline are visible to the left at about 11:00. Mexican Hat Rock is composed of lower slopes of Halgaito Shale, capped by

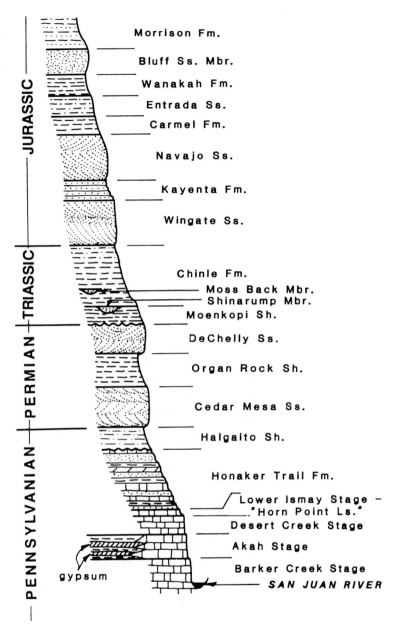

Figure 84. The stratigraphic section for the exposed rocks present on the Monument Upwarp. The entire section is not seen in any one place, but the rocks would appear in this order if the various exposures were to be put together.

a thin remnant of basal Cedar Mesa Sandstone, **Mile 26.5; Milepost 24**. The road to the left at **Mile 27.0** goes to Mexican Hat Rock. Alhambra Rock is on the skyline to the right. **Mile 27.5: Milepost 23:** The highway is again at the top of the Honaker Trail Formation.

Mile 28.5: Note the tiny, almost toy-like oil pumps on wells in the Mexican Hat Oilfield. The wells produce very little oil per day from very shallow depths. E. L. Goodridge noticed oil seeps beneath the bridge at Mexican Hat on a prospecting trip down the river in 1879. He filed placer claims on lands adjacent to the seep in 1882 and later drilled a shallow well that he claimed to be a "gusher" on March 4, 1908. Thus, the "San Juan Oilfield" was discovered, leading to the drilling of dozens of holes in later years. Each produces one or two barrels of oil a day at discovery. Most oil is produced from anticlines (upfolds), but production here is from the bottom of the Mexican Hat syncline. Because of the dry climate, the water table is very deep and any oil in the rocks trickles down into the syncline (downfold) in the dry rocks above the water table.

Mile 28.7: Entering Mexican Hat, Utah. Bridge over the San Juan River at **Mile 29.5.** This is a third-generation bridge, as the first two collapsed over the years. The road enters the Navajo Indian Reservation at the center of the river. The road is in the dark brown Halgaito Shale, just above the top of the gray limestone of the Honaker Trail Formation, as it begins to climb the east flank of the Cedar Mesa anticline. (It is important to stay on the highway, as it is not legal, or wise, to stray onto "private property" on the Navajo Reservation.)

Alhambra Rock, a diatreme-related igneous dike, is ahead and to the right at **Mile 30.0. Mile 30.5: Milepost 20.** The road to the left at **Mile 31.2** goes to Halchita, which has an elementary school, Indian services, and health services and was formerly the site of the Mexican Hat uranium processing mill. **Mile 31.5: Milepost 19**: The road climbs the steep east flank of the Cedar Mesa anticline at the Honaker Trail–Halgaito contact. Halgai Spring is in the valley to the left at **Mile 31.9** and is the origin of the term *Halgaito Shale* (pronounced "hal-guy-*toe*" and meaning "spring in the valley" in Navajo). Alhambra Rock is close on the right at **Mile 32.2. Mile 32.5: Milepost 18**: The highway is at or near the Honaker Trail–Halgaito contact.

Mile 34.5: Milepost 16: Beautiful view of Monument Valley to the left. The buttes and mesas are capped by remnants of the Shinarump Member of the Chinle Formation above thin beds of basal Moenkopi Formation, both Triassic in age. The prominent massive cliffs below are in the DeChelly (pronounced "day-*shay*") Sandstone with lower slopes of Organ Rock Shale, both of Permian age. To the left at about 3:00 the Three Sisters in Monument Valley are visible, and Rooster Rock is farther to the left. **Mile 35.5: Milepost 15**: Turnouts for photos here. Exposures of Organ Rock Shale, above a floor of Cedar Mesa Sandstone, are to the left at **Mile 39.0.**

A view of the northern buttes of Monument Valley is ahead at **Mile 40.6: Milepost 10.** The King on His Throne is straight ahead. The road to the left at **Mile 41.4** goes to Redlands Viewpoint, offering a grand view into Monument Valley. **Mile 41.6: Milepost 9**: Great view on the left, but one has to circumnavigate the bead shops.

Exposures of the pink, gypsiferous facies of the Cedar Mesa Sandstone are at **Mile 42.0**, and the northern buttes of Monument Valley are ahead. **Mile 42.7: Milepost 8**: The highway is near the top of the pinkish Cedar Mesa Sandstone, capped by the dark reddish brown slopes of Organ Rock Shale below massive cliffs of the orange-colored DeChelly Sandstone, all capped by thin remnants of the Moenkopi Formation and the Shinarump Member of the Chinle Formation. The top of the Cedar Mesa Sandstone is at road level at **Mile 43.8**, and good roadcuts of Organ Rock Shale are ahead.

Mile 44.6: Milepost 6: Monument Pass. The highway is in the Organ Rock Shale. Views to the south include Black Mesa on the far skyline, Agathla Peak in the intermediate distance, and Monument Valley in the foreground. To the right is Hoskininni Mesa in the distance behind Goulding's Trading Post. The highway from here descends the west flank of the Cedar Mesa anticline. The large dome of Navajo Mountain is visible in the far distance to the right, or west, at **Mile 46.2**.

Mile 48.7: Milepost 2: Goulding's Trading Post and the Monument Valley Indian boarding school are straight ahead. **Mile 50.3**: Crossroads. The road to the right goes to Goulding's, the boarding school, a campground, and on to Oljeto Trading Post. The road to the left goes to the Monument Valley Tribal Park, 4 miles to the Visitor Center; a fee area. A dirt road, rough in places, provides a quick overview of the inner reaches of Monument Valley. The valley floor is made up of the topmost beds of the Cedar Mesa Sandstone (Early Permian in age), here in its pink, gypsiferous mode. Above the relatively flat Cedar Mesa floor the myriad buttes and mesas consist of lower slopes of the dark brown Organ Rock Shale that support massive cliffs of the DeChelly Sandstone, both of Permian age. The little caplets on many of the buttes consist of a very thin remnant of Moenkopi Shale protected from erosion by the Shinarump Member of the Chinle Formation, both of Triassic age (see Figures 81 and 85).

The magical scenery of Monument Valley was brought to the public's attention by John Ford's many western movies filmed here, starring the legendary John Wayne. In those movies Monument Valley was located anywhere between Mexico and Canada and stretched for hundreds of miles in all directions. It was not uncommon for horsemen to travel for several weeks, all within the confines of the valley. Monument Valley still hosts numerous television commercials, keeping the majesty of the region in the public eye around the world.

Mile 50.7: Milepost 0: Arizona state line. The highway is still U.S. 163, heading south to Kayenta, Arizona. A distant view of Agathla Peak, a diatreme, is ahead.

Figure 85. The Totem Pole *(right)* and the Yei Bi Chei Dancers in the southern part of Monument Valley. The spires and buttes are in the DeChelly Sandstone, rising above lower slopes of the Organ Rock Shale, both Permian in age.

Mile 51.5: Milepost 416: The orange-red cliffs all around are remnants of the DeChelly Sandstone. The rocks are dipping toward the west, away from the Cedar Mesa and Douglas Mesa anticlines and toward the Oljeto sag (syncline) to the west. Hoskininni Mesa is to the west on the Organ Rock anticline; it was named for a Navajo headman who hid Navajos and their livestock in 1864, protecting them from capture and deportation to Fort Sumner, New Mexico, during the Long Walk.

Mile 53.5: Milepost 414: Agathla Peak is at 10:00; the cliffs to the right are in the DeChelly Sandstone. A gradational contact between the Organ Rock Shale and the overlying DeChelly Sandstone is exposed on the right at **Mile 54.6.** The highway is at about the Organ Rock–DeChelly contact, as seen in scattered exposures near **Mile 55.6.** The highway is at about the top of the DeChelly Sandstone at **Mile 56.4**; the Monument Upwarp is plunging southward here toward Kayenta, and the highway is slowly climbing up-section. The cliffs to the right are, from bottom to top, the dark brown Moenkopi Formation, the varicolored slopes of the Chinle Formation, and the brown massive cliffs of the Wingate Sandstone, topped by the ledgy cliffs of the Kayenta Formation and capped by the nearly white Navajo Sandstone on the skyline. **Mile 57.5: Milepost 410:** Small "boondocks" are numerous.

A small diatreme on the left at **Mile 58.6** has intruded the DeChelly-Moenkopi contact. The hill on the left at **Mile 59.1** is the DeChelly Sandstone at the base, beneath thin Moenkopi Shale slopes, capped by the Shinarump Member of the Chinle Formation, all dipping to the west. The west-dipping, light-colored ledge on the left at **Mile 61.1** is the Shinarump

Figure 86. Agathla Peak, also known as El Capitan, the neck of a volcanic blowout eruption, or diatreme. The rock consists of fragments brought up from great depth with the gaseous eruption, accompanied by fragments of strata near the surface at the time of eruption which fell into the vent and become incorporated in the volcanic rocks.

Member. The cliffs on the right consist of the Chinle through Navajo Formations.

Mile 62.2: Good view of Agathla Peak, also known as Agathlan by the Navajos, at about 10:00, with Owl Rock in the upper Chinle Formation at about 11:00. Owl Rock is directly ahead on the skyline at **Mile 62.5: Milepost 405.**

Mile 63.5: Milepost 404. Great view straight ahead of Agathla Peak (Figure 86). The Navajo word means "a place for the scraping of hides," a ritual sheep-shearing location. The monolith, called El Capitan by Kit Carson, is the solidified neck of an ancient volcanic blowout, or diatreme. Owl Rock is on the right at **Mile 64.5: Milepost 403** and actually looks like an owl from here; Agathla is on the left. The highway skirts the base of Agathla at **Mile 65.5.**

Three or four scattered small diatremes form dark-colored crags near **Mile 66.0.** Ahead and to the right is Black Mesa, capped by sandstone of Cretaceous age, forming the skyline. The ridge ahead at **Mile 68.0** is the Wingate Sandstone on the south-plunging nose of the Monument Upwarp. A small diatreme is on the left at **Mile 68.9. Mile 69.5: Milepost 398.**

To the right at **Mile 70.0** is the southward-dipping Wingate Sandstone with a ragged valley formed on the Kayenta Formation. This is the type section for the Kayenta Formation. **Mile 70.4**: The highway is going up-section into the nearly white Navajo Sandstone, dipping southward in the south-plunging nose of the Monument Upwarp. **Mile 70.5: Milepost 397.** The village of Kayenta is visible ahead. A bridge over Laguna Wash is at **Mile 71.3.**

Mile 71.7: Entering Kayenta, Arizona. Kayenta, a trading center on the

Navajo Indian Reservation, was first settled by John and Louisa Wetherill, who founded a trading post here in 1910. The original Navajo name means "a place where water runs like fingers out of a hill" and was later changed to Tyende, "where the animals bog down." With the proper Anglo corruptions of the Navajo language, the name became Kayenta. The town has been called "the farthest place from anywhere."

Mile 74.0: The end of U.S. 163 and the intersection with U.S. 160. (See Roadlog 27 to go east on U.S. 160 and Roadlog 28 to go west.)

37. Hanksville to Capitol Reef National Park and Torrey, Utah, via Utah 24

The route from Hanksville, Utah, westward on Utah 24 crosses the north end of the Henry Basin, first through rocks of Late Jurassic and Cretaceous age and then gradually going down-section into the Glen Canyon Group at the Waterpocket Fold. The Waterpocket Fold, one of the largest monoclines on the Colorado Plateau, is largely included in Capitol Reef National Park. Rocks of the Navajo Sandstone form mighty crags and cliffs along the monocline, and then the road proceeds down-section into rocks of Permian age near the crest of the Waterpocket Fold. The highway then follows the dipslope downward along the gentle west flank of the great anticline to Torrey, Utah.

Mile 0.0: Hanksville, Utah, at the intersection of Utah 24 and Utah 95: turn west onto the continuation of Utah 24 toward Capitol Reef National Park; Utah 95 goes toward Lake Powell. **Mile 1.0**: Heading west on a Scenic Byway. The red bluffs of siltstone in the Entrada Sandstone, the light-colored cliffs above of the Curtis Sandstone, overlain by the brown, thin-bedded siltstone of the Summerville Formation, are exposed across the river on the right (north); exposures of the brown mudstone of the Carmel Formation are to the left. A bridge crosses the Dirty Devil/Fremont River at **Mile 1.6**; the highway is in the silty Entrada facies.

Mile 2.0: Milepost 115. Good exposures of the brown Summerville Formation are all around. **Mile 3.3**: The contact of the Summerville Formation with the Curtis Sandstone is obvious on both sides of the road. The highway crosses upward through the Curtis Sandstone at **Mile 3.6.** Exposures of the Carmel Formation are on the right at **Mile 3.8**. Strata are dipping gently eastward into the Henry Basin. **Mile 4.3: Milepost 113**.

The Morrison Formation, consisting of varicolored mudstone and sandstone, is on the right and ahead at **Mile 4.5**. The Morrison Formation on the right at **Mile 6.0** is capped by thin Dakota Sandstone, a ledge-forming shoreline sandstone of the westward advancing Late Cretaceous seaway. Exposures of the Dakota Sandstone are on the right at **Mile 6.5.** The highway crosses upward into dark gray, slope-forming mudstone of the Tununk

Member of the lower Mancos Shale at **Mile 6.8.** The Cretaceous Ferron Sandstone Member of the Mancos Shale forms the cliffs ahead at **Mile 7.0.**

Mile 7.3: Milepost 110. The Mancos Shale is capped by Mesaverde cliffs on the right.

Mile 8.3: Milepost 109: The highway is in the Tununk and Ferron Members of the Mancos Shale. Fresh exposures of the very black Tununk Shale Member are on the right at **Mile 9.7.**

Mile 10.3: Milepost 107: The highway heads into a small canyon cut in the Ferron and Tununk Members. The highway crosses upward through the Ferron Sandstone Member of the Mancos Shale at **Mile 10.7.** Dirt roads are to right and left at **Mile 11.1**; the road is on top of the Ferron Sandstone, and slopes of the Blue Gate Shale Member of the Mancos Shale, capped by the Emery Sandstone, are all around. **Mile 12.4: Milepost 105.**

Mile 15.4: Milepost 102: Entering Caineville, Utah. Boulder Mountain is visible in the distance ahead. First settled in 1882 by Elijah C. Behunin for the Mormon church, Caineville was named for the Utah Territory's representative to Congress, John T. Caine. Periodic flooding by the river caused the farmland and village to be largely abandoned. **Mile 17.4: Milepost 100**: The highway is in the nearly black Blue Gate Member of the Mancos Shale. **Mile 18.3**: The Caineville Inn is on the right.

Cretaceous strata are sharply upturned along the east flank of the Caineville anticline at **Mile 18.8.** At **Mile 19.0** the Ferron Sandstone is sharply upturned above the Tununk Shale Member; the varicolored Morrison Formation and Cedar Mountain Formation (Lower Cretaceous) are visible ahead in the valley. **Mile 20.3: Milepost 97:** The highway parallels hogbacks of steeply east-dipping Ferron Sandstone. **Mile 21.3: Milepost 96:** The highway is following a racetrack valley in the Tununk Shale Member at **Mile 21.5**; the Morrison and Cedar Mountain Formations are exposed on the right. A bridge over the Fremont River is at **Mile 22.1** with a sharp turn to the left. **Mile 22.3: Milepost 95**: Sleepy Hollow Campground and Store is on the left.

The Waterpocket Fold in Capitol Reef National Park, one of the great monoclines on the Colorado Plateau, is obvious ahead at **Mile 24.3.** The highway is back in the varicolored Morrison Formation at **Mile 25.0**; strata dip to the west across the crest of the Caineville anticline. At **Mile 26.7** the Morrison Formation is dipping eastward along the east flank of the Waterpocket Fold. The highway is in the Salt Wash Member (sandstone) of the Morrison Formation at **Mile 27.5**, and good exposures of the Brushy Basin Member, varicolored mudstone, are ahead; the caprock is a thin Dakota Sandstone with beds dipping to the east.

Mile 28.3: Milepost 89. The road to left at **Mile 28.5** goes toward Notom,

Figure 87. Utah 24 crosses the Waterpocket Fold through jagged peaks and ridges eroded from the Navajo Sandstone of Jurassic age. The underlying stream-deposited Kayenta Formation is visible here beneath the rounded white cliffs of Navajo Sandstone. The feature directly above the road is Fern's Nipple.

Figure 88. The exposed layered rocks seen near the Visitor Center in Capitol Reef National Park. The dark-colored rocks at the base of the section are in the Early Triassic Moenkopi Formation. The lighter-colored slopes above are in the Upper Triassic Chinle Formation, and there is no basal Shinarump Member separating the two slope-forming formations. The high, jagged cliffs are in the Wingate Sandstone of Early Jurassic age.

following the syncline of the Waterpocket Fold. (Eventually, it connects with the Burr Trail and Bullfrog, Utah, but as of 2001 the road is paved for only the first 5 miles and can become a morass after rain.) Entering Capitol Reef National Park. Cliffs of the nearly white Curtis Sandstone are on the left at **Mile 29.9. Mile 30.3: Milepost 87:** The highway is still in the canyon of the Fremont River, going down in the section into the Navajo Sandstone. A small waterfall is in the canyon to the right at **Mile 30.9** with massive cliffs

Figure 89. Hilltop viewpoint along the Scenic Drive in Capitol Reef National Park, looking north. The lower dark-colored slopes are in the Moenkopi Formation, the upper lighter-colored slopes are in the Chinle Formation, and the cliffs are the Wingate-Kayenta-Navajo Sandstones.

of thick Navajo and Page Sandstones all around along the Waterpocket Fold. **Mile 31.4: Milepost 86**: The road is in the deep canyon of the nearly white, highly cross-bedded Navajo Sandstone (Figure 87).

Behunin Cabin is on the left at **Mile 31.6.** Grand Wash is on the left at **Mile 33.0.** At **Mile 34.0** a massive collapse of Navajo Sandstone into the canyon is on the left. The top of the Kayenta Formation is in the canyon walls at **Mile 34.5. Mile 35.4: Milepost 82.** The top of the brown Wingate Sandstone is at **Mile 35.5.**

Mile 36.1: Entering Fruita Historic District with an orchard on the right. Petroglyphs on the right at **Mile 36.5** are Fremont-type figures just above the Chinle-Wingate contact. The highway is now in varicolored mudstone of the Chinle Formation. Historic Fruita School and lush orchards are at **Mile 36.8.** There are good exposures of the Chinle Formation to the left and right at **Mile 37.1.**

Mile 37.4: Milepost 80: The Chinle-Moenkopi contact is on the right (Figure 88); the base of the Chinle is light gray above the dark brown Moenkopi. No Shinarump Member is present here, and it is only locally present elsewhere in this part of the Waterpocket Fold.

Mile 37.6: The Visitor Center and Scenic Drive are to the left. The scattered large black boulders are mostly of vesicular basalt, presumably having come from the High Plateaus lava flows. Turn south from the Visitor Center on the Scenic Drive.

A blacksmith shop and Mormon homestead are at **Mile 38.5** on the Scenic Drive. Crossing the little creek of the Fremont River at **Mile 38.7.** The Historic Gifford Farmhouse is at **Mile 38.8.**

The Fruita campground is at **Mile 38.9.** The hillside behind the campground is covered with large boulders of igneous rocks. **Mile 40.1**: A scenic turnout is at the top of the hill (Figure 89).

Figure 90. A bit farther along the Scenic Drive through Capitol Reef, the Shinarump Member of the Chinle Formation appears in the section. It is the bright white, thin bed just above the lower dark-colored slopes of the Moenkopi Formation. The cliffs at the skyline are in the Wingate Sandstone of Early Jurassic age.

Grand Wash Road is to the left at **Mile 41.1.** Channels of thin sandstone of the Shinarump Member of the Chinle Formation are exposed above the dark brown Moenkopi Formation on the left. **Mile 43.6**: Slickrock Divide. The road to this point has been in the Moody Canyon Member of the Moenkopi Formation, with good examples of rippled bedding planes along the way (Figure 90).

Mile 44.8: Egyptian Temple. The road crosses upward at **Mile 45.2** through the Shinarump Member of the Chinle Formation, dipping westward across the Waterpocket Fold. There are excellent exposures of the Shinarump Member, a stream-deposited sandstone at the base of the Chinle Formation, in a narrow little gully at **Mile 45.4.** Exposures of gray and purplish beds in the Chinle Formation are along the road and to the left at **Mile 45.5.**

Mile 45.7: Crossroads: Pleasant Creek is to the right, and Capitol Gorge is to the left with a picnic area. The pavement ends here, and the road ahead goes into Capitol Gorge. We'll turn around here and return to the Visitor Center.

Mile 53.8: Intersection at the Visitor Center: Turn left, or west, on Utah 24. The highway is in the Moody Canyon Member of the Moenkopi Formation. The deep canyon to the left at **Mile 54.2** exposes the Sinbad Limestone Member of the Moenkopi Formation.

Mile 54.5: The highway climbs the east flank of the Waterpocket Fold in the basal Moenkopi Formation (Figure 91). The high country to the left is capped at or near the top of the Virgin Limestone Member in the middle

Figure 91. A view to the east across the Waterpocket Fold in Capitol Reef National Park. The lower dark-colored slope is in the Moenkopi Formation, and the upper slope is in the Chinle Formation, both of Triassic age. The upper cliff is the Wingate Sandstone, with a crag of the Navajo Sandstone peeking through the notch; both are of Jurassic age.

Figure 92. The Goosenecks Point viewpoint is on the Sinbad Limestone Member of the Moenkopi, the upper cliff, overlying a thin red slope of the Moenkopi shale (the Black Dragon Member). The "Kaibab Limestone" (Black Box Dolomite) is the lower cliff, and exposures of the White Rim Sandstone (sometimes incorrectly called the Coconino Sandstone), the oldest rocks exposed in Capitol Reef National Park, are at creek level.

Moenkopi Formation. The road to the left at **Mile 56.2** goes to Panorama Point, with excellent views back toward the east across the Waterpocket Fold. Panorama Point is at about the crest of the anticline on a hard sandstone ledge, the Torrey Member of the Moenkopi Formation.

Mile 56.4: Take a right turn onto a gravel road toward the Goosenecks.

The turnaround and parking for Goosenecks Point are at **Mile 57.2**. The overlook, accessible from an easy walk of about 600 feet, is a view into the deep canyon of Sulfur Creek (Figure 92).

Mile 58.1: Back at Utah 24, turn left, or west, toward Torrey, Utah. Chimney Rock turnoff is to the right (north) at **Mile 58.7**. Orientation Turnout is on the left at **Mile 60.7**; good exposures of the Shinarump and Moenkopi are on the right. Leaving Capitol Reef National Park. Black vesicular basalt boulders are common across the valley floor and in roadcuts at **Mile 62.1**.

Mile 62.5: There is a good view of Boulder Mountain to the left, or south, capped by Tertiary basalt flows, undoubtedly the source of the black basalt boulders of the Capitol Reef area. Capitol Reef Resort is on the right at **Mile 63.3**. The Holiday Inn is on the right at **Mile 63.9**. The highway is on the Shinarump bench. **Mile 64.7: Milepost 71.**

Mile 65.5: Entering Torrey, Utah. The town was named for Colonel Torrey, who fought in the Spanish-American War. **Mile 65.7: Milepost 70.**

38. Torrey to Boulder, Utah, via Utah 12

Geologically speaking, the route from Torrey to Boulder on Utah 12 is a rather monotonous high traverse of the eastern side of Boulder Mountain. This faulted segment of the High Plateaus subprovince is capped by lava flows of Tertiary age, the weathered debris from which covers nearly the entire route. Boulder Mountain was certainly well named! (Some locals claim, however, that Boulder Mountain and Thousand Lake Mountain to the north had their names switched in an early cartographic error.) The saving graces are the numerous good views to the east across Capitol Reef National Park and the Henry Mountains beyond.

Mile 0.0: Intersection of Utah 24 and Utah 12 in Torrey, Utah. Head south toward Boulder, Utah, on Highway 12, a Scenic Byway.

Mile 0.8. The Shinarump Member, at the base of the Chinle Formation, caps buttes to the left and right; Boulder Mountain is obvious ahead, capped by Tertiary volcanic rocks. There are sharp curves ahead. The road crosses a creek at **Mile 2.7**; the cliffs to the left are in the Moody Canyon, the Sinbad Limestone, and the Black Dragon Members of the Moenkopi Formation (Early Triassic). **Mile 3.1**: The valley floor is filled with black vesicular basalt boulders. A sharp ridge of white Navajo Sandstone is on the right at **Mile 3.7**, apparently standing on end owing to displacement along a fault. The highway is still in the lower Moenkopi Formation. Basaltic boulders are in the roadcut on the right at **Mile 4.4**. The road to the right at **Mile 4.7** goes to Teasdale, Utah. **Mile 4.8: Milepost 119**. North Slope Road is on the right at **Mile 4.9**. Roadcuts are in basaltic boulders to the right and left at **Mile 5.2**.

Mile 5.8: Milepost 118: The mountainside is covered with basaltic boulders. Boulder Mountain, part of the Aquarius Plateau, is appropriately named. **Mile 6.1**: Grover, Utah, with massive cliffs of Navajo Sandstone in the distance. The road crosses a north-south fault at **Mile 7.2**; dark brown beds of the Moenkopi Formation are exposed north of the fault, and massive cliffs of white Navajo Sandstone are ahead and to the left. Boulder Mountain forms the skyline ahead. **Mile 7.6**: Entering Dixie National Forest. Exposures of the Navajo Sandstone are to the right and left at **Mile 8.2**. The mountainside at **Mile 8.6** is covered with basaltic boulders right and left.

Massive exposures of Navajo Sandstone are ahead and to the left at **Mile 9.1**. The highway is heading up into a densely vegetated ponderosa pine forest, floored by extensive lava boulders. **Mile 10.0**: There are good exposures of brown mudstone and siltstone of the Carmel Formation high on the left above the Navajo Sandstone. **Mile 10.6**: Elevation 8,000 feet. **Mile 10.9: Milepost 113**: Roadcuts on the left are in the Carmel Formation. The highway crosses a saddle at **Mile 11.5**, and the mountainside is covered with basaltic boulders.

Mile 11.9: Milepost 112: Dixie National Forest Campground is on the left. A good view of the Henry Mountains is to the southeast at **Mile 12.4**. Entering Garfield County at **Mile 12.5**.

The highway crosses a high saddle at **Mile 13.5** with good views to the east across the Waterpocket Fold to the Henry Mountains. The Cretaceous buttes and mesas in the northern Henry Basin are visible in the vicinity of Hanksville, Utah. The boulders strewn across the side of Boulder Mountain consist of basalt and andesite.

The scenic overlook of Larb Hollow, with educational displays, is on the left at **Mile 14.3**. Visible is the full length of Capitol Reef and the Waterpocket Fold with the Henry Mountains to the east. Volcanic boulders, mostly andesitic, cover the mountainside, hosting a dense forest of ponderosa pine, mountain hemlock, and aspen. **Mile 15.0: Milepost 109**. A rest area is on the right at **Mile 16.4**. An information display for Dixie National Forest is at **Mile 16.5**, along with access to the Great Western Trail. **Mile 17.2**: Pleasant Creek Campground is on the left. The road to Lower Bowens Reservoir is to the left at **Mile 17.4**. The mountainside is still covered by volcanic boulders. The road to the left at **Mile 18.4** goes to Oak Creek Campground.

Mile 20.2: Elevation 9,000 feet and climbing. Aspens grow densely here. At **Mile 22.3** there is another good view of the Henry Mountains to the east; a private road goes to the right. **Mile 23.1: Milepost 101**: View ahead to the west or southwest into the Kaiparowits Basin. Thick aspen stands cover the mountainside. A national forest overlook is at **Mile 24.0**. The view to the south is of rugged exposures of Navajo Sandstone with Navajo Mountain in the distance; Cretaceous and Tertiary sedimentary rocks in the Kaiparowits Basin are visible on the low skyline to the southwest.

Mile 24.1: Milepost 100: The mountainside is still covered with volcanic boulders. **Mile 25.4:** Homestead Overlook, on the left, offers a good view to the south across heavy aspen growth. A vast area of deeply eroded Navajo Sandstone is low to the left at **Mile 26.0. Mile 26.1: Milepost 98.** There is a good view of the Kaiparowits Plateau and Basin to the southwest from **Mile 26.7. Mile 27.1: Milepost 97.**

Mile 27.8: The road is in heavy aspen and scrub oak thickets on the boulder-covered mountainside. **Mile 29.1: Milepost 95.** The elevation is 8,000 feet at **Mile 30.0** as the road goes down a steep grade. **Mile 30.1: Milepost 94.** The highway is down into a pine forest with scrub oak at **Mile 30.5**; volcanic boulders are everywhere.

The light-colored sandstone exposure on the left at **Mile 34.8** is the Middle Jurassic Page Sandstone above the top of the Lower Jurassic Navajo Sandstone. The Page Sandstone is identical in appearance to the Lower Jurassic Navajo Sandstone but is separated from the Navajo by a regional erosion surface (unconformity). The hillside on the left at **Mile 35.1** is in the varicolored Carmel Formation. **Mile 35.2: Milepost 89.** The low knobs ahead at **Mile 35.3** are Page Sandstone, capped by the Carmel Formation. **Mile 35.8:** Leaving Dixie National Forest.

Mile 36.3: Entering Boulder, Utah. The town of Boulder is nestled in open valleys eroded from the Navajo and Page Sandstones, which are characterized by well-developed, large-scale cross bedding. The Carmel Formation caps the hills and ridges above town.

Mile 37.1: Intersection: Burr Trail is to the left, and Escalante is to the right. (See Roadlog 39 for the Burr Trail.)

39. Boulder to Bullfrog, Utah, via the Burr Trail

This beautiful drive climbs the gentle western flank of the Waterpocket Fold through the Grand Staircase–Escalante National Monument and across Capitol Reef National Park. The road begins in exposures of the Page and Navajo Sandstones, capped with beds of the Carmel Formation, and proceeds down-section into the Moenkopi Formation near the crest of the giant uplift. The breathtaking descent across the Waterpocket Fold and its spectacular monocline ends in the western Henry Basin in Glen Canyon National Recreation Area at Bullfrog, Utah. The road is now completely paved except for a short distance where it crosses the steep descent of the Waterpocket Fold.

Mile 0.0: Intersection of Utah 12 and the Burr Trail Scenic Backway in Boulder, Utah. Head east on the Burr Trail. The cliffs in the area are formed by the Navajo and Page Sandstones, capped by the Carmel Formation on most of the mesas. The Navajo and Page Sandstones are nearly identical in appearance: almost white, highly cross-bedded, and massive. The Page Sandstone, however, rests on the Navajo Sandstone above a regional erosional surface, or unconformity, that makes a nearly horizontal bedding plane. The Carmel Formation here consists of pink and gray mudstone containing thin beds of gypsum.

Mile 1.5: Entering Grand Staircase–Escalante National Monument. Volcanic boulders are scattered about the cliffs and slopes of the Navajo-Page Sandstones on the right at **Mile 2.3** with a caprock of Carmel Formation. **Mile 3.2**: The landscape consists of thick Navajo Sandstone.

The creek bottom at **Mile 4.2** is filled with thick deposits of volcanic boulders. At **Mile 4.7** volcanic boulder deposits are perched on terraces along the wash on the Navajo Sandstone, and they fill the valley floor at **Mile 5.0**. The volcanic boulders were undoubtedly derived from the volcanic rocks of nearby Boulder Mountain.

The Navajo Sandstone abruptly changes along the outcrop near **Mile 5.5** from almost pure white to light red. Deer Creek Campground is on the left at **Mile 6.3**. The top of the Page Sandstone on the left at **Mile 8.6** is capped by the Carmel Formation. The road is here following a small fault: note that the Navajo-Page Sandstones are higher on the right than on the left. Navajo Mountain is visible in the far distance ahead. **Mile 9.1**: The Navajo-Kayenta contact is in the gully on the right.

Mile 9.4: As the road comes to a rise in the hills, the Page Sandstone–Carmel contact is ahead, and a thick exposure of Kayenta Formation is in the wash to the left above cliffs of the Wingate Sandstone. The road now is in the Kayenta Formation, a stream-deposited sandstone that forms ledges between the Navajo and Wingate windblown sandstone cliffs. The entire Glen Canyon Group is red in color here. The Gulch is at **Mile 10.2**. The contact of the Kayenta Formation and the Wingate Sandstone is at road level. The canyon now is in high Wingate cliffs. The Wingate-Chinle contact is in the canyon floor at **Mile 12.1**. Typical exposures of varicolored, slope-forming mudstone of the Chinle Formation are ahead at **Mile 13.3**, heavily masked by Wingate talus from above. Highly fractured Wingate Sandstone is ahead and to the right at **Mile 14.8**, above slopes of the Chinle Formation.

Mile 16.0: The Wingate Sandstone overlies the Chinle Formation in Long Canyon. The Wingate changes abruptly from red to white here, a really photogenic area with almost no turnoffs (Figure 93).

Mile 16.8: Divide: the top of the Waterpocket Fold can be seen ahead, and the Henry Mountains are on the far skyline. The road is in the Chinle

Figure 93. A prominent butte in Long Canyon along the Burr Trail: cliffs of the Wingate Sandstone of Jurassic age overlie colorful slopes of the Triassic Chinle Formation.

Formation, and Shinarump-capped upper Moenkopi slopes (ledge-forming stream-deposited sandstone at the base of the Chinle Formation above dark brown mudstone of the Moenkopi) are visible in the valleys below between here and the Henry Mountains. The Wingate Sandstone cliffs to the east are in the Waterpocket Fold.

The mesas to the right at **Mile 17.1** are the Shinarump Member of the Chinle Formation, capping slopes of the dark brown Moenkopi Formation. The road is on top of the Shinarump Member at **Mile 17.3**, and the Shinarump-Moenkopi contact is at road level at **Mile 17.5**.

Mile 18.5: The dirt road to the right goes to Horse Canyon, Wolverine Canyon, and Little Beth Hollow; Capitol Reef is 11 miles ahead, and Bullfrog Marina is 52 miles ahead. The road is in the Moenkopi Formation. The Moenkopi through Wingate cliffs are visible toward the west, behind us, at **Mile 18.7** (Figure 94). We have just come out of Long Canyon and are heading up onto the Flats, upheld by the Torrey Member (sandstone) of the Moenkopi Formation.

Mile 19.7: The yellowish-colored sandstone in the gulch to the right and roadcuts to the left is in the oil-stained Torrey Member of the Moenkopi Formation. The red beds of the upper Moenkopi are bleached to a tan color owing to the hydrocarbon gas in the Torrey sandstone. Bleached red beds of the Moenkopi Formation are in roadcuts ahead at **Mile 20.0** with exposures of the tar sand in the gully to the right. The bleached Moenkopi is evident for the next 2.5 miles. **Mile 22.3-22.5**: Sandstone of the Torrey Member is

Figure 94. Layered rocks exposed along the Burr Trail in the Grand Staircase–Escalante National Monument. The lower, darker slopes are in the Moenkopi Formation below lighter-colored slopes of the Chinle Formation, both of Triassic age. The Shinarump Member of the Chinle Formation forms a prominent ledge that separates the two formations. The cliffs are in the Jurassic-age Wingate Sandstone.

broadly exposed. The sandstone does not appear to be petroliferous at this location. The mesas to the left at **Mile 23.8** are Moenkopi Formation, capped by the Shinarump Member of the Chinle Formation. The Wingate Sandstone cliffs beyond are in the Waterpocket Fold. **Mile 25.0: Milepost 40.**

The road at **Mile 25.9** is in the Moenkopi Formation. The cliffs ahead are thick Shinarump. Torrey Member sandstones are to the right and left in the little canyon at **Mile 27.6**, within the Moenkopi. **Mile 28.2: Milepost 45.** The road to the right at **Mile 29.1** is the Wolverine Loop Road. **Mile 29.8**: Divide. The road is in bleached beds of the Moenkopi.

Mile 30.1: Cattle guard at the crest of the Waterpocket Fold. There is a good view of the Henry Mountains in the eastern distance across cliffs of upturned Wingate Sandstone in the monocline. The road is in bleached mudstone of the Moenkopi Formation. Entering Capitol Reef National Park, leaving Grand Staircase–Escalante National Monument.

Mile 30.7: The pavement ends ahead, and the dirt road starts its descent from the crest of the Waterpocket Fold across the ever-steepening dipslopes of the Wingate Sandstone (Figure 95). The road continues to descend through sharp switchbacks on the steep dipslopes of the Kayenta Formation with jagged, massive exposures of Navajo Sandstone right and left. At the base of the monocline the road crosses the ragged hogbacks in the Navajo Sandstone and about 100 feet of the Page Sandstone in a sharp declivity and then crosses steeply dipping strata of the Carmel Formation in the valley floor. Scattered exposures of Entrada Sandstone are apparent in the wash on

Figure 95. The Burr Trail, here in Capitol Reef National Park, begins its descent across the Waterpocket Fold. The slopes in the foreground are in the Moenkopi Formation, and the upper slopes beyond are in the Chinle Formation, both of Triassic age. The prominent near buttes are eroded from steeply dipping exposures of the Wingate Sandstone, Jurassic in age, as the formation crosses the monocline. The white cliffs in the middle ground are the Navajo Sandstone, with glimpses of the underlying Kayenta Formation, also dipping steeply across the fold. In the distance are the Henry Mountains, formed by igneous laccolithic intrusions.

the east flank of the monocline. The Curtis and Summerville Formations are covered by the wash.

Mile 38.7: Big T intersection; the sign says Boulder 42 miles. The road to the left (north), County Route 1670, goes to Utah 24 (34 miles), Hanksville (64 miles), and Capitol Reef Visitor Center (48 miles). Reset odometer to 0.0.

Mile 0.0: Turn right to Bullfrog. The road follows an open valley adjacent to the Waterpocket Fold. The cliffs to the left are Morrison Formation (Brushy Basin Member varicolored mudstone), overlain by thick Cedar Mountain Formation; the Dakota Sandstone forms the crest of the hogback. The Cedar Mountain and Dakota Sandstones are very similar, both being ledge-forming, cross-bedded sandstones of Cretaceous age. Hogbacks of the Navajo and Page Sandstones are on the right, topped by beds of the lower Carmel Formation, along the steeply dipping east flank of the Waterpocket Fold (monocline).

Exposures of Brushy Basin Member of the Morrison Formation are on the left at **Mile 1.0**, capped by Cedar Mountain and Dakota ridges. Slopes of the dark gray Mancos Shale are at **Mile 1.2**, and cliffs of Mesaverde Sandstone are high on the left. **Mile 1.5: Route 1668, Milepost 60(?)**. Exposures in a draw to the left at **Mile 1.6** are of Morrison through the Mesaverde-equivalent Tarantula Mesa Sandstone. Ahead is a massive, white sandstone cliff, mapped as the Salt Wash Member of the Morrison Formation. Surprise Canyon Trail is on the right at **Mile 1.7**. The Salt Wash Member cliff is on the left. The red, rounded-weathering sandstone hogbacks on the right are the Entrada Sandstone.

Mile 2.3: The road to the right goes to Headquarters Canyon and Lower Muley Twist trailhead. The road turns left and climbs upward through the Salt Wash sandstone and on up-section. The road crosses the Brushy Basin Member of the Morrison Formation at **Mile 2.4**, dipping gently eastward away from the monocline, toward the Henry Basin (Rock Springs syncline). The base of the Cedar Mountain Formation (Lower Cretaceous) is at road level at **Mile 2.5.** The Cedar Mountain–Dakota contact, a regional erosional surface, is at road level at **Mile 2.6.**

The road crosses the top of the Dakota Sandstone (Upper Cretaceous) at **Mile 2.8.** The Tununk Member of the Mancos Shale is on the left with exposures on up-section to the Mesaverde-equivalent Tarantula Mesa Sandstone. The road, in the base of the Mancos Shale, can be very muddy.

Mile 3.2: Leaving Capitol Reef National Park. On the left are exposures of five members of the Mancos Shale: the Tununk, a low cliff of Ferron Sandstone, the Blue Gate Shale, the Muley Canyon Sandstone, and the Masuk, capped by the thin Tarantula Mesa Sandstone. The Ferron Sandstone Member forms the low cliffs in the Mancos Shale to the left at **Mile 3.3**, and the Tarantula Mesa Sandstone forms the top of the cliffs. The eastward-dipping bedding plane on top of the Dakota Sandstone forms the hogback to the right for some distance ahead.

The road at **Mile 6.5** is on top of the Dakota Sandstone dipslope, crosses the hogback at **Mile 6.7,** and is in the Brushy Basin Member of the Morrison Formation, a varicolored mudstone, at **Mile 6.8.** The Dakota Sandstone is much thinner here than at the last crossing of the hogback. The Dakota–Cedar Mountain Formation is on the left at **Mile 7.1**, and the Salt Wash Member of the Morrison, stream-deposited sandstone beds, is on the right. **Mile 8.2**: The monocline of the Waterpocket Fold is obvious ahead and to the right. The sandstone draping across the monocline is the Wingate. The road is in the Brushy Basin Member. The road again crosses the Dakota hogback at **Mile 9.5**, heading east.

Mile 9.9: The top of the Dakota Sandstone and base of the Mancos Shale are at road level. There is a good view of the complete Mancos Shale and Mesaverde-equivalent section ahead below the Henry Mountains. **Mile 10.9: Milepost 75.**

At **Mile 11.0** the road to the right goes to Bullfrog, and the road ahead goes to Starr Spring. Turn right. Pavement begins. The road heads back across the Dakota Sandstone hogback. The road at **Mile 11.4** is in a valley of the Brushy Basin Member of the Morrison Formation with the Dakota Sandstone hogback on the left. The road to the right at **Mile 11.8** goes to the

Halls Creek Overlook. The top of the Salt Wash Member of the Morrison Formation is on the left at **Mile 12.3**, below the Brushy Basin Member and Dakota Sandstone. **Mile 13.9, Milepost 80**: The road is at the top of the Salt Wash Member of the Morrison Formation.

Mile 14.7: The valley on the left is becoming a sharp canyon, exposing a thick Salt Wash Member. The road is in the Brushy Basin Member. **Mile 16.0**: Long Canyon, on the left (a fork of Bullfrog Creek), exposes a very thick section of the Salt Wash Member, the brown Summerville Formation, and in the bottom of the canyon, the silty facies at the top of the red Entrada Sandstone. The road is at the top of the Salt Wash Member, and the canyon rim is in the Dakota Sandstone.

Navajo Mountain can be seen ahead (south) at **Mile 16.5** with the Waterpocket Fold to the right (southwest) and the Henry Mountains to the left (east). The road is still at or near the top of the Salt Wash Member of the Morrison Formation. The Waterpocket Fold is obvious on the right at **Mile 17.4**, plunging gently to the south.

Mile 19.0: The road runs down the axis of the Balanced Rock anticline; the Piute Mesa syncline and Halls Creek are on the right (west), and the Rock Spring syncline and Bullfrog Creek are on the left (east). **Mile 20.2: Milepost 90.** Bullfrog Basin on Lake Powell can be seen ahead at **Mile 20.3**. A drilling location is on the right at **Mile 20.4.** The road descends steeply, on top of the Salt Wash Member, into Bullfrog Basin in the southern Henry Basin. Views to the right are of Lake Powell. **Mile 22.4**: Another view is ahead toward Bullfrog Marina: the thick Salt Wash Member is on the left, and the red, silty facies of the Entrada Sandstone, capped by the Summerville Formation, is in cliffs on the right. The Curtis Formation is not present here. Bullfrog is just about at the top of the Entrada Sandstone.

Mile 22.7: Entering Glen Canyon National Recreation Area. Newly paved road begins. There is a good view to the left at **Mile 23.2** of the Entrada, Summerville, and Morrison Formations.

Mile 23.3: Milepost 95. A steep and crooked grade descends through the Salt Wash Member at **Mile 23.8** with the bright red Entrada Sandstone, capped by the brown Summerville and Morrison Formations, in the canyon on the left. The top of the Summerville Formation is exposed on the left and after the hairpin curve on the right at **Mile 24.1.**

Midway up the cliff across the canyon at **Mile 24.5** is an exposure of Pleistocene terrace gravels above the red crinkled beds of the Entrada Sandstone. At **Mile 24.7** the road descends through the Summerville Formation–Entrada Sandstone contact with no white Curtis Sandstone present. Terrace gravels cap the low benches to the right and left. The road crosses a rocky wash of Bullfrog Creek at **Mile 25.4.** Exposures of gently folded

Entrada Sandstone are all around at **Mile 25.6. Mile 26.4: Milepost 100**: The Bullfrog Arm of Lake Powell is down to the right.

A dirt road to the right at **Mile 26.9** goes to Bullfrog primitive camping beaches. Another road to primitive camping is at **Mile 28.4. Mile 29.5: Milepost 105.**

Mile 30.5: Milepost 110: Intersection with the Bullfrog-Hanksville highway (see Roadlog 17).

40. Arizona 98 from U.S. 160 to Page, Arizona

Arizona 98 goes north from U.S. 160 west of Kayenta to Page, Arizona. The road generally follows the contact of the Page Sandstone below and red beds of the Carmel Formation above for most of the route. Exposures are very poor owing to a thick cover of windblown sand. Navajo Mountain, a prominent laccolithic mountain, is occasionally visible toward the north.

Mile 0.0: Intersection of U.S. 160 with Arizona 98 heading north to Page, Arizona. Turn north onto Highway 98, crossing the railroad tracks of the electric coal train, the Black Mesa & Lake Powell Line. The intersection is in the middle of the broad, relatively flat Klethla Valley, which has a great deal of sand cover. Black Mesa is behind, and Navajo Mountain is ahead at about 1:00.

In a shallow wash at **Mile 1.0** the top of the Page Sandstone is exposed. The Page Sandstone looks for all the world like the underlying Navajo Sandstone, but an erosional surface separates and distinguishes the two windblown sandstone beds. There are scattered exposures of the top of the Page Sandstone for the next 7 miles, peeking through a plateau that is heavily covered with sand. **Mile 3.5: Milepost 358.**

The road to the right at **Mile 6.0** goes to Shonto, Arizona (7 miles), meaning "water on the sunny side of a rock wall" in Navajo. Roadcuts at **Mile 8.1** are in the Carmel Formation. Scattered exposures ahead for the next 5 miles are in the Page Sandstone, mostly covered with sand. **Mile 9.5**: Entering Coconino County. **Mile 11.6: Milepost 350.**

Mile 12.3: The road to Navajo Mountain and Inscription House, Navajo Route 16, is to the right. The red soils at **Mile 13.7** appear to be from the Carmel Formation. Roadcuts for the next 3 miles are in the Carmel Formation with widely scattered exposures of the Page Sandstone. **Mile 16.6: Milepost 345**: Elevation 6,687 feet.

At **Mile 17.0** the mesa on the left and the prominent butte on the right are in the Navajo Sandstone. Roadcuts at **Mile 18.0** are in the Page Sandstone, topped in places with basal Carmel Formation. There is a scenic view turnout at **Mile 18.5** with a good view of Navajo Mountain to the right and exposures all around of the Page/Navajo Sandstone. The prominent butte

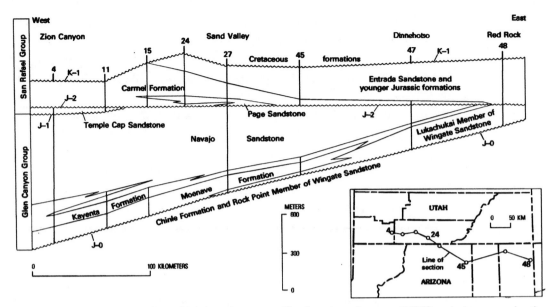

Figure 96. It has taken years of hard work on the part of USGS geologists to resolve the regional relationships of the many formations of Jurassic age on the Colorado Plateau. The recognition of regional unconformities (erosional breaks in the geologic record) was the key to unraveling an otherwise baffling story. The wavy lines in the diagram indicate the unconformities that can be recognized on a regional basis, thus placing the various formations into an understandable context. The unconformities have been given code names, such as J-2 and J-5, shown on the right side of the drawing. The chart above is for a particular line of cross section, as indicated, but it is also useful for understanding the complexities of the Jurassic System across much of the province. From Peterson and Pipiringos 1979.

ahead is Navajo Sandstone. **Mile 19.7: Milepost 342:** Square Butte, close by on the left, is the Entrada Sandstone, including the Cow Springs Member, mostly white sandstone with occasional red bands. The Carmel Formation is exposed in the deep gully here. White Mesa to the southwest is the Entrada Sandstone, with a thin Romano Sandstone, capped with the Dakota Sandstone. The countryside opens up at **Mile 21.3**, and cliffs of Navajo Sandstone are to the left. The road must be near the top of the Kayenta Formation. **Mile 21.7: Milepost 340:** There are no rock exposures across the plain on the Kaibito Plateau for several miles.

Mile 26.7: Milepost 335: Kaibito, Arizona, is visible at about 10:00. Red rocks are exposed in a gully at **Mile 28.9** with sandstone and shale beds that may be in the Carmel Formation. The road to the left at **Mile 29.3** goes to Kaibito, a Navajo word meaning "willows at a spring." **Mile 30.0:** Red beds in roadcuts and in rolling hills are in the Carmel Formation(?).

Mile 30.7: Milepost 331: The main road to Kaibito (misspelled on the sign) is to the left in poor exposures of the Carmel(?) Formation. The road is in a structural sag with Carmel exposed in the low part of the syncline. **Mile 32.7: Milepost 329:** Roadcuts are in the Carmel Formation with expanses of the Page/Navajo Sandstone ahead and to the left (Figure 96). There are poor exposures of the top of the Page Sandstone around the country at **Mile 34.3**, and a low mesa at about 1:00 is Page Sandstone with a thin Carmel Formation cap.

Mile 36.8: Milepost 325: The low mesa on the right is Page Sandstone with a thin cap of Carmel Formation dipping gently to the east. There are pretty good exposures of the Page Sandstone at **Mile 38.5. Mile 41.8: Milepost 320:** The contact between the Page Sandstone and the Carmel Formation is exposed on the right. The road crosses upward through the Page-Carmel contact at **Mile 43.7.** The broad plain ahead is on top of the Page Sandstone.

Mile 44.8: Milepost 317: Navajo Mountain is prominent at 3:00. The Kaiparowits Plateau is visible at about 2:00 in the distance. Shallow roadcuts at **Mile 48.0 to Mile 50** are in the Carmel Formation. **Mile 53.9: Milepost 308:** The prominent butte ahead consists of the Navajo Sandstone over the Kayenta Formation with lower red, rounded cliffs of Wingate/Moenave Sandstone. The Wingate Sandstone grades from windblown deposits to the east to a stream-deposited sandstone and mudstone toward the west. This region is about in that transition zone.

Mile 55: Visible ahead are Lake Powell and the coal-burning Navajo Power Plant near Page, Arizona. At **Mile 55.9** the butte on the right, Leche-e Rock ("Red" Rock in Navajo), has Carmel Formation at the base, then Entrada Sandstone, then the Romano Sandstone (a lateral equivalent of the Summerville Formation), capped by the Salt Wash Member of the Morrison Formation. There is no Tidwell Member here. Lake Powell is visible ahead at **Mile 56.5.** The road ahead is in the Carmel Formation.

Mile 58.9: Milepost 303: The Navajo coal-powered generator is prominent ahead. The roads to the right at **Mile 61.1** go to the power plant. Roadcuts in red beds at **Mile 61.2** are in the Carmel Formation, and the road is heading down into the Page Sandstone. **Mile 61.8: Milepost 300:** The road to the right goes to the Navajo Power Plant. The road to the right at **Mile 62.4** goes to Antelope Point.

Mile 63.3: Entering Page, Arizona. The townsite was originally Navajo land, but the federal government traded federal acreage within the Aneth Oilfield in southeastern Utah for this land to provide homesites and shopping facilities for workers to build Glen Canyon Dam. The town now caters mostly to tourists and other visitors to Lake Powell. Page is located on an erosional bench on the Page Sandstone, this being the type area for the formation. **Mile 64.7:** Stoplight at the intersection with Copper Mine Road on the south edge of Page. Turn right to downtown Page and proceed straight ahead to the intersection with U.S. 89. **Mile 67.0: Milepost 295.**

Mile 67.3: Intersection with U.S. 89: to the right is Page, and to the left is Flagstaff. (To go south on U.S. 89, see Roadlog 41. To go north on U.S. 89, see Roadlog 42.)

41. Page to Bitter Springs, Arizona, on U.S. 89

From Page, Arizona, U.S. 89 follows exposures of the Navajo and Page Sandstones to the Echo Cliffs monocline and descends the western cliffs of the north-south fold through Triassic rocks to the Bitter Springs Junction of U.S. 89A.

Mile 0.0: Intersection of Arizona 98 and U.S. 89 at the edge of Page, Arizona. Head southwest on U.S. 89 toward Bitter Springs.

Mile 1.2: Milepost 545: In the distance to the right are the Vermilion Cliffs, consisting of Navajo Sandstone in the upper white cliffs, the Kayenta Formation forming ledgy cliffs below, and the Moenave Formation forming the lower red cliffs (all Jurassic). The varicolored slopes at the base are in the Chinle and Moenkopi Formations (Triassic). The road is about at the Page-Navajo contact. The Page and the Navajo Sandstones are nearly identical in appearance, but a widespread erosional surface (the J-2 unconformity shown on Figure 96) forms a prominent bedding plane between the two formations, thus justifying the distinction of separate stratigraphic units. This area, of course, is the type section of the Page Sandstone.

Mile 2.2: Milepost 544: The upturned Navajo Sandstone beds on the right in the middle distance are along the east flank of the Echo Cliffs monocline. The road crosses Water Holes Canyon at **Mile 4.2**. The road is in the upper Navajo Sandstone at **Mile 4.6**, and the Page Sandstone above forms the skyline cliffs: the contact is a prominent marker bed. **Mile 6.2: Milepost 540**: The contact between the Navajo and Page Sandstones is obvious ahead. **Mile 7.2**: Elevation 5,000 feet. **Mile 8.2: Milepost 538**: The road crosses the reddish-colored contact from the Navajo Sandstone upward into the Page Sandstone.

Mile 8.6: Prominent cliffy exposures of the Page Sandstone are seen to the left above the top of the Navajo Sandstone. The marker bed, just above the highway, is here a red mudstone/siltstone layer about a foot thick with an inch or two of conglomerate at the top of the red layer. This is typical of the J-2 unconformity. The highway at **Mile 8.8** is following the contact between the Page and Navajo Sandstones. The beds ahead begin to climb steeply up to the west.

The roadcut at **Mile 9.4** is in the Page Sandstone. **Mile 10.3: Milepost 536**: The road is on a broad flat surface on top of the Page Sandstone. The buttes and snags ahead at **Mile 16.5** are the Navajo and Page Sandstones along the Echo Cliffs monocline, a sharp north-south-trending fold bounding the east side of the Marble Plateau. A safety pulloff is to the right at **Mile 18.5.** Ahead is a sharp, deep roadcut through steeply dipping Navajo Sandstone along the Echo Cliffs monocline. In the next half-mile the road quickly crosses the Navajo, Kayenta, and Moenave Formations and emerges from the deep roadcut.

Mile 19.2: Viewpoint turnout is to the right. The road is crossing the Echo Cliffs monocline; high cliffs of the Moenave through Navajo Sandstone are high on the left. The Marble Platform is obvious to the right, and Marble Canyon is visible at about 3:00. Beyond the Marble Platform to the right, the Kaibab Plateau and uplift is visible. The Vermilion Cliffs are to the right rear at about 4:00. The road begins a steep descent through the varicolored Chinle Formation (Triassic), heavily masked by large talus, as it crosses the monocline. Exposures of the Chinle Formation are in roadcuts at **Mile 20.3. Mile 20.4: Milepost 526**: Bitter Springs, Arizona, is visible low on the right. **Mile 21.3: Milepost 525.**

Mile 22.3: The intersection of U.S. 89 and U.S. 89A. Turn left for Flagstaff (join Roadlog 31, Flagstaff to Marble Canyon, at **Mile 105.5**); turn right for Marble Canyon and Kanab (see Roadlog 42).

42. Marble Canyon Bridge, Arizona, to Kanab, Utah, on U.S. 89A

U.S. 89A heads west from Marble Canyon, Arizona, along the Vermilion Cliffs, consisting of the Kaibab through Navajo Formations. The highway crosses the East Kaibab monocline about at the top of the Kaibab Limestone, reaches the Jacob Lake intersection at the crest of the uplift, and then descends the western flank of the large uplift to Fredonia, Arizona. A sidetrip to the North Rim of Grand Canyon National Park can be made by following Arizona 67 south from Jacob Lake.

Mile 0.0: North end of Marble Canyon Bridge at **Milepost 538** on U.S. 89A. Head west toward Kanab, Utah. Leaving the Navajo Indian Reservation.

The road to the right at **Mile 0.2** goes to Lees Ferry Historical Site and Campground and the boat ramp for Grand Canyon river trips. Just beyond the side road is Marble Canyon Lodge. The Vermilion Cliffs (Figure 97) are to the north (right) for several miles along the highway. This area was designated the Vermilion Cliffs National Monument in the year 2000.

Mile 1.9: Milepost 540: The highway is near the Kaibab-Moenkopi contact. Vermilion Cliffs Lodge is on the right at **Mile 3.6. Mile 6.5**: Elevation 4,000 feet. **Mile 7.0: Milepost 545**: The highway is at the top of the Kaibab Limestone. A bridge across Soap Creek is at **Mile 8.7.** Well-developed hoodoos are just beyond on the right. **Mile 8.0: Milepost 547**: Cliff Dwellers

Figure 97. The Vermilion Cliffs, now a national monument, west of Marble Canyon. The highway here is at about the top of the Kaibab Limestone (Permian), the lower brown slopes near the center of the photo are in the Moenkopi Formation, and the low cliff above is the Shinarump Member of the Chinle Formation, overlain by the upper varicolored slopes of the upper Chinle Formation. The red rocks of the Moenave Formation form much of the high cliffs, capped by the high white cliffs of the Navajo Sandstone.

Lodge is on the right. Hatch River Expeditions warehouse is on the right at **Mile 9.1** in the lower Moenkopi Formation. **Mile 10.5**: The Kaibab Plateau and uplift is ahead on the skyline across the East Kaibab monocline.

Mile 14.0: Milepost 552: The Shinarump cliff is now near the base of the lower slope of the Vermilion Cliffs on the right. **Mile 15.0: Milepost 553**: The East Kaibab monocline and Kaibab uplift are visible to the left on the skyline. **Mile 16.0**: The highway is here on the top of the Shinarump Member of the Chinle Formation. The highway turns to the north toward the East Kaibab monocline. **Mile 18.1: Milepost 556** is in rolling exposures of the Petrified Forest Member of the Chinle Formation. **Mile 18.5**: Elevation 5,000 feet. The East Kaibab monocline is obvious ahead. **Mile 22.1: Milepost 560**.

Mile 25.1: Milepost 563: The road here is paralleling the East Kaibab monocline. The gray rock layers rolling over to the east across the monocline on the left are the Kaibab Limestone. The Moenkopi Formation and the Shinarump Member are covered in the valley floor. The Vermilion Cliffs to the right consist of the upper Chinle varicolored slopes up through the red Moenave Formation, capped by the white Navajo Sandstone. **Mile 27.1: Milepost 565.** A chain-up parking area is on the right at **Mile 27.2.**

Mile 28.2: Milepost 566: The highway begins climbing the flank of the East Kaibab monocline on the Kaibab Limestone (Permian). Entering Kaibab National Forest at **Mile 28.3**; good exposures of the Kaibab Limestone are in the roadcut to the right. **Mile 28.7**: Red beds are between limestone beds in roadcuts on the right. **Mile 29.1: Milepost 567**: The highway is climbing across the East Kaibab monocline; red beds of the Moenkopi Formation are in the roadcuts. Moenkopi red beds are in the roadcuts at **Mile 29.5.**

A viewpoint is on the left at **Mile 29.9.** The road is definitely on the

Kaibab Limestone here. There is apparently some faulted duplication of section across the monocline. There are numerous exposures and roadcuts in the Kaibab Limestone as the road winds its way up across the fold.

Mile 31.2: Milepost 569: The road is in a gully that is at right angles to the monocline, well down into the Kaibab Limestone; the rocks are beginning to flatten near the structural crest. The Kaibab is very nodular in roadcuts at **Mile 31.7**, where white chert (siliceous) nodules are common through the section. **Mile 32.2: Milepost 570.** Elevation is 6,000 feet at **Mile 32.3.** The road is coming into a heavily forested area at **Mile 32.8**, climbing into a juniper-pinyon environment, with ponderosa pine becoming predominant as the road climbs higher.

Mile 35.1: Milepost 573: There are poor, scattered exposures of Kaibab Limestone capping the Kaibab Plateau. **Mile 38.2: Milepost 576**: The road is at the crest of the Kaibab monocline in the Kaibab National Forest, heavily wooded with ponderosa pine. **Mile 41.2: Milepost 579.**

Mile 41.5: Entering Jacob Lake, Arizona, junction of Arizona 67 and U.S. 89A. Reset your odometer to continue on U.S. 89A or to begin Roadlog 43, Jacob Lake to the North Rim of Grand Canyon.

Mile 0.0: Jacob Lake. Intersection of U.S. 89A and the road to the North Rim of Grand Canyon; head west on U.S. 89A toward Kanab, Utah. The highway is near the crest of the Kaibab uplift in densely forested terrain.

Mile 0.7: Milepost 580: The highway is presumably on top of the Kaibab Limestone, although the rocks are heavily obscured by dense forest. Occasional roadcuts ahead are in the Kaibab Limestone for the next 20 miles. **Mile 5.7: Milepost 585.** The road is descending the western flank of the Kaibab uplift at **Mile 8.4**, pretty much on the Kaibab Limestone. **Mile 10.7: Milepost 590.** A scenic view is on the right at **Mile 11.2**, elevation 5,700 feet.

Mile 14.1: Leaving Kaibab National Forest. There are good views to the north into the "Grand Staircase" country, as the road is well down the flank of the Kaibab uplift. **Mile 15.7: Milepost 595.** The Grand Staircase (Figure 98) is obvious to the north (right) at **Mile 16.3**, showing the stratigraphic section up into the Tertiary volcanic caprock of the High Plateaus (the Paunsaugunt Plateau). **Mile 19.4**: Elevation 5,000 feet. **Mile 20.7: Milepost 600.** Roadcuts at **Mile 21.8** are in the lower Moenkopi Formation (Triassic). **Mile 26.7: Milepost 606**: A thin limy bed (Virgin Limestone Member?) in the Moenkopi Formation.

Mile 29.3: Entering Fredonia, Arizona. **Mile 30.0**: Intersection of Arizona 389 and U.S. 89A. Arizona 389 heads west and then north, becoming Utah 59 at the border and terminating at Hurricane, Utah (see Roadlog 51).

Mile 31.8: Milepost 611: Leaving Fredonia. The low hills to the right and left are the Shinarump Hills (*shinar* is the Paiute Indian word for "wolf," and

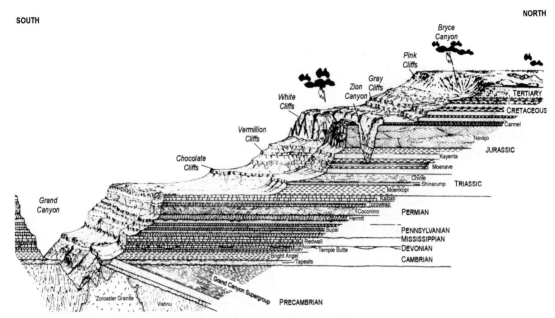

SOUTH

NORTH

Figure 98. Schematic block diagram of the rock record in the Grand Staircase, from
Grand Canyon on the south to Bryce Canyon on the north. Formation names, shown on
the right, indicate the rocks exposed in the various regional cliffs that are given color
names, shown on the left of the drawing. Modified from W. K. Hamblin in Hintze 1988.

rump means "rump"), consisting of red beds of the Moenkopi Formation,
capped by the Shinarump Member of the Chinle Formation in the type area
of the Shinarump Member. The Shinarump Member consists of stream-deposited sandstone, conglomeratic in places, that fills erosional channels
carved into the top of brown mudstone of the Moenkopi Formation.

Mile 32.8: Milepost 612: There are good exposures of the Shinarump
Member above road level to the right and left. The Shinarump Member is at
road level at **Mile 33.7** to the right and left. **Mile 33.9:** Entering Utah, Kane
County.

Mile 34.0: Entering Kanab, Utah. The valley broadens, and the town of
Kanab is visible off to the left. The name *Kanab* means "willow" in the
Paiute Indian language. The village was settled in 1864, abandoned because
of problems with the Indians, and resettled in 1871. John Wesley Powell's
Colorado River expedition of 1871-72 overwintered in Kanab. The Kanab
airport is to the left at **Mile 34.5.** Kanab is situated in the varicolored Petrified Forest Member of the Chinle Formation with red cliffs above and
around the town of the Moenave Formation (Jurassic). **Mile 36.8:** Stoplight
at the intersection of U.S. 89 and 89A; this is the end of U.S. 89A.

43. Jacob Lake to the North Rim of Grand Canyon via Arizona 67

Mile 0.0: Intersection of U.S. 89A and Arizona 67 in Jacob Lake. Head south on Arizona 67 to the North Rim of Grand Canyon. The road is at or near the top of the Kaibab Limestone in heavily forested terrain. **Mile 0.4:** Campground to the right in a ponderosa pine forest.

Mile 0.6: Roadcuts in the Kaibab Limestone. The road is basically on top of the Kaibab Limestone for the next 30 miles. The route is along the crest of the Kaibab uplift with heavy forest cover that makes for a beautiful drive but very few rock exposures. **Mile 5.7: Milepost 585. Mile 10.7: Milepost 590. Mile 13.5:** Elevation 8,840 feet.

Mile 15.7: Milepost 595: Roadcuts in the Kaibab Limestone ahead. The road comes out of heavy forest into several little alpine glades. **Mile 20.7: Milepost 600.** Kaibab Lodge is to the right at **Mile 25.6.** Deer Lake crossroads is at **Mile 26.5.**

Mile 31.0: Entering Grand Canyon National Park; the entry pay station is at **Mile 31.2.** Roadcuts are still mainly in the Kaibab Limestone. **Mile 40.9:** The road to the left goes to Point Imperial. Turn left. As in most national parks, the road is narrow and crooked and is potentially hazardous. **Mile 44.7:** Picnic area is on the right in a mixed aspen forest.

Mile 46.3: Y in the road: to the left is Point Imperial, and to the right is Cape Royal. Turn left. Roadcuts and exposures ahead are in the Kaibab Limestone.

Mile 48.9: Enter a parking circle for Point Imperial. The view from Point Imperial is down into the lower part of Marble Canyon. Navajo Mountain is visible in the far distance to the east beyond the Marble Plateau. The East Kaibab monocline is to the south, just downriver from the Little Colorado River confluence with the Colorado. Low in the immediate foreground are down-faulted exposures of the Late Precambrian Chuar Group, dipping eastward, in angular unconformity with the Tapeats Sandstone at the base of the Paleozoic section. Just up from the Colorado River is Kwagunt Creek and Chuar Butte. Make the parking loop and head back to the Y in the road.

Mile 51.8: The Y. Straight ahead goes to Cape Royal. Turn right. **Mile 57.2:** Stop sign at the intersection with the main road. The Visitor Center is to the left. Turn left. Roadcuts ahead are in the Kaibab Limestone. ·

Mile 58.1: The road on the left goes to the Kaibab Trail. The road to the right at **Mile 58.9** goes to a campground and ranger station. The road to the right at **Mile 59.1** goes to a campground. **Mile 59.3:** Picnic area on the left. **Mile 60.2:** End of the road at the lodge and North Rim Visitor Center. Turn around in the parking area and return to Jacob Lake or visit other viewpoints.

44. Kanab, Utah, to Utah 12 via U.S. 89

Kanab, Utah, is along the western extension of the Vermilion Cliffs, in the Chinle Formation (Triassic), with red cliffs of the Moenave Formation (Lower Jurassic) glorifying the northern cliffs. After leaving the canyon of Kanab Creek, U.S. 89 generally follows the Sevier fault north to the intersection with Utah 12, which leads to Bryce Canyon National Park. The fault zone is beautifully exposed in some stretches of the highway but is obscured from view much of the way.

Mile 0.0: Intersection at the stoplight of U.S. 89 and 89A in Kanab, Utah. Take U.S. 89, a Scenic Byway, north toward Bryce Canyon National Park. Kanab is in the Petrified Forest Member of the Chinle Formation (Triassic), surrounded to the north by red bluffs of the Moenave Formation (Early Jurassic).

Mile 1.2: Leaving Kanab and heading up Kanab Creek Canyon, carved into red, interbedded fluvial overbank and lake sandstone and siltstone/mudstone beds of the Moenave Formation. The road is at about the top of the Chinle Formation. **Mile 1.9: Milepost 66.** The light-colored sandstone capping the cliff on the left at **Mile 2.9: Milepost 67** is the Springdale Sandstone Member of the Moenave Formation. The Moenave Formation here consists of the lower Dinosaur Canyon Member, the middle Whitmore Point Member, and the upper Springdale Sandstone Member. The Whitmore Point Member contains palynomorphs (plant spores) that are Early Jurassic in age. The view ahead at **Mile 4.3** is of the thick Moenave Formation and the reddish-colored Kayenta Formation, capped by the Lamb Point Tongue of the Navajo Sandstone (Figure 99). The road crosses Kanab Creek at **Mile 4.6.**

Mile 4.7: The contact between the lower red Kayenta Formation and the upper white Lamb Point Tongue of the Navajo Sandstone is at road level. The Lamb Point Tongue is highly cross-bedded, appearing to be windblown in origin, as is the main body of the Navajo Sandstone. The highway at **Mile 5.2** leaves Kanab Creek Canyon, heading up a tributary, Three Lakes Canyon. The road is in the main body of the Kayenta Formation, capped by the Tenny Canyon Tongue of the Kayenta Formation. The road to the right at **Mile 5.6** goes up Kanab Creek. The Tenny Canyon Tongue has a windblown origin in this area, appearing more like the main body of the Navajo Sandstone. **Mile 5.9: Milepost 70.**

Mokie (or Moqui) Cave is to the right at **Mile 6.4.** The term *Moqui* was formerly used to refer to the Hopi Indians and is a common name for ruins or caves. The top of the Tenny Canyon Tongue of the Kayenta Formation and base of the Lamb Point Tongue of the Navajo Sandstone are ahead at **Mile 6.6.** The canyon rims are here capped by basal Navajo Sandstone. At

Figure 99. The stratigraphic relationships of the Lower Jurassic formations between Kanab Creek Canyon on the west and areas to the east such as Canyonlands National Park. The Kayenta and Wingate Formations grade westward into the Moenave Formation, with its three members, and the Navajo Sandstone of the eastern Colorado Plateau contains tongues of stream and tidal flat deposits to the west. Note that there are two separate Kayenta Formations, apparently only loosely related to each other. Drawn from written communication with Fred "Pete" Peterson, 2001.

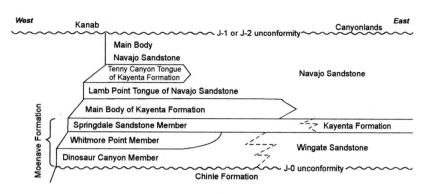

Figure 99. The stratigraphic relationships of the Lower Jurassic formations between Kanab Creek Canyon on the west and areas to the east such as Canyonlands National Park. The Kayenta and Wingate Formations grade westward into the Moenave Formation, with its three members, and the Navajo Sandstone of the eastern Colorado Plateau contains tongues of stream and tidal flat deposits to the west. Note that there are two separate Kayenta Formations, apparently only loosely related to each other. Drawn from written communication with Fred "Pete" Peterson, 2001.

Mile 7.1 the contact between the Kayenta Formation and the Navajo Sandstone is near road level. The Navajo is highly cross-bedded, as is typical for the windblown sandstone. The road to the left at **Mile 8.2** goes to the Coral Pink Sand Dunes State Park; the loose sand cover in this area is "coral pink." The countryside at **Mile 8.6** is heavily covered with windblown pink sand.

Mile 9.9: Milepost 74: The high cliffs and mesas to the right consist of thick exposures of Navajo Sandstone. The Navajo Sandstone cliffs to the right at **Mile 10.9: Milepost 75** are capped by the Temple Cap Sandstone (Middle Jurassic, in the San Rafael Group) and the basal Carmel Formation with a prominent treeline at the contact. Numerous "boondocks" and a lot of brush are at road level.

Mile 12.8: Milepost 77: The Kolob Plateau is visible to the left at about 10:00 with widespread exposures of the Navajo Sandstone in the Zion Canyon country. The highway crosses a divide and begins its descent toward the East Fork of the Virgin River. There is a good view to the north of the Markagunt Plateau, capped by volcanic rocks of Tertiary age. There is another road to the left at **Mile 13.6** to the Coral Pink Sand Dunes State Park.

Mile 13.8: Milepost 78: As the highway crosses the Sevier fault zone, cliffs of the Lower Jurassic Navajo Sandstone are to the right and the Middle Jurassic Carmel Formation is to the left; the down-thrown side of the fault is to the left, or west. **Mile 14.4**: The Paunsaugunt Plateau is visible in the right distance; this is the High Plateau that hosts Bryce Canyon National Park. The highway wanders through hills of the Carmel Formation at **Mile 14.7** with light gray marl beds, thin limestone beds, and an occasional gypsum bed. Here in the type locality of the Carmel the formation consists of a lower limestone and marlstone unit, the Co-op Creek Member, overlain by red sandstone and siltstone beds of the Crystal Creek Member. Massive gypsum occurs in the Paria River Member, which is overlain by sandstone of the Winsor Member at the top.

At **Mile 15.1**, the region around Zion National Park, the southern Kolob Terrace is visible in the left distance and the Markagunt Plateau can be seen in the high country ahead. **Mile 15.8: Milepost 80:** The road is in a deep gully carved into the rather thick Carmel Formation, here largely a light tan color, although it also contains red beds.

Mile 17.2: Entering Mount Carmel Junction, Utah. Utah 9 heads west toward Zion National Park and Hurricane, Utah (see Roadlog 48). The Carmel Formation is widely exposed here in its type area. The high cliffs to the right are in the Navajo Sandstone, capped up high by the Carmel Formation, east of the Sevier fault. Go straight ahead on U.S. 89 toward Bryce Canyon National Park. Leaving Mount Carmel Junction at **Mile 17.7.**

At **Mile 18.5** a thick bed of gypsum of the Paria River Member of the Carmel Formation is above road level; it comes down to road level at **Mile 18.8.** The road crosses Muddy Creek and enters Mount Carmel at **Mile 18.9.** The townsite is situated very near the Sevier fault zone. To the right are high massive cliffs of the Navajo Sandstone, capped by the Temple Cap Sandstone and basal Carmel Formation; the fault is at the base of the cliffs. The road is in the Carmel Formation along the western down-thrown side of the fault. **Mile 20.0: Milepost 84**: Leaving Mount Carmel village.

Mile 20.5: The road enters exposures of massive gray shale of the Tropic Shale, the equivalent of the Mancos Shale (Late Cretaceous) farther east, capped by sandstone of the Straight Cliffs Formation, visible on the left. The Straight Cliffs Formation is in the same stratigraphic position as the Mesaverde to the east but is somewhat older here. Huge cliffs of Navajo Sandstone are to the right (east) across the Sevier fault. **Mile 20.7: Milepost 85.**

Mile 21.0: Entering the village of Orderville, Utah. The Sevier fault runs about through town, with Tropic Shale, capped by beds of the Straight Cliffs Formation, on the left and Navajo Sandstone cliffs on the right. At **Mile 22.0**, leaving Orderville, the highway crosses the Virgin River. The highway is on the Sevier fault at **Mile 22.3.** The rocks on the right of the fault dip steeply toward the Virgin River. The Navajo Sandstone dips steeply to the left toward the fault at **Mile 22.9**; the highway is on the fault, and sandstone of the Straight Cliffs Formation is on the left.

At **Mile 23.6** a sign says "Entering Glendale" with no human habitation in sight. The Tropic Shale is faulted against the Navajo Sandstone on the right at **Mile 24.5.** The highway enters Glendale at **Mile 25.4. Mile 25.7: Milepost 90.** The cliff ahead at **Mile 29.4** is a lava flow above rocks of the Straight Cliffs Formation. There is a rest stop on the left at **Mile 30.6: Milepost 95.** The thick conglomerate here is thought to mark the top of the Straight Cliffs Formation below the Wahweap Formation, both Late Cretaceous in age. There are deeply weathered exposures of the orange and pink Claron Formation off to the left at **Mile 31.8.** The hill ahead at **Mile 32.0** is capped by volcanic rocks overlying rocks of the Wahweap Formation. Roadcuts at **Mile 32.5** are in rocks of the basal Tertiary, with lava in the roadcut at **Mile 32.7.** The road to the right at **Mile 35.0** goes to Alton, Utah. The soil all around is pink in color, suggesting that the highway is in the Claron Formation (Tertiary). **Mile 35.6: Milepost 100**: The valley and the highway are still in the Claron Formation. The highway climbs through the orange and pink member of the Claron Formation at **Mile 37.5** into light-colored rocks, largely freshwater limestone, of the middle Claron.

Mile 39.7: Long Valley Junction, where Utah 14 heads west toward Cedar Breaks National Monument and Cedar City, Utah (see Roadlog 49). The highway is still in nearly white beds of the middle Claron Formation. **Mile 40.6: Milepost 105.** Entering Garfield County at **Mile 43.8. Mile 45.6: Milepost 110**: The Paunsaugunt Plateau is across the valley to the right (east), and exposures of the orange and pink Claron Formation are in the cliffs. The highway is here still in the white upper member of the Claron Formation. The highway crosses Asay Creek at **Mile 46.8**. At **Mile 47.7** there are good exposures of the Claron Formation in the distance at 2:00 in the Red Canyon vicinity. Roadcuts at **Mile 48.7** are in white marl of the upper Claron Formation. At **Mile 50.0** the highway crosses through a basaltic lava flow. **Mile 50.5: Milepost 115.**

Dark gray shale of the Wahweap(?) Formation is exposed on the left at **Mile 51.1**, and exposures of the pink Claron Formation are ahead. Entering the village of Hatch, Utah, at **Mile 51.7;** leaving Hatch at **Mile 52.6.** A basaltic lava flow caps the low bluffs on the left and in roadcuts ahead at **Mile 54.3. Mile 55.5: Milepost 120**: The dark gray shale is again exposed on the left, and obvious exposures of the orange and pink Claron Formation are in the valley walls to the right.

Mile 60.2: Intersection with Utah 12, which heads east through Red Canyon toward Bryce Canyon National Park (see Roadlog 47).

VI

The High Plateaus

West of the San Rafael Swell and Circle Cliffs Uplift, the western margin of the Colorado Plateau is broken into northerly oriented fault-block uplands, the High Plateaus subprovince. Faults that bound the individual High Plateaus are generally north-south normal faults of major proportions. Because the faults are much like the structures that bound the many mountain ranges of the Basin and Range Province to the west, they are often referred to as "Basin and Range faults." The overall effect is that the western margin of the Colorado Plateau grades westward into the mountains of western Utah and Nevada through intermediate steps; relative movement along many of the normal faults is down to the west, although a few are downthrown to the east. For no apparent reason, the faults step one at a time farther to the west toward the south along the ancient Wasatch Line. The fault that marks the western boundary of the Colorado Plateau to the north is at Salina, Utah, whereas the western-bounding fault to the south is as far west as St. George, Utah, and the Grand Wash Cliffs, which form the west end of Grand Canyon.

Because most of these Basin and Range faults drop the margin of the Colorado Plateau toward the west, the rocks exposed in the High Plateaus are younger than those farther to the east, being almost entirely of Cretaceous and Tertiary age. The rocks of Cretaceous age become much thicker in the west and much coarser in grain size as the source area for the sand and gravel, the Sevier Orogenic Belt, is approached. The source area consists of highlands formed by eastward-moving thrust-faulted blocks in Cretaceous time, remaining as highlands throughout Cretaceous time in western Utah. Thus, shale beds thin and sandstone beds thicken and become conglomeratic toward the west across the margin of the plateau country. Thicknesses of the deposits vary from about 6,000 feet to the east on the Colorado Plateau to as much as 15,000 feet to the west. The Cretaceous section here includes, from the bottom up, the Mancos Shale, Star Point Sandstone, Blackhawk Formation, Castlegate Sandstone, Price River Formation, and North Horn Formation; the North Horn straddles the Cretaceous-Tertiary boundary.

The thick Cretaceous section is overlain by mudstone and siltstone of Tertiary age along the High Plateaus. These sedimentary rocks are best displayed in the colorful countryside around Bryce Canyon National Park and Cedar Breaks National Monument, Utah. The high plateau country, especially to the south, is capped by lava flows, mostly emanating from the Marysvale area, although several lesser volcanic centers are to be found. Very fresh lava flows of Quaternary age are frequently encountered as well.

The subprovince lies generally along the ancient Wasatch Line, which may reflect the western margin of North America in Late Precambrian time. The Wasatch Line is dissected at the surface by the Basin and Range normal faults. Originally, the highly generalized trend was recognized to mark the transition from the thin sedimentary deposits of the oceanic shelf regions of the Colorado Plateau westward into the very thick rocks of the Cordilleran seaway. That interpretation is certainly true as seen today, for the entire Paleozoic section thickens westward by several fold as it crosses beneath the High Plateaus. The trend is documented in only a very few deep wells, since the Paleozoic rocks are buried deeply beneath the High Plateaus. Exposures in the Virgin River canyon between the Nevada state line and St. George, Utah, provide a view at the surface of the dramatic changes at the Colorado Plateau border. Continental deposits of the Triassic Moenkopi and Jurassic Carmel Formations thicken and grade westward into marine rocks along the zone of transition, and the Cretaceous section becomes much thicker and coarser and almost entirely nonmarine. The Wasatch Line is truly a zone of drastic change throughout the Paleozoic and Mesozoic Eras.

These changes are apparent in the exposed sections along Interstate 70 as it crosses the San Rafael Swell between Green River and Salina, Utah. To the east, along the San Rafael "Reef," the stratigraphic section is typical of Colorado Plateau geology, but west of the crest of the broad structure the formations thicken and begin to change. At the entrance to Salina Canyon on the gentle west flank of the immense fold, Mesaverde-equivalent rocks thicken and individual formations become nearly indistinguishable. Emerging from the canyon at Salina, the Jurassic section includes thick marine evaporites, as seen in salt domes, and the Paleozoic section in the mountains to the west is amazingly thick.

Especially impressive are the radical changes in the Moenkopi Formation as seen between St. George and Kanab, Utah, along the highway through Zion National Park. To the west the section is much thicker and contains members that are more marine. Lower members of the Middle Jurassic Carmel Formation openly display marine characteristics in Cedar Canyon just east of Cedar City, Utah, where thick gypsum and massive limestone deposits replace the red bed sequence so commonly seen to the east. Here the Carmel section is truncated by an unconformity at the base of the Dakota Sandstone, and younger Jurassic formations have been removed by erosion.

Rocks of Tertiary age are best appreciated in the bright orange and pink mudstone and marl that forms the amazing landscapes of Bryce Canyon National Park and Cedar Breaks National Monument. Formerly known as the Wasatch Formation, the colorful rocks are now called the Claron Formation of Early Tertiary age. The formation's distinctive color forms a convenient bookmark in an otherwise complex part of the stratigraphic section. Volcanic rocks of Tertiary age cap the High Plateaus in the region.

45. Green River to Salina, Utah, via Interstate 70

Heading west from Green River, Utah, Interstate 70 first encounters the San Rafael Reef, where the Jurassic and Triassic formations are crossed in rapid succession in spectacular exposures along the prominent monocline. The oldest rocks exposed along the highway on the San Rafael Swell are in the Black Box Dolomite (formerly known as the "Kaibab Limestone") of Permian age. From the crest of the huge asymmetrical anticline, the highway proceeds down the gentle west flank of the swell through beautiful exposures of progressively younger Mesozoic strata. Near Fremont Junction a complex series of north-south faults cross the route in deep canyons, exposing a mixture of Cretaceous and Tertiary formations that are difficult to distinguish. The western margin of the Colorado Plateau Province is evident where the highway crosses the Sevier fault zone at Salina, Utah. (For I-70 east to Grand Junction, Colorado, see Roadlog 1.)

Mile 0.0: Begin on the west on-ramp (158) onto Interstate 70 from Green River, Utah, and head west on I-70 toward Salina.

Mile 1.4: Milepost 157: The highway is in the lower Mancos Shale, the Book Cliffs and exposures of the Mesaverde Group (Upper Cretaceous) are high on the right, and the San Rafael Swell is ahead. **Mile 2.1: Exit 156:** U.S. 6 and 191 head north toward Price and Salt Lake City (and connect with Roadlog 52, Green River to Helper, Utah). **Mile 3.3: Milepost 155.**

Mile 8.3: Milepost 150: The highway is at the base of the Mancos Shale with the San Rafael Swell obvious ahead. Roadcuts are in the Mancos Shale at **Mile 8.8.** The highway at **Mile 9.8** descends through a thin Dakota Sandstone (Upper Cretaceous) into the Lower Cretaceous Cedar Mountain Formation and Brushy Basin Member of the Morrison Formation (Upper

Figure 100. Interstate 70 enters the rugged, upturned rocks of the Page and Navajo Sandstones that form the San Rafael Reef, or monocline. The sharp fold formed where sedimentary rocks were draped across a deep-seated normal fault.

Jurassic). At **Mile 10.3** the top of the Salt Wash Member of the Morrison is at road level.

Mile 10.8: Exit 147: Utah 24 heads south to Hanksville and Lake Powell. The intersection is in the Salt Wash Member of the Morrison Formation, and the ominous San Rafael "Reef" looms ahead. At **Mile 11.6** roadcuts are in the Salt Wash Member, which is thick and well developed here.

At **Mile 12.3** the highway crosses down through red beds of the Summerville Formation (Middle Jurassic). The white hogbacks ahead in the San Rafael Reef are in thin beds of the Page Sandstone and much thicker beds of the Navajo Sandstone (Lower Jurassic). Early pioneers crossing the region likened their mode of travel to ocean voyages. They called their wagons "prairie schooners," and relatively impassable topographic features, such as this monoclinal fold, "reefs." Thus, the San Rafael Reef.

Mile 12.6: Crossing the San Rafael River. **Mile 13.1: Milepost 145**: Exposures of the dark reddish brown, thin-bedded Summerville Formation are to the right and left. A road to a view area is on the right at **Mile 13.9** in the very axis of the syncline adjacent to the San Rafael Reef. The highway is in the Carmel Formation (Jurassic), against the steeply dipping beds of the Navajo and Page Sandstones (Figure 100).

The highway at **Mile 14.5** enters very deep roadcuts into the Navajo Sandstone as the road crosses the magnificent monocline. The highway crosses the Navajo-Kayenta contact at **Mile 14.8** and descends into the massive, steeply eastward-dipping Wingate Sandstone (Jurassic) at **Mile 14.9.** At **Mile 15.0** the highway is in the top of the Chinle Formation (Triassic). **Mile 15.4**: The top of the Moenkopi Formation (Triassic) is at road level. The lower part of the Moenkopi is here a tawny yellowish color, caused by the

Figure 101. View back toward the east across the San Rafael monocline from the viewpoint at the top of the Sinbad Limestone Member of the Moenkopi Formation of Early Triassic age. The upper ledgy slopes are in the Late Triassic Chinle Formation, overlain by sandstone formations of the Glen Canyon Group of Early Jurassic age.

bleaching of the usual red beds, but the upper Moenkopi is still red. The middle member of the Moenkopi Formation, the Sinbad Limestone Member, is at road level at **Mile 15.7**.

Mile 16.0: The top of the Black Box Dolomite (Permian), formerly referred to as the "Kaibab Limestone," forms the caprock to the eastward-dipping San Rafael monocline. The roadcuts at **Mile 16.7** are near the top of the Sinbad Limestone Member of the Moenkopi Formation (Triassic). The road to the right at **Mile 17.5** goes to a viewpoint for a look back eastward across the San Rafael Reef (Figure 101).

Mile 18.2: Milepost 140: There are good exposures of the Black Box Dolomite ("Kaibab") in gullies to the right. The highway is in the lower Moenkopi Formation, the Black Dragon Member. The rimrock here is the Sinbad Limestone Member of the Moenkopi. The roadcuts at **Mile 19.0** are in the Black Box Dolomite ("Kaibab"). In the bottom of the deepest canyon the White Rim Sandstone, an eastern equivalent of the Toroweap Formation, is exposed, formerly incorrectly called the "Coconino Sandstone."

Mile 19.4: The highway is now on top of the Black Box Dolomite with good exposures in gullies below. This prominent limestone/dolomite unit was originally called the "Kaibab" because its appearance and stratigraphic position were similar to those of the Kaibab Limestone in the Grand Canyon. A study of the rare fossils found here suggested that this unit is somewhat younger than the true Kaibab and is better correlated with the Park City Formation in western and northern Utah. The highway runs near the top of the Black Box Dolomite for the next 2 miles.

Mile 23.1: Milepost 135: The highway is still ascending the dipslope toward the crest of the San Rafael Swell, here climbing up-section into the lower Moenkopi Formation (the Black Dragon Member), and is in the Sinbad Limestone Member at **Mile 23.9**. Roadcuts at **Mile 24.3** are in the Sinbad Limestone Member.

Mile 24.9: This is about the crest of the San Rafael Swell, where rocks ahead dip gently westward. The San Rafael Swell is obviously a very large asymmetrical anticline. The cliffs ahead are in the Glen Canyon Group, consisting in ascending order of the Wingate Sandstone, the Kayenta Formation, and the Navajo Sandstone, all of Early Jurassic age. The Wasatch Plateau is visible ahead and to the right in the far distance. From here to the west the highway crosses up-section into progressively younger rocks, as erosion has beveled the westward-dipping strata. **Mile 28.0: Milepost 130.**

Mile 28.5: Exit 29 is a ranch road. The highway is in the upper Moenkopi Formation, the Moody Canyon Member. The cliffs to the right are Chinle Formation, capped by the basal Wingate Sandstone. The highway crosses up-section at **Mile 31.4** from the Moenkopi Formation into the Chinle Formation (both Triassic). The cliffs above are capped by lower Wingate Sandstone.

Mile 33.0: Milepost 125: Exposures of the upper Chinle Formation and Wingate Sandstone form the landscape here; the white cliffs ahead are in the Navajo Sandstone. At **Mile 33.6** the highway is crossing up-section into the Wingate Sandstone, here rather thin-bedded; the Kayenta Formation is seen in roadcuts at **Mile 34.1**. Good exposures of Navajo Sandstone are to the right and ahead at **Mile 34.3**. The highway is in the Kayenta Formation, here dipping gently toward the west. The highway passes up-section at **Mile 36.1** from exposures in the Kayenta Formation into the Navajo Sandstone. The road on the right at **Mile 37.2** goes to a view area in the Navajo Sandstone where the High Plateaus subprovince can be seen in the distance toward the west and northwest.

Mile 38.0: Milepost 120: The beautiful canyons below to the right and left are in the Navajo Sandstone. The exposures at **Mile 38.2** are in the Page Sandstone. The westward dip of the strata is now more exaggerated. The view ahead at **Mile 39.0** is to the High Plateaus, mainly the Wasatch and Fish Lake Plateaus. The highway is back into the Navajo Sandstone here. The highway is at the top of the Kayenta Formation at **Mile 39.7** with cliffs of Navajo Sandstone all around. The Kayenta Formation is here rather thick and well developed. The highway climbs back up into the Navajo Sandstone at **Mile 41.6**. The hills at **Mile 42.2** are capped by the Carmel Formation on the rather thin Page Sandstone. **Mile 42.8: Milepost 115**: The highway climbs up-section into the Carmel Formation from the Page Sandstone. The road to the right at **Mile 43.4** goes to a view area in the Carmel Formation with scattered beds of gypsum.

Mile 44.5: Grand views of the High Plateaus are to the west beyond red rocks of the Jurassic section and gray rocks of the Cretaceous section in the middle distance. The highway descends across a dipslope in the Carmel Formation, and roadcuts are in the lower Carmel at **Mile 47.8: Milepost 10.** At **Mile 49.7** the highway crosses up-section into the Entrada Sandstone, which is here in the soft red silty facies. At **Mile 51.4** there are beautiful exposures of the Entrada Sandstone all around, capped in the higher buttes by greenish-gray sandstone beds of the Curtis Formation. This area is known as Sinbad Valley. **Mile 52.8:** Exit 105.

The light-colored cliffs around **Mile 53.8** are in the Curtis Formation, a marine sandstone, with exposures of the dark brown Summerville Formation beyond.

A road to the right at **Mile 54.4** goes to a view area. The highway at **Mile 55.0** is crossing upward through the Curtis Sandstone. Red beds of the Summerville Formation are exposed ahead and to the right. The skyline ahead is capped by the Morrison Formation. **Mile 56.2:** The Summerville-Morrison contact is visible in the cliffs to the right.

The highway at **Mile 57.0** crosses upward through the basal Morrison Formation, the Tidwell Member, and into the Salt Wash Member of the Morrison. At **Mile 57.3** the Brushy Basin Member of the Morrison Formation is exposed to the right and left, and at **Mile 57.8: Milepost 100** the highway is in the Brushy Basin Member.

At **Mile 58.4** a thin bed of the Buckhorn Conglomerate Member is exposed at the base of the Cedar Mountain Formation (Early Cretaceous). The highway at **Mile 58.7** crosses upward through the Dakota Sandstone and into the Tununk Member (lower) of the Mancos Shale (Upper Cretaceous), and high cliffs of Upper Cretaceous Ferron Sandstone beds are all around. **Mile 62.7: Milepost 95:** The highway is in the Lower Blue Gate Shale Member of the upper Mancos Shale with cliffs of the Emery Sandstone Member of the Mancos Shale ahead. The highway crosses the Lower Blue Gate–Emery contact at **Mile 63.4.**

The highway crosses the top of the Emery Sandstone into mudstone and sandstone beds of the Upper Blue Gate Member at **Mile 63.7** and then climbs up-section into the Star Point Sandstone of the Mesaverde Group at **Mile 64.0.**

The hillsides at **Mile 64.6** are covered to the right and left with black volcanic boulders, which came down from the high plateau to the south. The covered hillsides are on the Blackhawk Formation. The highway at **Mile 64.9** is in light-colored mudstones of the North Horn Formation of latest Cretaceous and early Paleocene age. **Mile 65.8:** Entering Sevier County. Volcanic boulders shroud a light gray soil (apparently in the North Horn Formation) in the hillsides all around.

Mile 67.8: Milepost 90: Ahead is Exit 89, Fremont Junction; Utah 10 goes north to Price, Utah, and Utah 72 goes south to Fremont, Utah. The highway is on a broad plain covered with volcanic boulders over gray shale of the Upper Blue Gate, or Masuk, Member of the Mancos Shale. The highway has crossed a major north-south Basin and Range–type fault; this area is a northern extension of the Paunsaugunt fault zone. At **Mile 68.9** the highway is in the Upper Blue Gate Member of the Mancos Shale, heading into the canyon and crossing the south end of the Wasatch Plateau leading down to Salina, Utah.

Note: From here to Salina, numerous north-south normal faults cross the canyon, making identification of the various Cretaceous formations very difficult. The Mesaverde-equivalent and younger sandstone formations become thicker and much coarser in grain size closer to the source of the sediments. This complexly faulted zone marks a gradation from Colorado Plateau structure and stratigraphy to the east to the thicker and more complex stratigraphic relationships in the Basin and Range Province to the west.

The highway at **Mile 70.8** is back into the Mesaverde Group sandstone section. This is a rusty yellow, ledgy sandstone section (the Star Point Sandstone?). The highway goes up-section at **Mile 72.2** from the rusty yellow sandstone (Star Point?) into light gray shale (Blackhawk?). **Mile 72.4:** Entering Fish Lake National Forest. **Mile 72.6: Milepost 85**: Light gray mudstone forms the canyon walls (North Horn?), capped by the Castlegate Sandstone cliffs.

Mile 74.3: The highway is at the top of the light gray shale and the base of a massive sandstone. For the next 6+ miles the highway is crossing through a north-south fault zone. *Note:* From here to the west, the terminology changes to that of the central Utah Cretaceous section. The rocks exposed here are largely within the Indianola Group. The gray shale section should now be called the Allen Valley Shale, and the overlying light-colored sandstone should be called the Funk Valley Formation of the Indianola Group, of Late Cretaceous age. **Mile 74.8:** The highway is at the top of the light-colored sandstone and the base of the gray shale, North Horn(?) of Paleocene age. **Mile 76.2:** Sandstone ledge within the gray shale (North Horn?). **Mile 77.6: Milepost 80**: Alternating sandstone and gray shale (North Horn?).

At **Mile 81.5** there is a nice view ahead of the high rims of the Wasatch Plateau, capped by Late Cretaceous–Tertiary sedimentary rocks. The highway is in interbedded sandstone and mudstone. The white cliffs at the top of the plateau may be the Green River (Tertiary-Eocene). **Mile 82.5: Milepost 75.** Nearly white mudstone and interbedded sandstone are at road level at **Mile 83.7.**

Mile 85.5: Exit 72. The section here is nearly white, getting down into pink mudstone of the North Horn Formation. Ledgy sandstone cliffs are high ahead. The canyon ahead at **Mile 86.7** narrows and steepens as the road climbs into ledgy sandstone and mudstone cliffs of the Green River Formation. From **Mile 87.5: Milepost 70** to **Mile 92.4: Milepost 65** the road is in the very thick section of ledgy sandstone and some mudstone.

Mile 96.1: The exit goes to Gooseberry Road and Campground. The highway is in the top of the ledgy sandstone and entering into pink mudstone (Colton Formation?). The high plateau is capped with very light-colored beds of the Green River Formation (Tertiary). **Mile 97.4: Milepost 60**: The highway is in dark red beds, capped by almost white mudstone high above. These red beds are dark red to pink mudstone and sandstone of the Middle Tertiary Crazy Hollow Formation. **Mile 99.0**: There are good exposures and roadcuts in pink beds of the Crazy Hollow Formation with light gray sandstones above. **Mile 100.6**: Leaving Fish Lake National Forest. The highway is near the top of the red beds and the base of the very light gray mudstone and sandstone beds of the Jurassic Arapien Shale, exposed ahead.

Mile 102.3: Milepost 55: The highway is crossing a major Basin and Range fault zone. This is the Sevier fault and the Pavant thrust-fault belt, marking the eastern edge of the Great Basin Province and the western margin of the Colorado Plateau Province.

Mile 102.6: Take exit 54 to Salina, Utah. **Mile 103.0**: Intersection and stoplight; turn right to Salina on U.S. 89.

46. Salina, Utah, to Utah 12 via Interstate 70 and U.S. 89 (plus Utah 20)

U.S. 89 follows the Sevier fault zone and Pavant thrust belt from Salina, Utah, southward. The fault-bounded valleys are considered the western margin of the Colorado Plateau in Basin and Range–type fault zones.

Mile 0.0: Intersection of U.S. 89 and Interstate 70 at the south edge of Salina, Utah; head south on I-70. Salina is definitely in the Basin and Range Province at the western boundary of the Colorado Plateau Province. The Spanish name *Salina* comes from the occurrences of salt and salt mines near the town in the Arapien Shale (Middle Jurassic), now called the Twelvemile Canyon Formation by the U.S. Geological Survey. The village was settled in 1863 and abandoned three years later because of clashes between whites and Indians. It was resettled in 1871.

Volcanic rocks in the Pahvant Range are to the right (west), and light-colored mudstone exposures of the Arapien Shale (Carmel-Entrada-Summerville equivalent) are along the slopes of the southern High Plateaus to the left. The rolling light gray hills along the valley are spangled with shiny flakes of alabaster gypsum from the Arapien Shale outcrops. The highway follows a broad valley, floored by Quaternary sediments, along the Sevier fault zone.

Mile 5.0: Milepost 50: Volcanic rocks in the Tushar volcanic field are ahead; there are four volcanic calderas in the Tushar Mountains. **Mile 7.4**: Exit 48 goes to Utah 24, which heads east toward Capitol Reef National Park. The Fish Lake Plateau is obvious high and to the left. Light gray to dark red sedimentary rocks of Tertiary age are randomly mixed with the volcanic rocks in the Pahvant Range to the right. **Mile 10.7: Milepost 45**: The highway is following a wide-open valley, filled with Tertiary and Quaternary sediments, along the Sevier fault zone. **Mile 15.9**: Exit 40, to Richfield, Utah. **Mile 16.4: Milepost 40.** The highway follows the west flank of the Sevier Valley, and Tertiary red beds are along the base of the Pahvant Range to the right. **Mile 19.0**: Exit 37, to south Richfield.

Mile 21.7: Milepost 35 (these mileposts are not very consistent with odometer readings): Tertiary volcanic rocks, ash flows, and ubiquitous boulders are to the right. **Mile 24.4**: Exit 32, to Elsinore and Monroe, Utah. Volcanic rocks are to the right, and a jumbled mass of volcanic rocks is on the left side of the valley. **Mile 26.7: Milepost 30**: There are volcanic rocks to the right and in low ridges midvalley to the left. Black volcanic boulders cover the countryside.

Mile 30.3: Exit 26, to Joseph, Utah. Volcanic rocks are to the right and left. **Mile 31.7: Milepost 25.**

Mile 32.6: Milepost 24: Take Exit 23 to U.S. 89. I-70 continues west through Clear Creek Canyon, which divides the Pahvant Range to the north from the Tushar Mountains to the south. Head south on U.S. 89 toward Bryce Canyon. Exposures of nearly white mudstone (fault gouge?) are to the right. The road to the right at **Mile 34.2** goes to Fremont Indian State Park.

Mile 34.9: Crossing the Sevier River. The road heads into the canyon of the Sevier River, a jumbled mess of volcanic rocks on both sides. This is the edge of the Marysvale volcanic field, a volcanic complex of up to 15,000 feet of rhyolitic and basaltic rocks and one of the largest volcanic fields in Utah. Here in the Tushar Mountains volcanism began about 30 million years ago (Oligocene) and continued until about 4 or 5 million years ago (Late Miocene). The volcanic rocks of the Tushar Mountains have produced significant amounts of gold, silver, mercury, lead, zinc, uranium, manganese, iron, and alunite, a source of aluminum, potash, and sulfur (Stokes 1986). Although the volcanic field is still considered prospective, the mining districts are largely abandoned.

Mile 35.7: Milepost 190: Volcanic rocks make up the canyon walls to the rim. Volcanic agglomerates are exposed to the right and left at **Mile 37.5**. The road crosses a volcanic blowout center at **Mile 40.3**. **Mile 40.7**: Thunderbird Trading Post and Big Rock Candy Mountain Resort. The road leaves the volcanic blowout center and heads into dark-colored and jumbled volcanic rocks. **Mile 41.1**: Entering Piute County. A rest area is on the right at **Mile 41.7.**

The road at **Mile 44.0** leaves the narrow Sevier River canyon and enters a broad valley along the Sevier fault zone. **Mile 44.4:** Crossing Beaver Creek. Volcanic rocks are exposed along the valley walls.

Mile 45.5: Entering Marysvale, Utah. The origin of the name is problematic, but the village was known as Merry Vale to Mormon visitors in the late 1800s. It was settled in 1863, abandoned because of Indian problems, and later resettled by Mormon pioneers. Leaving Marysvale at **Mile 46.1** through thick roadcuts in river terrace gravels.

Mile 46.6: Milepost 180: Volcanic rocks are high on the left and right. The road is on a broad, flat valley on top of a gravel terrace. **Mile 51.5: Milepost 175**: This Basin and Range–type valley along the Sevier fault zone is bordered on the right and left by volcanic rocks. **Mile 54.5**: The road to the left goes to Piute Lake State Park. Volcanic mountains are to the left and right at **Mile 56.5: Milepost 170**. Entering Junction at **Mile 59.0**, elevation 6,000 feet.

The road to the left at **Mile 62.5** goes to Kingston, Otter Creek Reservoir, and Capitol Reef National Park. The road crosses the Sevier River at **Mile 63.7.** Volcanic rocks are exposed to the right and left in the mountains. **Mile 65.7: Milepost 160**: Still in the broad valley with volcanic mountains to the right and left.

Mile 66.0: Recrossing the Sevier River. Entering Circleville, Utah, named for Circle Valley, meaning a valley encircled by mountains, and settled in 1864 by Mormon pioneers; the home of Butch Cassidy. Butch Cassidy's hideout is on the right at **Mile 66.8. Mile 67.5**: Leaving Circleville.

At **Mile 68.5** the valley along the Sevier River narrows and heads into volcanic rocks once again. **Mile 68.7**: Entering Garfield County, named for President James A. Garfield in 1882.

Mile 70.7: Milepost 155: The road is in a narrow canyon surrounded by Tertiary basalt flows. Terrace gravel deposits form the valley floor. **Mile 75.6: Milepost 150.** The valley opens up at **Mile 78.8** with volcanic rocks forming the surrounding hills. **Mile 83.7**: Intersection with Utah 20, which heads west (see sidetrip below); continue south on U.S. 89. **Mile 85.5: Milepost 140.**

Volcanic rocks of Miocene age form the east valley wall at **Mile 86.2**, but exposures of the orange and pink Claron Formation (Paleocene–Eocene) are visible at about 10:00. Tertiary-age volcanic rocks still form the valley wall to the right. **Mile 90.4: Milepost 135.**

Entering Garfield County again at **Mile 91.8.** The orange and pink Claron Formation is widely exposed in the hills to the left (east).

Mile 92.6: Entering Panguitch, Utah. The name is Paiute, meaning "fish" and "water." The village was settled in 1866 and abandoned because of clashes with Indians; it was resettled in 1871. Stop sign at **Mile 93.7.** Intersection of U.S. 89 and Utah 143, which heads west toward Cedar Breaks National Monument and Brian Head. Turn left on U.S. 89. Leaving Panguitch and heading east at **Mile 94.5**; the Tertiary Claron Formation is visible ahead, capped by volcanic rocks. **Mile 95.4: Milepost 130.**

The road to the right at **Mile 99.0** goes to a gravel pit of volcanic material. The colorful Claron Formation is widely exposed ahead and to the left, and volcanic rocks form the hills to the right. **Mile 100.3: Milepost 125.**

Mile 100.5: Intersection with Utah 12, which goes to Bryce Canyon National Park. (Connect with Roadlog 47, Bryce Canyon via Utah 12, or Roadlog 44, Kanab, Utah, to Utah 12.)

Sidetrip: Utah 20 at U.S. 89, heading west to Interstate 15

Mile 0.0: At the intersection of U.S. 89 and Utah 20, head west on Utah 20 for a drive of 20 miles across volcanic rocks.

Volcanic rocks (Miocene) are exposed to the right and left at **Mile 1.2.** **Mile 1.5: Milepost 19.** Entering Iron County at **Mile 3.1.** Exposures are agglomerates, or volcanic breccias. **Mile 5.5: Milepost 15.** The road crosses a divide at **Milepost 10,** still in volcanic rocks. **Mile 20.2:** Intersection with I-15.

47. Bryce Canyon National Park via Utah 12

Mile 0.0: Intersection of U.S. 89 and Utah 12, a Scenic Byway. Turn east on Highway 12 toward Red Canyon and Bryce Canyon.

Mile 3.0: Entering Red Canyon. A well-exposed fault (the Sevier fault or fault zone) is obvious at the entrance to the canyon, separating a lava flow of Tertiary age from the Claron Formation, Early Tertiary orange and pink mudstone and marl, which forms the scenery at Bryce Canyon. There is a scenic turnout at **Mile 3.2** and a Forest Service Ranger Station for the Dixie National Forest at **Mile 3.3.** A Forest Service campground is at **Mile 3.8.** **Mile 4.0: Milepost 4.**

At **Mile 4.6** there is a short tunnel and then a second tunnel in a tenth of a mile, with another scenic turnout at **Mile 4.8.** There are beautiful exposures of the Claron Formation at **Mile 5.0: Milepost 5.**

The road reaches the head of Red Canyon at **Mile 7.2** and enters a broad grassy plain with a few scattered pine trees, at the top of the Claron Formation. Roadcuts along the grassy plain are mostly in gravel deposits. At **Mile 9.0: Milepost 9** the Escalante Mountains, another of the High Plateaus, are visible to the left at about 10:00 to 11:00. The elevation at **Mile 9.2** is 7,619 feet. Bryce Canyon Pines Motel is on the left at **Mile 10.0: Milepost 10,** and a rest area is on the left at **Mile 10.5.**

The road crosses the East Fork of the Sevier River at **Mile 11.6**, and there is another scenic turnout on the left at **Mile 12.6.** A road to the left at **Mile 13.0**, and again at **Mile 13.2,** goes to the Bryce Canyon Airport. The High Plateau ahead and to the left is the Table Cliff Plateau at the south end of the Aquarius Plateau; the Escalante Mountains are visible to the left (north).

The turnoff to Bryce Canyon National Park, Utah 63, is at **Mile 13.6.** Reset trip odometer to **Mile 0.0** and turn south toward Bryce Canyon National Park on Utah 63. Entering Ruby's Inn village at **Mile 1.2.** For some unknown reason, **Milepost 1** is at **Mile 1.7.** Entering Dixie National Forest at **Mile 1.8**, with nice ponderosa pine stands; a scenic turnout is at **Mile 2.2.**

Mile 2.7: Entering Bryce Canyon National Park on a heavily wooded plateau. Bryce Canyon was named for Ebenezer Bryce, who homesteaded at the mouth of the canyon in 1876 and ran a cattle ranch there. The scenic cliffs along the east-facing facade of the Paunsaugunt Plateau were designated a national park in 1928. The park entry gate and pay station is at **Mile 3.7**, and the North Campground is on the left and the Visitor Center on the right at **Mile 3.9.**

The road to the left at **Mile 4.2** goes to Sunrise Point; at **Mile 4.8** the road to the left goes to the Bryce Canyon Lodge, with another road to the left to Sunset Point at **Mile 5.0.** Sunset Campground is to the right at **Mile 5.1.** The bright orange and pink lower member of the Claron Formation, with its complex and grotesque erosional forms, is the focus of all the viewpoints. Exposures and roadcuts along the route are all in the Claron Formation of Early Tertiary age.

Mile 5.7: Milepost 3 (measured from the national park boundary): The road is in a heavily wooded forest with scattered grassy glens and pink-colored soils. The Back Country Trailhead is at **Mile 8.9** with exposures of the Claron Formation below. There are numerous turnouts and viewpoints along the route for the next 10 miles.

Mile 20.7: Milepost 18: Turnaround and campground at the end of the road at Rainbow Point, elevation 9,115 feet. As often happens, an outdated exhibit at Rainbow Point presents an incorrect stratigraphic section for the park. According to current knowledge, the Tropic Shale is at the base of the cliffs below, overlain by the Straight Cliffs Formation and then the Wahweap Formation (all of Late Cretaceous age), capped by the Claron Formation, not the "Wasatch Limestone" as stated. These are the "Pink Cliffs" at the top of the "Grand Staircase" of southern Utah (see Figure 98, Roadlog 42). Return to the park entrance.

48. Mount Carmel Junction, Utah, to Interstate 15 via Utah 9 through Zion National Park

The trip begins in Mount Carmel Junction, Utah, the type area for the well-exposed Carmel Formation of Middle Jurassic age. In a short distance the road climbs into the Dakota Sandstone and Tropic Shale of Late Cretaceous age before dropping into a magnificent canyon carved into the Navajo Sandstone in Zion Canyon National Park. From the tunnel near the park entrance the route descends down-section through the Kayenta and Moenave Formations into the Chinle Formation through a ruggedly beautiful canyon to the entrance to Zion Canyon on the Virgin River. Then, gradually but persistently, the highway continues down-section through a very thick Moenkopi Formation and onto the Permian-age Kaibab Limestone as the road crosses the Hurricane fault near Hurricane, Utah.

Mile 0.0: Intersection of U.S. 89 and Utah 9 at Mount Carmel Junction, Utah. Head west on Utah 9 toward Zion National Park. Great exposures of the Carmel Formation are all around this area, which is the type section, here subdivided into five members. They are, in ascending order, the Co-op Creek Member, which is the basal limestone unit; the Crystal Creek Member, a fine-grained sandstone unit formerly mapped as the Entrada Sandstone; the Paria River Member, which contains considerable amounts of gypsum; and the Winsor Member, a sandstone unit at the top. There is no Entrada Sandstone here, as it has been removed by erosion at the base of the Cretaceous section, the Dakota Sandstone.

Mile 2.7: Milepost 54: There is a good view ahead of the Kolob Terrace on the skyline. The gray rocks in the hills ahead at **Mile 4.5** are in the Cretaceous Dakota Sandstone, the dark gray Tropic Shale (Mancos), and thick shoreface sandstone beds of the Straight Cliffs Formation. **Mile 4.7: Milepost 52.**

Mile 6.0: The hills have been heavily covered for the last 3 miles, but the road is probably on the Carmel Formation. The gray cliffs ahead are in rocks of Cretaceous age. Roadcuts and exposures to the left at **Mile 6.7: Milepost 50** are in the Tropic Shale (Mancos Shale equivalent), and then the road is on brush-covered terrain with no rock exposures. The canyon to the left at **Mile 8.4** has good exposures of the Tropic Shale and the Straight Cliffs Formation.

Ahead at **Mile 9.9** are exposures of the Temple Cap and Navajo Sandstones of Jurassic age. The Temple Cap Formation consists of a lower thin Sinawava Member (red mudstone) and an upper cross-bedded sandstone member that resembles the Navajo, the White Throne Member. The road here is in the Temple Cap Formation, just above the Navajo Sandstone. The road starts down into a canyon at **Mile 10.4** at about the top of the Navajo

Figure 102. Checkerboard Mesa Viewpoint near the east entrance to Zion National Park. Large-scale cross bedding is quite apparent in the Navajo Sandstone of Early Jurassic age, punctuated by near-vertical fractures in the rock that provide a checkerboard effect.

Sandstone with the Temple Cap forming the cliffs above. There are beautiful exposures of the Navajo Sandstone ahead. **Mile 11.7: Milepost 45:** The road descends through cliffs of Navajo Sandstone.

Mile 12.3: Entering Zion National Park in a sharp canyon bounded by Navajo Sandstone cliffs. **Mile 13.0:** Pay station. **Mile 13.2:** Checkerboard Mesa viewpoint (Figure 102). Then down, down, down through the Navajo Sandstone, with magnificent eolian cross bedding in a very rugged, narrow canyon.

The top of the Kayenta Formation, mostly a tidal flat and windblown dust deposit here, is exposed at road level at **Mile 17.2.** There are huge cliffs of beautifully cross-bedded Navajo Sandstone above. Canyon overlook parking is at **Mile 18.1.**

Mile 18.2: Entering a long tunnel with one-way traffic and a stoplight. The tunnel exit at **Mile 19.3** is at about the Kayenta-Navajo contact with high, massive cliffs of Navajo Sandstone above.

At **Mile 21.7** the top of the Moenave Formation, the stream-deposited Springdale Sandstone Member, forms red, ledgy sandstone cliffs (Figure 103). At **Mile 22.4** the road is in varicolored shale slopes of the upper, Petrified Forest Member of the Chinle Formation (Triassic).

Mile 22.7: Intersection with the Zion Canyon road, going north along the Virgin River for 6.2 miles to the Zion Lodge and trail. This is a most beautiful drive and hike and highly recommended.

Figure 103. Zion National Park's east entrance road, just below the tunnel, winds its way down from massive cliffs of the Navajo Sandstone through thinner-bedded rocks of the Kayenta and the red Moenave Formations, all of Early Jurassic age.

Mile 23.4: The Visitor Center is on the right. South Campground is on the left at **Mile 23.8** in the Petrified Forest Member of the Chinle Formation. Watchman Campground is on the left at **Mile 24.1**. **Mile 24.2**: Park entry fee station. Leaving Zion Canyon National Park. **Mile 24.5: Milepost 32**.

Mile 25.0: Entering Springdale, Utah. The sandstone ledges to the right and left at **Mile 27.4** are in the Shinarump Member of the Chinle Formation, stream-deposited sandstone filling erosional channels cut into the top of the Moenkopi Formation. The road at **Mile 27.6** is in the top of the "upper red member" of the Moenkopi Formation (Triassic), beneath the Shinarump cliffs.

Mile 28.5: Milepost 28: Entering Rockville, Utah. The white beds at **Mile 30.0** are in the Shnabkaib Member of the Moenkopi Formation, as are the white beds at **Mile 30.6**. The view ahead down the valley at **Mile 31.0** is of the Shinarump rim on the left above Moenkopi red beds with a lava flow capping the hill on the right.

Mile 31.4: Milepost 25: There are volcanic rocks on the mountainside to the right and ahead. The Moenkopi Formation here consists of an "upper red member," the Shnabkaib Member (white, containing some gypsum, limestone, and mudstone), and the "middle red member." The Moenkopi red beds at **Mile 33.0** are covered with lava boulders from the lava flow above. The road has just passed the west end of the lava flow at **Mile 36.5: Milepost 20**. Hurricane Mesa, ahead, is capped by the Shinarump Member above slopes of a very thick Moenkopi Formation. A water tower and a test track, used for astronaut acceleration experiments in the early training days, are located on the mesa.

Mile 38.0: Entering the village of Virgin, Utah, elevation 3,552 feet. **Mile 41.5: Milepost 15**: The Virgin Limestone Member of the Moenkopi Formation is low and above to the right. The road is in the "lower red member" of the Moenkopi. The Pine Valley Mountains are visible ahead in the distance.

Roadcuts at **Mile 42.5** are in gray limestone of the Kaibab Formation (Permian), rolling over very sharply to the west toward the Hurricane fault. Exposures and roadcuts in the Kaibab Formation at **Mile 43.2** are still dipping strongly westward into the fault. The road crosses the Hurricane fault zone at **Mile 43.8**. Volcanic lava flows and cinder cones are visible ahead and to the right.

Mile 44.1: Entering La Verkin, Utah, with a good view of the fault zone to the right. Intersection at **Mile 44.3**: Utah 9 turns left, and Utah 17 turns right. Turn left on Highway 9. Heading south toward Hurricane, the highway is along the down-thrown side of the Hurricane fault, which is to the left. The Hurricane fault scarp to the left at **Mile 45.5** is rimmed with a basaltic lava flow. **Mile 45.7**: Bridge across the Virgin River and into a canyon of Quaternary basalt that also caps the rim to the left atop the fault escarpment.

Mile 46.0: Entering Hurricane, Utah. **Mile 46.5: Milepost 10. Mile 49.0**: Leaving Hurricane. (To go south on Utah 59, Hurricane to Fredonia, Arizona, see Roadlog 51.) **Mile 50.4: Milepost 6**: A lava flow on the Moenkopi Formation is to the right and left. A view of the Virgin anticline is ahead. **Mile 51.5: Milepost 5**: The northeast flank of the Virgin anticline is obvious from left to right ahead. The road passes through a thin lava flow at **Mile 52.1** and another at **Mile 52.5**.

The road crosses the northeast flank of the Virgin anticline at **Mile 53.0** in a lower member of the Moenkopi Formation, dipping steeply toward the east. The southwest limb of the Virgin anticline, visible ahead at **Mile 53.4,** is capped by the Shinarump Member of the Chinle Formation. **Mile 53.6: Milepost 3**: The Shinarump Member caps the "upper red member," the Shnabkaib Member, and the "middle red member" of the Moenkopi Formation ahead. The road is about at the crest of the Virgin anticline at **Mile 53.8.** The Shinarump forms the circular hogbacks that surround the fold. The road climbs the west flank of the Virgin anticline at **Mile 54.8**, and the Moenkopi-Shinarump contact is at road level.

Mile 55.0: The road crosses the Shinarump hogback, dipping westward from the anticlinal crest, and climbs into the varicolored Petrified Forest Member of the Chinle Formation (Triassic). **Mile 55.6: Milepost 1.** Roadcuts are in the Chinle Formation at **Mile 55.9.** Interstate 15 is ahead with the red cliffs of the Moenave Formation (Jurassic) exposed beyond the interchange.

Mile 56.5: The road passes beneath I-15. (Connect with Roadlog 50, **Mile 44.8.**)

49. Cedar City to Cedar Breaks National Monument and Long Valley Junction, Utah, via Utah 14

Leaving Cedar City, Utah 14 almost immediately begins to cross a 10-mile-wide series of north-south faults that constitute the Hurricane fault zone and the western boundary of the Colorado Plateau. The rocks are mostly Jurassic in age, including a well-exposed intrusive plug of gypsum of the Carmel Formation. Rather poor exposures of the Upper Cretaceous Dakota, Tropic, and Straight Cliffs Formations cap the Jurassic rocks in the upper canyon. Breaking out of the fault zone on the high plateau, the road is in latest Cretaceous rocks, capped by the bright orange and pink marl and mudstone of the Claron Formation of Eocene (Early Tertiary) age, which typifies the beauty of Cedar Breaks National Monument. The Claron is exposed most of the way down the east side of the Markagunt Plateau, punctuated by a Quaternary lava flow.

Mile 0.0: Intersection of Utah 14, a Scenic Byway, and Utah 130 in Cedar City, Utah. Head east on Utah 14.

Mile 0.5: The road crosses the Hurricane fault into a section of red mudstone interbedded with limestone and gypsum of the Moenkopi Formation (Triassic), all dipping steeply into the mountains toward the east. The gypsum-bearing section appears to be duplicated by faulting. The road enters Cedar Canyon, a sharp canyon with walls of the Moenkopi Formation. **Mile 1.1**: The massive white Navajo Sandstone is faulted against red beds of the Jurassic Carmel Formation. Several faults cross the canyon. At **Mile 1.4** massive pink sandstone of the Crystal Creek Member of the Carmel Formation is dipping very steeply toward the east. The road crosses another fault at **Mile 1.5** with the Crystal Creek Member repeated.

The road crosses another fault at **Mile 1.9**, and massive gypsum, of the Carmel Formation of Jurassic age, is exposed to the east. Although the gypsum appears to be an intrusive plug, much like a salt dome, it is mapped as vertical beds of gypsum plus some folding due to drag on a fault. **Mile 3.0: Milepost 3.** Pink mudstone of the Winsor Member is mixed in with the gypsum at **Mile 4.0** along the eastern margin of the gypsum "plug." The road to the right at **Mile 4.8** goes to the Kolob Reservoir. **At Mile** 5.4 the eastern margin of the gypsum "plug" is badly covered with brush and talus; massive rocks high above the gypsum dip steeply to the north on the left and to the south on the right, thus forming an anticline, or dome.

Mile 6.0: Milepost 6: Massive carbonate rocks of the Co-op Creek Member of the lower Carmel Formation (Jurassic) are just above the highway to the left and right. The massive limestone of the Co-op Creek Member is down to road level at **Mile 6.5.** The narrow, steep-walled canyon at **Mile 7.3** is in rocks of the Co-op Creek Member from here to **Mile 8.0: Milepost 8.** The road takes a sharp right turn into a narrow tributary canyon. **Mile 9.9: Milepost 10.** The top of the massive limestone of the Co-op Creek Member is at **Mile 10.5.** A major north-south fault is crossed about here. The valley opens up but is heavily covered with grass and trees. The previous 10 miles have been a remarkable cross section of the Hurricane fault zone, the transition between the Colorado Plateau Province to the east and the Great Basin, or Basin and Range Province, to the west.

Entering Dixie National Forest at **Mile 11.5.** The countryside is heavily wooded. **Mile 11.9: Milepost 12.** The road to the left at **Mile 12.1** goes to Cedar Canyon Campground. At **Mile 12.3** the orange and pink mudstone and marl of the Early Tertiary Claron Formation is visible high and ahead. **Mile 12.9: Milepost 13:** The section here is light gray mudstone and interbedded sandstone of the Late Cretaceous Straight Cliffs Formation. Good exposures of the pink Claron Formation are high ahead. Good exposures of the Kaiparowits and Claron are ahead at **Mile 13.3.** The countryside is heavily wooded with ponderosa and aspen forest at **Mile 13.9: Milepost 14.** Large roadcuts at **Mile 14.1** are in the Straight Cliffs Formation.

The road to the right at **Mile 14.6** goes to Webster Flat; roadcuts are in the Straight Cliffs Formation. There is a nice view to the south of the Zion Canyon country. **Mile 14.9: Milepost 15:** Densely wooded country is mostly aspen. The road at **Mile 15.2** is in the heavily covered base of the Claron Formation. **Mile 15.9: Milepost 16:** Pink mudstones of the Claron Formation are poorly exposed. Zion Overlook is on the right at **Mile 16.4** with good views of the Zion Canyon country in the low distance. The roadcuts on the left are in the basal member of the Claron Formation. **Mile 16.9: Milepost 17:** The road tops out here and begins a descent to the east.

Mile 17.8: Intersection with Utah 148, a Scenic Byway. Turn left on 148 to Cedar Breaks National Monument, heading north across a wooded surface on the Isom Breccia (Oligocene), with Blowhard Mountain to the left. **Mile 20.3:** Entering Cedar Breaks National Monument. **Mile 21.4:** Visitor Center at Point Supreme, elevation 10,350 feet. Cedar Breaks consists of rugged erosional terrain in the Claron Formation (Paleocene and Eocene). Turn around and return to Utah 14.

Back at the intersection with Utah 14 at **Mile 24.9**, turn left and proceed to the east toward U.S. 89. The valley opens into a broad, flat-bottomed meadow at **Mile 25.5** as the road crosses the Markagunt Plateau. Markagunt is a Paiute word meaning "highland of trees." A very fresh Quaternary lava flow is visible to the left at **Mile 28.0.** Roadcuts are in a thin limestone at **Mile 29.2**, probably basal Claron Formation (Flagstaff Limestone, Paleocene). **Mile 29.5**: Entering Kane County.

Mile 29.9: Roadcuts and poor exposures of light gray, thin-bedded limestone are in the lower Claron Formation. There is an overlook on the right to Navajo Lake at **Mile 31.0.** The roadcut on the left is a light gray limestone in the Claron Formation. **Mile 31.8: Milepost 25.** A very recent (Quaternary) blocky lava flow is on the left and right at **Mile 32.1**; it emanated from volcanic centers a short distance to the north. The recent lava is along the road for the next 2.5 miles. The road to the right at **Mile 32.3** goes to Navajo Lake. **Mile 34.7**: Spruces Campground is to the left at the edge of the lava flow. **Mile 34.9: Milepost 28**: Small patches of lava are scattered about.

Duck Creek Village is on the left at **Mile 36.6. Mile 37.9: Milepost 31**: Crossroads. Another crossroads at **Mile 39.5** goes to a summer-homes development to the right. Lava flows cap the hills to the left at about 8:00. Roadcuts at **Mile 40.9: Milepost 34** are in a chalky limestone above pink mudstone of the Claron Formation. The road to the right at **Mile 41.6** goes to Swayns Creek. The road to the right at **Mile 43.9** goes to Stout Canyon.

Mile 44.9: Milepost 38: The view to the left down a canyon is of the Canaan Peak Formation of latest Cretaceous to earliest Tertiary age. Leaving Dixie National Forest at **Mile 45.1.** The soils are still pink, so the road must be in the Claron Formation. At **Mile 46.0** the road is down into the gray sandstone and mudstone section, probably the Canaan Peak Formation. **Mile 46.9: Milepost 40**: The road is now in a valley formed in light brown clastics (sandstone and mudstone).

Mile 47.3: Milepost 41: Stop sign at Long Valley Junction, Utah; intersection with U.S. 89 (connect with Roadlog 44 at **Mile 39.7**). Long Valley was formed by erosion of the Sevier River along the extensive Sevier fault zone of Tertiary age.

50. Arizona-Nevada State Line to Hurricane, Utah, via Interstate 15 and Utah 9

To capture the western margin of the Colorado Plateau Province, this route begins at the Arizona-Nevada state line and proceeds through the narrow, rugged canyon of the Virgin River in massive exposures of lower Paleozoic rocks. The canyon separates the Beaver Dam Mountains to the north from the Virgin Mountains to the south; both comprise the eastern margin of the Basin and Range Province. At about midcanyon, where the highway crosses the Grand Wash fault zone, rocks of Pennsylvanian and Permian age mark the boundary of the Colorado Plateau Province. The highway skirts the beautiful Virgin anticline near St. George, Utah, and then crosses the structure between St. George and Hurricane, Utah. At the town of Hurricane the highway crosses the Hurricane fault.

Mile 0.0: Interstate 15 at the Arizona-Nevada border; head east toward St. George, Utah. The border is immediately east of Mesquite, Nevada, in a broad intermontane basin, definitely in the Basin and Range Province.

Mile 4.3, Milepost 5: To the right are the Virgin Mountains, a Basin and Range mountain complex. The highway crosses the Virgin River at **Mile 8.5**. **Mile 9.3: Milepost 10.**

Entering the deep gorge of the Virgin River at **Mile 12.0**, crossing a Basin and Range Tertiary-age fault zone. Massive carbonate cliffs at road level appear to be Redwall Limestone of Mississippian age; the Callville and Pakoon Formations (Pennsylvanian and Permian) are high on the left. The rocks exposed here are in the greatly thickened sections of the Basin and Range Province and are difficult to identify, especially when one is being bombarded by heavy, high-speed traffic. The gorge effectively divides the Virgin Mountains to the south from the Beaver Dam Mountains to the north, both Basin and Range structures.

Rocks of Devonian(?) age appear below the Redwall Limestone at **Mile 13.0**; this would be the Muddy Peak Dolomite, equivalent to the Temple Butte Formation of Grand Canyon. A massive dolomite below the Devonian must be the Nopah Dolomite of Late Cambrian age. The beds here are dipping strongly toward the west, in the direction of the bounding fault. **Mile 13.5**: Elevation 2,000 feet. There is a great thickness of massive carbonate rocks here, largely of Cambrian age(?), beneath the Redwall Limestone.

Mile 14.3: Milepost 15: The rocks are plunging strongly northward from the Virgin Mountains anticline. A short distance to the south, rocks of Precambrian age are exposed in the core of the anticline. The highway is well below the Redwall Limestone in carbonate rocks of Devonian and Cambrian age.

Mile 16.2: Milepost 17: The highway breaks out of the very deep, narrow gorge into a broad canyon formed along a complex fault zone associated with the Grand Wash fault. Pennsylvanian and Permian-age red beds of the Supai Group are visible ahead in the distance.

Mile 17.3: Exit 18: Take exit to Cedar Pocket, Arizona. The Virgin River Canyon Recreation Area, a campground and viewpoint, is along the complex Grand Wash fault zone. A short walk leads to a viewpoint at the top of a knoll, where informative plaques explain the geology in simple terms. The Grand Wash fault is up the canyon a short distance to the east, separating the Colorado Plateau from the Basin and Range Provinces. East of the Grand Wash fault are "layer-cake" exposures of the Pennsylvanian and Permian Callville and Pakoon Formations, overlain by the Permian Supai Group, here consisting of the Queantoweap Sandstone, with Toroweap and Kaibab limestone cliffs capping the section.

To the west the view is of the Basin and Range Province: the Callville through Kaibab Formations are visible to the south, and the Redwall and older formations can be seen to the west across a fault complex. The Beaver Dam Mountains are to the north, and the Virgin Mountains are to the south. The Mississippian Redwall Limestone is 700 to 900 feet thick; the Pennsylvanian Callville, equivalent to the Hermosa Group in the Four Corners area, is 1,500 to 2,000 feet thick; the Permo-Pennsylvanian Pakoon Dolomite, equivalent to the Elephant Canyon Formation in Canyonlands, is 700 to 900 feet thick; and the Permian Queantoweap is 1,500 to 2,000 feet thick. The Queantoweap is equivalent to the Esplanade Sandstone in Grand Canyon and the Cedar Mesa Sandstone in Canyonlands. There is no Coconino Sandstone present here. The Permian Toroweap Formation, consisting of massive limestone and gypsum, is 450 to 850 feet thick and grades northeastward into the White Rim Sandstone. The Kaibab limestone and gypsum is 330 to 650 feet thick. This point is truly a well-defined western margin of the Colorado Plateau Province.

Mile 18.4: Leaving the recreation area at I-15, heading east. Exposures of the Callville Limestone are to the left and right, and red beds of the Queantoweap, Toroweap, and Kaibab Formations are above and ahead. **Mile 19.8: Milepost 20**: The highway crosses the Grand Wash fault with deep roadcuts ahead in the Callville Formation and ledgy slopes above of Pakoon Dolomite. Like the equivalent Elephant Canyon Formation, the lower Pakoon is probably of late Virgilian (Bursum) age (latest Pennsylvanian) and the upper part is probably of earliest Permian (Wolfcampian) age.

There are deep roadcuts at **Mile 21.4** in alluvium and Queantoweap sandstone beds. Roadcuts ahead at **Mile 21.9** are in the Queantoweap Sandstone. There are good exposures at **Mile 23.0** of the Queantoweap in the canyon to the right and left. Its bedding features are identical to those of the Esplanade Sandstone in Grand Canyon, but the section here is much thicker. Note that the Coconino Sandstone and the Hermit Shale are missing in this area.

Mile 24.3: The lower cliff is the Brady Canyon Member of the Toroweap Formation, and the upper cliff is in the Fossil Mountain Member of the Kaibab Formation. Intervening slopes consist of gypsum and gypsiferous mudstone members of the two formations. **Mile 24.8: Milepost 25**: The highway is here along a bench of old river terrace gravel. The highway at **Mile 28.6** is in red beds of the lower Moenkopi Formation (Triassic), just above the Kaibab Formation. The Pine Valley Mountains can be seen ahead to the north of St. George, Utah.

Mile 29.0: Milepost 0: The Arizona-Utah state line. The hill on the right at **Mile 31.0** is the southwest-plunging nose of the Virgin anticline. The gray rocks holding up the rounded hill are the Virgin Limestone Member of the Moenkopi Formation. The Shinarump Member of the Chinle Formation is visible ahead in hogbacks along the flank of the anticline. The highway at **Mile 32.3** is in the "middle red member" of the Moenkopi Formation, dipping northward away from the Virgin anticline. At **Mile 32.8** the highway is crossing the lighter-colored, banded Shnabkaib Member of the Moenkopi Formation.

Mile 34.0: The contact between the Shnabkaib Member and the "upper red member" of the Moenkopi is at road level, with the Shinarump Member of the Chinle Formation above the "upper red member," still dipping northward. The highway crosses up-section through the Shinarump Member at **Mile 34.5**. The highway at **Mile 34.8** crosses the top of the Shinarump Member onto varicolored shale slopes of the Petrified Forest Member of the Chinle Formation. The plateau on the left is capped by a Tertiary lava flow.

Mile 35.1: Milepost 6: Entering St. George, Utah. **Mile 37.0**: The red rocks to the north across St. George are in the Moenave Formation of Jurassic age. The high bench on the right is capped by a lava flow. The cliffs to the left at **Mile 38.7** are in the Moenave Formation, and the northern flank of the Virgin anticline, capped by the Shinarump Member, is obvious to the right.

Mile 39.1: Milepost 10. Very young lava flows and cinder cones are visible to the left at **Mile 40.0**. The highway is in the Moenave Formation. **Mile 41.3**: Leaving St. George.

At **Mile 43.4** an inverted valley-fill of basaltic lava flow is incised into the Moenave Formation. The original valley was filled with the lava flow, but since the lava was more resistant to erosion, further down-cutting caused the valley to move adjacent to the present lava ridge. **Mile 44.0: Milepost 15**: There are good exposures of the red Moenave Formation on the left and the Virgin anticline on the right.

Mile 44.8: Take Exit 16 to Hurricane, Utah, on Utah 9, heading onto the flank of the Virgin anticline, rimmed by the Shinarump Member of the Chinle Formation. The low hogback on the right at **Mile 45.7** is the Shinarump Member, dipping strongly to the northwest. The road crosses the hogback of the Shinarump Member at **Mile 46.5**. There is a good view of the Virgin anticline (Figure 104). The ridge to the left is held up by the Shinarump, above the upper red and Shnabkaib Members of the Moenkopi Formation, with lower members of the Moenkopi in the valley. To the right (south) across the axis of the Virgin anticline, the section goes back up through the Moenkopi to the Shinarump hogback, which dips strongly toward the southeast. The high country ahead is in the Zion country, where the massive light-colored Navajo Sandstone is exposed.

Figure 104. The Virgin anticline between St. George and Hurricane along Utah 9. The cliff-forming sandstone on the left is the Shinarump Member of the Chinle Formation, overlying the slopes of the Moenkopi Formation, both of Triassic age. The hogbacks in the right center of the photograph are the same rocks dipping in the opposite direction across the axis of the anticline.

Mile 46.8: Milepost 2: The road is at the top of the Shnabkaib Member of the Moenkopi Formation, heading down toward the axis of the Virgin anticline. The road at **Mile 48.0** goes up into the "upper red member" of the Moenkopi Formation on the southeastern flank of the anticline. **Mile 48.2:** The road crosses through the Shinarump Member into the varicolored Petrified Forest Member of the Chinle Formation, dipping strongly southeastward off the Virgin anticline.

A lava flow is to the right at **Mile 49.0** and in the roadcut, and the northeasterly plunging nose of the Virgin anticline is visible to the left. The road crosses upward through the second lava flow at **Mile 49.4** and then runs along the top of the lava flow. **Mile 49.9: Milepost 5.** There are more lava flows to the left and right at **Mile 50.8.**

Entering Hurricane, Utah, at **Mile 52.8.** The Hurricane fault scarp is ahead just beyond town. (Connect with Roadlog 48 or Roadlog 51.)

51. Hurricane, Utah, to Fredonia, Arizona, via Utah 59/ Arizona 389

Leaving Hurricane, Utah, Utah 59 first crosses the obvious Hurricane fault zone and associated volcanic rocks and then climbs to a broad plateau that is capped by limestone of the Permian Kaibab Formation. Becoming Arizona 389 at the state line, the highway follows the Arizona Strip all the way to Fredonia, Arizona, basically at the top of the Kaibab Formation. Exposures of colorful rocks of Mesozoic age form cliffs to the north for the entire route.

Mile 0.0: Intersection of Utah 9 and Utah 59 in Hurricane, Utah; take Utah 59. The road turns to the right at one block and then left at stop signs.

Mile 0.2: Leaving Hurricane. The road crosses the Hurricane fault and begins the ascent of the fault scarp, with a lava flow capping the low cliff on the right. The road crosses up through exposures and roadcuts in lava flows at **Mile 0.5**, and the Hurricane fault is visible to the left (north). A small oil-field was discovered on the visible anticlinal drag fold along the up-thrown side of the fault in the middle distance.

Mile 1.2: Milepost 21: The road is climbing through exposures and road-cuts in limestone of the Kaibab Limestone (Permian) for the next 1.5 miles. The Kaibab here dips to the west toward the Hurricane fault. The road at **Mile 2.6** is on top of the Kaibab Limestone on a broad plateau surface. The red beds to the left are exposures of all the members of the Moenkopi Formation (Triassic), with a small butte capped by the Shinarump Member of the Chinle Formation (Triassic), seen from here to **Mile 3.5**. There are road-cuts and exposures at **Mile 4.5** of the "lower red member" of the Moenkopi Formation.

There is a lava flow at **Mile 6.7**, and a couple of cinder cones are to the right in the middle of the valley. **Mile 7.1: Milepost 15**. The road is in the Shnabkaib Member of the Moenkopi at **Mile 8.0**. The high white butte ahead at **Mile 8.4** is capped by the Navajo Sandstone (Jurassic) of Zion Canyon country. **Mile 11.8**: Passing Little Creek village on the left. A cinder cone, Little Creek Mountain, is on the right at **Mile 12.9**; the cliffs to the left expose rocks up through the Navajo Sandstone. The Moenkopi Formation is poorly exposed on the left at **Mile 16.9**.

Mile 17.1: Milepost 5: Poor exposures of the lower Moenkopi and Kaibab Formations are visible to the right. There are beautiful exposures in the cliffs to the left at **Mile 20.0** of the Chinle, Moenave, Kayenta, and Navajo Formations. **Mile 21.7**: Entering Hilldale, Utah, which straddles the Arizona-Utah border.

Mile 23.0: Milepost 1 in Arizona (the highway is now Arizona 389). The road left at **Mile 23.4** goes to Colorado City, Arizona, part of the so-called Arizona Strip, the section of Arizona between the Utah border and Grand Canyon. The villages in the vicinity are well-known fundamentalist Mormon settlements, where polygamous relationships are tolerated though illegal. The road to the left at **Mile 26.4** goes to Cane Beds, Arizona. **Mile 27.0: Milepost 5. Mile 28.5**: Elevation 5,000 feet. The cliffs to the left at **Mile 31.0** are the varicolored Chinle Formation, capped by the red Moenave Formation.

Mile 32.0: Milepost 10: Exposures to the left are in the "upper red member" of the Moenkopi Formation with a low cliff of the Shinarump Member of the Chinle Formation (Triassic). Above that is the varicolored Petrified Forest Member of the Chinle, capped by red cliffs of the Moenave Formation (Jurassic). The view ahead and to the right in the distance is of the

Kaibab uplift, capped by the Kaibab Limestone. The Grand Canyon is carved through the Kaibab uplift farther south.

Mile 36.0: Entering the Kaibab Paiute Indian Reservation. **Mile 36.9: Milepost 15**. The road to the left at **Mile 41.0** goes to Pipe Springs National Monument. The name comes from an incident in 1858 in which William Hamblin, a well-known Mormon guide, reportedly shot out the base of another man's smoking pipe on a bet. The spring water was used for decades by Indians, travelers, and settlers.

Mile 41.9: Milepost 20: The north-plunging nose of the Kaibab uplift is obvious ahead and to the right. The low cliff on the left is the Shinarump Member, and the high cliffs beyond are in the Moenave Formation. The road is on the poorly exposed Moenkopi Formation.

Mile 46.1: The dirt road to the right goes to Toroweap, about 60 miles to the south, where Recent cinder cones, especially Vulcans Throne, line the bench above Grand Canyon and where lava flowed down the walls of the canyon to Lava Falls Rapids on the Colorado River. There is a campground at the rim of the canyon. **Mile 46.8: Milepost 25. Mile 51.8: Milepost 30.**

Mile 52.3: Leaving the Kaibab Paiute Indian Reservation. **Mile 53.7**: Entering Fredonia, Arizona, and Coconino County, elevation 4,671 feet. Stop sign at **Mile 54.4**: Intersection of Arizona 389 and U.S. 89A in Fredonia. (Join Roadlog 42 at **Mile 30.0**.)

VII

The Uinta Basin
and Dinosaur National Monument

We complete our clockwise, circuitous journey through the geology of the Colorado Plateau with a tour of the Uinta Basin in the northern extremity of the province. The Uinta Basin lies mostly in northeastern Utah, where a gigantic lake formed in Early Tertiary time. It was part of a complex of lake basins that formed around the uplifted Uinta Mountains: the Uinta Basin to the south in Utah, the Piceance (the polite map spelling of the original "Piss Ants") Basin in northwestern Colorado, and the Green River Basin of southwestern Wyoming. The lakes (really one large, partially segmented lake) formed as the Colorado Plateau was regionally uplifted in the south and tilted toward the north. Runoff from the tilted land flowed northward into the lake complex, bringing sediments eroded from higher regions. Thick sand was deposited along the southern and western margins, changing gradually to silt and mud near the farther, deeper reaches of the lake surrounding the rising Uinta Mountains.

This process, of course, took place later and stratigraphically above the complex interfingering of nearshore marine sandstone formations of the Cretaceous section. The region is sometimes called "the dinosaur triangle" for the abundant fossil dinosaur bones in the Jurassic Morrison Formation in scattered quarries near Grand Junction, Colorado, and Price, Utah, as well as in Dinosaur National Monument near Vernal, Utah.

As mentioned in the previous section, a highland arose in western Utah from which vast quantities of coarse sand and gravel spilled eastward into the Cretaceous seaway. Thick sandstone near the source to the west grades into the distal open sea to the east, and no place shows the complexities of this pattern better than the Book Cliffs, which form the southern border of the Uinta Basin. As we head into the western Uinta Basin northward from Helper, Utah, the highway climbs up-section through a complex of interbedded nearshore sandstone, back-beach swamp deposits, and marine black shale of Mancos Shale affinities. The largely sandstone "Mesaverde" section is

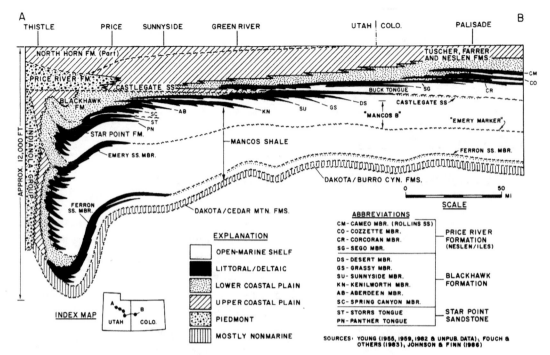

Figure 105. Schematic cross section showing stratigraphic changes in Cretaceous time from west to east across the southern Uinta Basin. The source of the sand and gravel was in mountains in western Utah that had formed by severe thrust-faulting in Cretaceous time. Thick conglomerates were deposited to the west that would grade eastward into nearshore and marine sandstone as the Cretaceous shoreline advanced and retreated with sea-level fluctuations. Tongues of sandstone, rocks of the Mesaverde Group, gradually pinched out eastward, away from the mountainous source areas and into open marine environments that received only gray and black mud of the Mancos Shale. The dark-colored mud deposits eventually took over most of the section toward the open seaway that lay to the east. From R. G. Young in Cole 1987.

complicated by numerous tongues of finer-grained deposits, replete with a bewildering set of local, complex, and discouraging member names. I have attempted to indicate the various named units in the appropriate roadlogs. As one heads eastward along the Book Cliffs toward Grand Junction, Colorado, sandstone tongues thin and pinch out and tongues of Mancos Shale thicken, largely taking over the section. The intertonguing relationships are too staggering to describe here in words but are apparent in Figure 105. In brief, sandstone dominates the Cretaceous section toward the source region at the west end of the Book Cliffs, and marine black shale becomes dominant eastward toward the open sea. At the close of Cretaceous time, when the last dinosaur staggered and fell into the bog, the landscape was a low plain across the Uinta Basin, and a regional unconformity developed as the last Cretaceous seaway slowly receded from the continental interior.

Tertiary lake sediments began to accumulate above the Late Cretaceous erosional surface in the Uinta Basin, and coarse sandstone and conglomerate of the North Horn Formation was restricted to the western margin of the basin. The coarse-grained sediments grade eastward into fine-grained mudstone and sandstone of the Price River Formation. The Flagstaff Limestone, freshwater algal carbonate rocks, is present above the North Horn only along the western margin of the basin. Above the Flagstaff, floodplain deposits of the Colton and Wasatch Formations spread across the basin, eventually grading upward into thick lake deposits of the Green River Formation. Delicately preserved fossil fish and other freshwater organisms are abundant in rocks of the Green River Formation, especially where it extends into southwestern Wyoming.

Separating the Uinta and Green River Basins along the Utah-Wyoming border is the maverick anticlinal uplift of the Uinta Mountains. It is the only east-west-trending mountain range in the Rocky Mountain region. An anticlinal (compressional) uplift in this orientation, especially of this magnitude, has mystified structural geologists for decades. The structure is at right angles to any possible scenario that can be applied to Laramide mountain-building processes, since Late Cretaceous–Early Tertiary compressional forces were west-to-east, in direct opposition to Utah's highest mountain range. But if we apply the global stress field that dominated the Precambrian through Paleozoic tectonic scheme, when compressional forces were from the north rather than from the west as in Laramide time, the Uinta Mountains fit well. We would expect east-west compressional structures like the Uinta Mountains to be common, as well as north-south extensional features such as the normal faults that underlie the great monoclines to the south. And this interpretation requires the presence of the northwest- and northeast-trending wrench fault zones that we have seen dominating the Colorado Plateau structure. The rejuvenation of the ancient force field in Laramide time indeed formed the surface structures we now see throughout the province, even the Uinta Mountains. This interpretation is further justified by the fact that nearly all formations of Paleozoic age thin or pinch out across the axis of the uplift, signifying a very long tectonic history for the highland. Smaller east-west anticlinal structures parallel the Uinta Mountain orientation to the southeast in the Split Mountain, Blue Mountain, and Rangely anticlines. So the origin of the Uinta Mountains is not so magical and mysterious after all.

Paleozoic and Mesozoic strata dip steeply southward from the Precambrian core of the Uinta Mountains anticline north of Vernal, Utah, exposing the entire stratigraphic section along the "Drive Through the Ages" highway. Similar steeply dipping strata occur along the southern flank of the

Figure 106. Aerial view of the Split Mountain anticline, aptly named by John Wesley Powell in his epic trip down the Green and Colorado Rivers in 1869. Here the Green River enters the east-west-trending fold, carving the magnificent canyon down the axis of the structure, only to leave the western plunging nose of the anticline downstream. The light-colored rock that caps the structure is the Weber Sandstone of Pennsylvanian-Permian age. Older rocks exposed in the canyon are of Pennsylvanian and Mississippian age. The upturned beds in the foreground are steeply dipping rocks of Mesozoic age exposed along the flanks of the fold, with the Green River in the lower right.

Split Mountain anticline just north of Jensen, Utah, in Dinosaur National Monument, where the excellent exposures of the Jurassic Morrison Formation contain abundant dinosaur fossil remains. In a remarkable display, dinosaur bones have been preserved in their natural setting in the rock, the steeply dipping sandstone bed that serves as the north wall of the Visitor Center. The bones have been carefully etched out of the rock wall by paleontologists and left in place for visitors to see. Dozens of nearly complete dinosaur skeletons were quarried from the area of the national monument by Earl Douglass, beginning in 1908, and sent to the Carnegie Museum in Pittsburgh, where they have been prominently displayed for decades. But all is not bones in Dinosaur National Monument, for the Green and Yampa Rivers have carved breathtaking canyons through the Uinta Mountains that expose the complete stratigraphic section from the 10,000-foot thickness of Precambrian quartzite of the Uinta Mountain Group upward through the entire Paleozoic section. These canyons are visited only by boat, and river trips are spectacular journeys through time.

The trip south from Vernal to Grand Junction, Colorado, begins with outstanding views of Split Mountain (Figure 106) and the Blue Mountain monocline and extends through the largest oilfield in Colorado, on the crest of the Rangely anticline. The oilfield produces vast quantities of oil from the deep Weber Sandstone of Pennsylvanian age and the shallow fracture zones of the Mancos Shale. South of Rangely, Colorado, Late Cretaceous and Tertiary rocks are well exposed in the canyons of Douglas Creek, along a structural high, the Douglas Creek arch, that separates the Uinta from the Piceance Basin. Stratigraphic changes from the western Uinta Basin are apparent, as the Cretaceous section is much more like the Mesaverde Group, which we have encountered to the south on the central Colorado Plateau. The serpentine road south from Douglas Pass, in rocks of Tertiary and Late Cretaceous age, provides excellent views of a more typical Upper Cretaceous section.

52. Green River to Helper, Utah, via U.S. 6 and 191

The highway between Green River and Helper, Utah, passes between the Book Cliffs to the north, with rocks of Late Cretaceous age, and the northern plunging nose of the San Rafael Swell to the west. The road crosses the Ferron anticline near Price, Utah.

Mile 0.0: Take Exit 156 on Interstate 70, 2 miles west of Green River, Utah, and head north on U.S. 6/191 toward Price, Utah.

Mile 0.5: Milepost 300: There are good views of the San Rafael Swell to the west and the Book Cliffs to the right. The Book Cliffs ahead wrap around the northern plunging nose of the San Rafael Swell. The highway is low in the dark gray, slope-forming Mancos Shale (Late Cretaceous). **Mile 10.4: Milepost 290.** An overpass at **Mile 12.6** crosses the railroad tracks (Denver & Rio Grande). **Mile 20.4: Milepost 280.**

Mile 21.3: Entering Woodside, Utah. Settled in 1881, it soon became a thriving farm village, but it is now largely abandoned. **Mile 21.4**: Crossing the Price River. There are roadcuts in the Mancos Shale at **Mile 22.7**, covered by pediment gravel.

Mile 25.3: Milepost 275: The north-plunging nose of the San Rafael Swell is obvious to the left. The light gray Roan Cliffs, formed in rocks of Tertiary age, are visible beyond the largely tan-colored Book Cliffs to the right. The small buttes around here are capped with pediment gravel.

Mile 30.2: Milepost 270: A well-developed pediment surface forms the skyline ahead, sloping down toward the left from the Book Cliffs. A *pediment* is a broad, gently sloping plain eroded on a bare rock surface, usually developed during previous times of erosion in a semiarid environment. Pediments form at the base of an abrupt or receding mountain front or plateau escarpment, in this case the Book Cliffs. The erosional surface is often veneered by alluvium from the upland masses, as seen here. The roadcuts at **Mile 31.4** are passing up through the pediment surface, which has a thick gravel mantle. The highway at **Mile 32.0** is on top of the pediment gravel layer.

The road to the right at **Mile 32.7** goes to Horse Canyon. The Wasatch Plateau is visible to the west at **Mile 33.5** beyond the northerly plunging nose of the San Rafael Swell. **Mile 35.2: Milepost 265:** The pediment surface is again obvious in the middle distance to the right at about 2:00. **Mile 36.0:** Exposures of a sandstone layer, the Ferron Sandstone Member of the Mancos Shale, are in a canyon to the left. **Mile 38.0:** Entering Carbon County, named for large deposits of coal in the area.

Mile 40.1: Milepost 260: The road is climbing to the pediment surface again. At **Mile 41.6** the Wasatch Plateau forms the skyline in the distance ahead and to the left. Roadcuts are capped by pediment gravel. Another overpass crossing the railroad is at **Mile 43.1.** The road to the right at **Mile 43.7** is Utah 123, going to Sunnyside, Utah, a coal-mining town established on the sunny side of the canyon in 1912. **Mile 45.1: Milepost 255.**

The ridge on the right at **Mile 46.1** is held up by the Ferron Sandstone Member of the Mancos Shale. The highway at **Mile 46.7** crosses the nose of the Ferron anticline. At **Mile 47.6** the highway starts down the northwesterly plunging nose of the Ferron anticline. **Mile 48.1:** The road crosses the outcrop of the Ferron Sandstone Member of the Mancos Shale. The structure approximates the northern plunge of the San Rafael Swell.

Mile 50.0: Milepost 250: Wellington, Utah. Settled in 1878, the Mormon settlement was named for Wellington Seeley, Jr., a justice of the Emery County Court. The road to the right at **Mile 50.3** goes to Nine-Mile Canyon. **Mile 53.5:** Leaving Wellington. **Mile 55.0: Milepost 245.**

The fork to the right at **Mile 57.0** goes to Price, Utah. **Mile 58.2:** Exit 241 goes to Castledale and Huntington, Utah. **Mile 58.7:** Crossing the Price River. **Mile 59.5:** Exit 240 goes to west Price.

Mile 62.2: Two thin sandstone beds of the Garley Canyon Member of the upper Mancos Shale are seen to the left and ahead. The road to the right at **Mile 63.4** goes to Carbondale, Utah. The two sandstone beds of the Garley Canyon Sandstone Member are on both sides of the road.

Recrossing the Price River at **Mile 64.0.** The road is now above the Garley Canyon Sandstone Member in the dark gray Blue Gate Shale Member of the Mancos Shale. Sandstone ledges of the Mesaverde Group are prominent ahead. **Mile 64.8: Milepost 235. Mile 65.0:** The Emery Sandstone Member of the Mancos Shale is poorly exposed in the canyon wall to the left, overlain by the Blue Gate Shale Member of the uppermost Mancos Shale.

Mile 65.9: Entering Helper, Utah, established in 1883 as Pratts Landing. By 1892 extra locomotive engines were being added to freight trains here to provide sufficient power to climb the steep grade to Soldiers Summit. The added locomotives were known as "helper engines," and thus the name *Helper* came into use. For years the town has been an important coal-mining and railroad center. Much of the coal mined nearby is shipped by rail to Salt Lake City.

Mile 67.7: Leaving Helper. The cliffs all around town are in the upper Mancos Shale (the Blue Gate Shale Member), overlain by the Panther and Storrs Members of the Star Point Sandstone of the Mesaverde Group. The upper cliffs are in the Spring Canyon, Aberdeen, and Kenilworth Sandstone Members of the Blackhawk Formation, Mesaverde Group (Late Cretaceous). **Mile 68.8**: The low cliffs above on the left are in the Panther Sandstone Member of the Star Point Sandstone. Port of entry is on the left at **Mile 69.3.**

53. Helper to Vernal, Utah, via U.S. 191 and U.S. 40

U.S. 191 heads to the north up Willow Creek Canyon through the Book Cliffs, composed of rocks of the Mesaverde Group of Late Cretaceous age. It then crosses the Roan Cliffs in the overlying sedimentary rocks of Tertiary age and heads down Indian Canyon to Duchesne, Utah. Stratigraphic relationships are complex, leading to often confusing formation and member names (see Figure 105). From Duchesne to Vernal, Utah, U.S. 40 runs along the deepest part of the Uinta Basin in rocks of Tertiary age, and the anticlinal Uinta Mountains are visible to the north.

Mile 0.0: Intersection of U.S. 6, heading west toward Salt Lake City, and U.S. 191, heading north toward Duchesne, Utah. Take U.S. 191 northbound. The two sandstone beds above the intersection are the Panther and Storrs Members of the Star Point Sandstone (Mesaverde Group, Late Cretaceous). Above that is the Blackhawk Formation, which has several named members. The Castlegate Sandstone forms the prominent cliffs in the distance up the Price River Canyon. On the right is the Carbon Power Plant, which generates electricity from coal mined from the Blackhawk Formation.

Mile 0.2: Bridge across the Price River. The Carbon Power Plant is ahead across the railroad tracks. The road passes through the plant at **Mile 0.3.** The sandstone to the right and left at **Mile 0.4** is the Spring Canyon Member of the Blackhawk Formation, overlain by the Aberdeen Member, a slope-forming unit. The road heads into Willow Creek Canyon. The road to the left at **Mile 0.9** goes to the Willow Creek Mine, which produces coal from the Blackhawk Formation. A cemetery is on the left at **Mile 1.2.** The red rocks on the right are clinker beds, the result of naturally burned coal.

Mile 1.3: Coal seams in the Kenilworth Member of the Blackhawk Formation are exposed across the river. The upper massive cliff is in the Castlegate Sandstone. The roadcuts on the right at **Mile 1.9** expose thin coal seams in the Blackhawk Formation, of swamp, shoreface, and terrestrial origin. **Mile 2.4**: This is the approximate top of the Blackhawk Formation and base of the stream-deposited Castlegate Sandstone. **Mile 2.8: Milepost 160.**

Mile 3.0: The road is about at the top of the Castlegate Sandstone and the base of the Price River Formation. The Price River is composed mainly of sandstone, with thin shale beds and an occasional coal bed deposited in floodplain environments exposed in one sandstone cliff after another; the four named members are difficult to distinguish.

Mile 5.2: This is the approximate top of the Price River Formation and the base of the North Horn Formation. The North Horn, a light gray to tan mudstone unit with a few scattered sandstone beds, was deposited in stream and lake environments and is about 1,000 feet thick in this area. The Cretaceous-Tertiary boundary is placed at about the middle of the formation, based on the highest occurrence of dinosaur fossils and the lowest occurrence of Paleocene fossils.

Mile 5.7: Milepost 163: Roadcuts are in a reddish-colored mudstone unit capped by a sandstone bed in the North Horn Formation. At **Mile 6.3** roadcuts are in a light greenish-gray mudstone unit. The road is back into reddish-colored mudstone and sandstone beds at **Mile 7.0.**

Mile 7.7: Milepost 165: Intersection in Emma Park: stay on U.S. 191, the road to the right.

Mile 8.0: The road is at the base of the Flagstaff Limestone (Tertiary); light-colored mudstone and thin limestone beds are overlain by the Colton Formation (Wasatch equivalent to the west).

The road at **Mile 8.3** is in pale red mudstone and thin sandstone beds of the Colton Formation (Tertiary). **Mile 8.7**: Entering Duchesne County. The low hills to the right are in the Colton Formation. The Paleocene-Eocene boundary is placed at about two-thirds of the way up into the Colton. The highway is now in the heart of the Roan Cliffs at **Mile 9.3.**

Mile 9.7: Milepost 167: This is the approximate base of the Green River Formation of Eocene age, freshwater deposits of Lake Uinta. Fossils found in the Green River include fish, plant fragments, clams, and ostracodes. Massive exposures in the cliff ahead at **Mile 10.1** are in the Green River Formation. **Mile 11.0**: Crossing Willow Creek. Exposures to the left at **Mile 12.0** are in the Green River Formation. **Mile 12.2**: The light-colored rocks along the road are marlstone beds in the Parachute Creek Member of the Green River Formation. **Mile 12.7: Milepost 170:** The road is still in the Green River Formation.

Mile 15.1: Timber Lane Junction. The road crosses a divide in the Roan Cliffs, elevation about 9,100 feet. **Mile 15.2:** Entering Ashley National Forest. Just across the divide at **Mile 15.3** there is a broad view of the western Uinta Basin. The countryside and roadcuts are in the Green River Formation. **Mile 15.7:** As it enters the left fork of Indian Canyon, the crooked road has an 8 percent grade ahead. The Green River Formation is exposed along the way. **Mile 17.7: Milepost 175. Mile 22.6: Milepost 180.**

Mile 24.3: Jones Hollow. Exposures of very light gray rocks of the Green River Formation dominate the walls of Indian Canyon. **Mile 25.4:** Exposures of light brown mudstone and siltstone beds above are in the Uinta Formation. The old road on the hillside to the left at **Mile 27.0** goes to an abandoned wurtzilite mine in the lower part of the Uinta Formation. Wurtzilite, which occurs in veins, is a black asphaltic residue similar to gilsonite, or uintaite.

Mile 27.5: Milepost 185: Leaving Ashley National Forest. Entering the Uintah and Ouray Indian Reservation. **Mile 28.7:** The approximate base of the Uinta Formation is at road level.

Mile 32.5: Milepost 190: The road is back into the very light gray to white mudstones of the Green River Formation. The road to the left at **Mile 33.4** goes to the right fork of Indian Canyon. **Mile 37.6: Milepost 195:** The road is still in the Green River Formation.

Mile 39.2: Minor east-west faults of the Duchesne fault zone disrupt the beds on the right side of the canyon. The road is back into light brown mudstone and siltstone beds of the Uinta Formation at **Mile 39.5.** The road to the right at **Mile 40.9** goes to Cottonwood Ridge and Sowers Canyon. The highway is in the limestone and sandstone facies of the Uinta Formation.

Mile 42.5: Milepost 200: A cemetery is on the right at the north end of Indian Canyon where the valley widens. The road crosses the Strawberry River and enters Duchesne, Utah, elevation 5,500 feet. Although the town was settled in 1904, when the Uinta Basin was opened to white settlers, the name *Duchesne* was not officially accepted until 1915. The town is in the Uinta Formation; the Uinta Mountains are visible ahead to the north.

Mile 43.3: Intersection of U.S. 191 and U.S. 40, which heads to the right to Roosevelt, Utah. Turn right on U.S. 40 toward Vernal, Utah. Crossing the Strawberry River again at **Mile 44.0.** Leaving Duchesne. **Mile 44.5: Milepost 89** on U.S. 40.

Mile 45.5: Milepost 90: Roadcuts are in a terrace gravel. The strata in this area are nearly horizontal, as the highway is near the structurally lowest part of the Uinta Basin. **Mile 50.5: Milepost 95**: The highway is in a wide-open valley; scattered hills of Uinta Formation are capped by terrace gravel. Crossing Antelope Creek at **Mile 54.0**. The Uinta Mountains, a 100-mile-long, east-west-trending anticline with Precambrian rocks exposed in the core, are well displayed to the left. **Mile 55.4: Milepost 100. Mile 60.4: Milepost 105.**

Mile 61.5: Entering Myton, Utah. **Mile 62.0**: Crossing the Duchesne River and leaving Myton. There is a nice view at **Mile 64.5** of the Uinta Mountains in the distance to the north. The higher mountains are Kings Peak, the highest mountain in Utah at 13,526 feet, and Mount Emmons, at 13,449 feet. **Mile 65.4: Milepost 110.**

Mile 68.0: Entering Roosevelt, Utah, elevation 5,100 feet, the oil capital of the Uinta Basin. The settlement was platted in 1905-6 and named for Theodore Roosevelt. The red sandstone cliffs at **Mile 69.6** are in the lower Brennan Basin Member of the Duchesne River Formation (Oligocene). **Mile 71.3**: Intersection in Roosevelt: turn right on U.S. 40/191.

Mile 71.9: Entering Uintah County. The highway traverses Roosevelt Valley for the next several miles. The red beds to the left are in the lower Duchesne River Formation. **Mile 75.4: Milepost 120**: Roadcuts are in the red Duchesne River Formation. The Bottle Hollow Resort, operated by the Ute Indians, is on the right at **Mile 77.0**. The road to the right at **Mile 78.0** goes to Fort Duchesne, the headquarters for the Uintah and Ouray Indian Reservation. **Mile 78.4**: Crossing the Uintah River. **Mile 80.4: Milepost 125**. The highway is in the Duchesne River Formation for the next 16 miles. **Mile 87.0**: Utah 88 goes to Ouray, Utah. **Mile 95.3: Milepost 140**.

Mile 96.6: View area on the right. To the east is the Split Mountain anticline, across the town of Vernal, Utah, with the Uinta Mountains to the north. South of Split Mountain is the Section Ridge anticline of the Blue Mountain Plateau. All these anticlinal structures are oriented east-west. To the northwest along Asphalt Ridge, oil-impregnated sandstone of the Mesaverde Group has been quarried for several decades for use in paving and maintaining the county road system.

Mile 98.1: Entering Vernal, Utah, and Ashley Valley, formed in the Mancos Shale. **Mile 100.5**: Junction in downtown Vernal with U.S. 191. (Connect with Roadlog 54 or Roadlog 57.)

54. Vernal, Utah, North on U.S. 191, the "Drive Through the Ages" Highway

U.S. 191, a Scenic Byway, heads north from Vernal, Utah, toward the south flank of the Uinta Mountain anticline, an east-west-trending structure more than 100 miles long. The route crosses down-section through rocks of Cretaceous age to the Precambrian quartzite core of the huge anticline. Road signs mark the various formation names along the route.

Mile 0.0: Intersection of U.S. 40 and U.S. 191 in the middle of Vernal, Utah. Head north on U.S. 191 toward Flaming Gorge Reservoir.

Crossing Ashley Creek at **Mile 1.2.** The Uinta Mountains are ahead; to the left at about 10:00 is Asphalt Ridge, where the rocks are obviously dipping strongly to the southeast into the eastern Uinta Basin. Petroliferous Mesaverde sandstone in Asphalt Ridge has been quarried for decades for use in paving the county roads. The road is along a hogback of Mesaverde sandstone at **Mile 3.5.**

Start the "Drive Through the Ages" at **Mile 3.6.** U.S. 191 heads north across the flank of the Uinta Mountains anticline (upfold). Unlike other mountain ranges on or near the Colorado Plateau, the enormous anticline trends east-west, thus marking the northern edge of the Colorado Plateau Province. The highway starts in the Upper Cretaceous Mancos Shale along the northern flank of the Uinta Basin and progressively traverses rocks of increasingly older age, going stratigraphically down-section as it crosses over the southward-dipping rock layers that are being eroded from the top of the anticline. The road encounters ever deeper and older strata as it approaches the structural axis of the Uinta arch. Signs along the road designate the names and ages of the formations crossed. The signs are usually placed in the middle of the formation exposures rather than at the formation contacts. Thus, the names designate the rocks both above and below the sign location, and no contacts are indicated. In some cases the formation names are obsolete, not having been updated in many years. Location by mileage of the signs marking the individual formations, shown in bold, are as follows:

Mile 3.7: *Mancos Shale*, a Late Cretaceous formation some 5,000 feet thick in this area. It is a dark gray shale that weathers to a tan or yellowish-buff color. It was deposited in deep, or at least stagnant, marine waters and commonly contains ammonitic cephalopods, large clams and oysters, shark teeth, and microscopic foraminifera. The estimated age is 95 to 75 million years before the present.

Mile 3.9: *Frontier Formation*, Late Cretaceous in age, composed of two sandstone layers separated by a shale. It forms a prominent hogback in this region, about 250 feet thick. The formation was deposited as beach and nearshore sand, containing small shark teeth, with coastal swamp deposits in the middle, including an occasional thin coal bed. The Frontier is about 95 million years old.

Mile 4.1: *Mowry Shale*, a dark gray, siliceous marine shale that weathers to a silver gray. It contains abundant fossil scales and bones of teleost fish. The approximate age is 105 million years.

Mile 4.2: Milepost 205: Steinaker Lake is to the left, and a beautiful view of the Uinta Mountains is ahead. ***Dakota Sandstone*** is a light-colored sandstone unit deposited as nearshore sand along the margin of the advancing Western Interior Seaway some 110 million years ago. The Dakota is a prominent bench- or ridge-forming unit that is widespread from Montana south to New Mexico along the ancient shoreline. It is about 50 feet thick here and contains conglomeratic layers, ripple marks, and some petrified wood.

Mile 4.8: ***Morrison Formation***, Late Jurassic in age (160 to 135 million years old on the sign, now considered to be 155 to 148 million years old), here about 800 feet thick. Colors of the mudstone, with occasional thin beds of sandstone and limestone, vary from light gray to green and lavender to red in ever-changing and mottled hues. The very widespread formation was deposited in lake (lacustrine) and stream (fluvial) environments in a semi-arid climate. The dinosaur quarry in nearby Dinosaur National Monument displays numerous partial skeletons that accumulated in a fluvial channel in the Morrison Formation.

Mile 5.3: Steinaker State Park is on the left. Another road to Steinaker Lake is to the left at **Mile 5.6. Mile 6.2: Milepost 207.** The road at **Mile 7.0** follows a valley between the Morrison Formation on the right and the Entrada Sandstone on the left; the valley is apparently on the Curtis Formation. **Mile 9.1: Milepost 210.**

Mile 9.8: ***Curtis Formation,*** now called the Curtis Member of the Stump Formation, varying in thickness up to 250 feet thick and composed of gray-green marine sandstone, shale, and limestone. It is of Middle Jurassic age, some 165 to 160 million years old on the sign (now considered to be around 158 million years old), and contains numerous belemnites (cigar-shaped internal skeletons of squid-like marine animals) and usually the mineral glauconite (a green-colored silicate formed only in warm, shallow seawater). The overlying Redwater Member of the Stump Formation (Upper Jurassic) is not indicated in the sign sequence. There is a great view ahead into the Uinta Mountains.

Mile 9.9: ***Entrada Sandstone***, a ridge-forming, light-colored sandstone formation, composed of mostly windblown sand. It is about 200 feet thick here; its estimated age on the sign is about 170 to 165 (now believed to be 159 to 160) million years before the present.

Mile 10.1: The road on the right goes to Red Fleet State Park, 2 miles. ***Carmel Formation***. Composed of usually reddish-brown mudstone and siltstone, the Carmel here contains a basal thin limestone. The formation probably was deposited on coastal mudflats in Middle Jurassic time. It commonly forms valleys or swails between the more resistant Entrada Sandstone above and the Page Sandstone below. The Page Sandstone, not recog-

nized on the roadside signs, is here about 30 feet thick, resting unconformably on the Nugget Sandstone.

Mile 10.5: *Navajo (Nugget) Sandstone,* sometimes called the Glen
Canyon Sandstone or, in northern Utah and Wyoming, the Nugget Sandstone. The Nugget and its equivalent, Navajo Sandstone, ranging in thickness from 700 to 1,000 feet, forms light-colored cliffs and rounded knolls
over nearly the entire Colorado Plateau, from Arizona to Wyoming. It was
deposited mainly as large windblown sand dunes, and its setting in Jurassic
time (195 to 170 million years ago) was much like the vast Sahara Desert of
North Africa today. It forms prominent cliffs, ridges, and hogbacks wherever it is exposed. The cross bedding indicates that the wind was blowing
from the northwest toward the southeast across the entire desert. The road
here crosses the Nugget hogback.

Mile 10.8: *Chinle Formation.* Late Triassic (205 to 195 million years old
on the sign, now considered to be 209 to 230 million years old), the Chinle
Formation is a varicolored mudstone unit, with thin interbeds of sandstone
and conglomerate, deposited in a fluvial (stream) and lake (lacustrine) net-
work that covered nearly the entire Colorado Plateau. It is here about 250
feet thick and contains some petrified wood, in lesser amounts than in the
Petrified Forest National Park in northern Arizona. Farther south the
Chinle is subdivided into several formal members, but in this region only
the lower unit, the Shinarump Member, is distinguished.

Mile 10.9: *Shinarump Member,* the lower sandstone and conglomerate
member of the Chinle Formation, known as the Gartra Grit in northern
Utah. The stream-deposited sandstone and conglomerate varies greatly in
thickness where it fills erosional channels in the top of the underlying
Moenkopi Formation. The Shinarump forms prominent benches and low
cliffs, as it is far more resistant to erosion than the overlying and underlying
mudstone units. A dinosaur track site is to the right at **Mile 11.1.**

Mile 11.3: *Moenkopi Formation,* a red to dark reddish brown, mostly
mudstone formation of the Early Triassic (225 to 210 million years old), deposited in a vast tidal flat that formed landward, or east, of the Triassic seaway in western Utah. It is here about 800 feet thick.

Mile 13.0: *Park City Formation.* The Park City is of Permian age (270 to
260 million years old) but is somewhat younger than the Kaibab Formation
of the southern Colorado Plateau. It is a three-part formation in this region
and would be better separated into three formations, as used in the phosphate mining operation here. A good view is to the left at **Mile 13.6** of a
phosphate strip-mining operation. **Mile 14.1: Milepost 215.** A huge phosphate open-pit mine is to the left at **Mile 14.3.**

Mile 14.5: A viewpoint to the left is an overlook of the phosphate mining operation. There is also a nice view to the south of the Uinta Basin. The Park City Formation is here designated the *Franson Formation*, with three members, overlying the *Phosphoria Formation*, in which the phosphate deposits occur, a much better way to designate this stratigraphic unit. Roadcuts are in the "Park City Formation" on the right at **Mile 16.2.**

Mile 16.8: Geological turnout on the left with a fine view to the south across the eastern Uinta Basin. This is still at about the top of the "Park City Formation." **Mile 17.3: Milepost 218**: Entering Ashley National Forest. Roadcuts are in the "Park City Formation" at **Mile 17.7** as the road climbs up the south flank of the Uinta Mountain arch.

Mile 18.3: *Weber Sandstone*. Although officially designated Pennsylvanian in age, the Weber Sandstone so closely resembles the Lower Permian Cedar Mesa Sandstone of Canyonlands country that the age designation seems questionable. It is a light-colored, highly cross-bedded sandstone unit that forms impressive scenic cliffs and canyons, such as the canyon to the left. It is here about 1,000 feet thick. The formation is usually interpreted to be a windblown dune sand deposit, but parts may be of nearshore marine origin. **Mile 19.3: Milepost 220.**

A picnic area is on the left at **Mile 19.8.** A steep grade is ahead at **Mile 21.4.** The road is in limestone with interbedded shale of the Morgan Formation.

Mile 21.5: *Morgan Formation*. About 1,400 feet thick in this area, the Morgan is of Middle Pennsylvanian age, the equivalent of much of the Hermosa Group in Canyonlands National Park. The Morgan, as used here, includes the Round Valley Limestone and consists of interbedded limestone, light-colored sandstone, and red mudstone. Some beds are quite fossiliferous, indicating a shallow marine environment of deposition. Crossing Little Brush Creek at **Mile 22.1.**

Mile 24.3: Milepost 225: The road to the right goes to Diamond Mountain. *Madison Limestone*. As indicated here, the Madison includes the entire Mississippian section, consisting of the Doughnut, Humbug, Deseret, and Lodgepole Formations. The generalized unit consists of limestone, dolomite, sandstone, and some shale of marine origin; fossils include corals, crinoids, brachiopods, and gastropods. The "Madison Limestone" is poorly exposed along the highway. There are no rocks of Ordovician, Silurian, or Devonian age in this area; the missing gap in time represents a major disconformity.

Mile 25.0: *Lodore Formation*. There are no rock exposures, just the sign to indicate its presence. The Upper Cambrian formation consists of light-colored sandstone and siltstone with interbedded green and red shale where exposed. In Lodore Canyon along the Green River, brachiopods and a few trilobites are present in the formation. It seems incongruous that the Madison and the Lodore are not better exposed, but both are hidden by high mountain weathering, soil cover, and alpine plants.

Mile 25.7: Entering Dagget County, elevation 8,428 feet at the crest of the Uinta Mountain anticline. A Forest Service access road is to the right at **Mile 26.4.** The highway starts down the north flank of the Uinta Mountain anticline. **Mile 29.2: Milepost 230**: No rock exposures in sight. **Mile 30.5**: Lodgepole Campground is on the right. The talus on the mountainside to the left is Uinta Mountain Quartzite, Precambrian in age.

Mile 31.2: Milepost 232: Precambrian *Uinta Mountain Group*, "core of the Uinta Mountains." Although exposures are nearly nonexistent here, the group consists of quartzite and some slate interbeds, perhaps as much as 25,000 feet thick, that were deposited in a tongue of the sea perhaps a billion years ago. The age and the great thicknesses of slightly metamorphosed sandstone are somewhat doubtful, but the unit is definitely very thick and old. Owing to the poor exposures, there is no use in proceeding farther than this end of the "Drive Through the Ages Highway."

Mile 32.7: Entering Flaming Gorge National Recreation Area. There is a nice view along the northern flank of the Uinta Mountains. There are poor exposures of the Uinta Mountain Group in roadcuts on the left. A good place to turn around.

55. Dinosaur National Monument Quarry

Mile 0.0: Intersection in Jensen, Utah, of U.S. 40 and Utah 149, which goes to Dinosaur National Monument. There are magnificent views of Split Mountain to the right at about 10:00.

Mile 1.4: Split Mountain is to the right (Figure 107). Rounded hills of Mancos Shale are to the left. The road is in flat farmland along the Green River. **Mile 2.0: Milepost 2**. Crossing Brush Creek at **Mile 2.7**. There are exposures to the left at **Mile 3.0: Milepost 3** of yellowish-weathering Mancos Shale. **Mile 5.0**: Entering Dinosaur National Monument. The Green River is to the right with Split Mountain beyond.

The hogback to the left at **Mile 5.2** is very steeply dipping Frontier Sandstone; the dark gray Mancos Shale is in the lower slopes, dipping strongly southward away from the Split Mountain anticline. **Mile 5.9**: Entry pay station; turn left immediately to the Dinosaur Quarry. The road crosses the hogback of Frontier Sandstone. **Mile 6.2**: Crossing the silvery gray Mowry Shale slope. The road at **Mile 6.3** crosses the Upper Cretaceous Dakota Sandstone hogback and Lower Cretaceous Cedar Mountain Formation and

Figure 107. Split Mountain, an east-west-trending anticline that has been dissected by the Green River northeast of Jensen, Utah. The light-colored caprock is the Weber Sandstone of Late Pennsylvanian, possibly Permian, age.

Figure 108. The Visitor Center in Dinosaur National Monument. The wall of the building on the right is a steeply dipping bed of sandstone of the Morrison Formation of Late Jurassic age in which numerous skeletal parts of dinosaurs have been partially excavated.

immediately enters exposures of the varicolored Morrison Formation of Late Jurassic age. The beds dip very steeply to the south.

Mile 6.6: The Visitor Center parking lot and Dinosaur Quarry are in the nearly vertical Morrison Formation. Cliffs of Navajo Sandstone are to the left; the Split Mountain anticline, with its white cliffs of Weber Sandstone, is to the east. The north wall of the Visitor Center is a nearly vertical wall of sandstone in the Morrison Formation (Upper Jurassic) (Figure 108). Numerous dinosaur bones have been painstakingly exposed by drill and chisel in their natural setting in the rock (Figure 109). Return to U.S. 40 in Jensen.

Figure 109. The wall of sandstone of the Morrison Formation inside the Visitor Center at Dinosaur National Monument, where paleontologists have partially excavated many fossilized bones of Late Jurassic dinosaurs. The bones have been left in place, allowing visitors to view how dinosaur remains are commonly found in nature.

56. Dinosaur, Colorado, to Harpers Corner, Dinosaur National Monument

The Harpers Corner road first crosses a sharp monocline just north of U.S. 40 at Dinosaur, Colorado, exposing steeply dipping strata of the Frontier Sandstone (Cretaceous) down-section into the Chinle Formation (Triassic). The road then tops out on the heavily covered Yampa Plateau, where structural complexities are obscured by grass and sagebrush. The Permian Park City Formation is exposed above views into deep canyons of the cliff-forming, white Weber Sandstone. The mile-long hike to Harpers Corner provides a spectacular view of Steamboat Prow, the Mitten Park fault, and Whirlpool Canyon, with its beautifully exposed section from the Precambrian Uinta Mountain Group up through the Pennsylvanian Weber Sandstone.

Mile 0.0: Turn north from U.S. 40, just east of Dinosaur, Colorado. Dinosaur National Monument Visitor Center is to the right. The road immediately crosses through a hogback of Frontier Sandstone and a swale of silvery-gray Mowry Shale, both Cretaceous in age.

Mile 0.2: The road crosses through the Dakota–Cedar Mountain hogback, dipping very steeply to the south. Following in short order are exposures of the Morrison Formation and underlying Redwater and Curtis Members of the Stump Formation, all of Jurassic age. The Jurassic Entrada Sandstone at **Mile 0.6** is on red beds of the Carmel Formation along the road. Roadcuts are in the Entrada Sandstone at **Mile 0.7**. The prominent cliffs ahead and to the left at **Mile 2.0** are capped by Entrada Sandstone, overlying slopes of Carmel Formation and high cliffs of the massive, white

Nugget Sandstone. The roadcuts at **Mile 2.5** are in the Nugget Sandstone (Jurassic).

The road is near the top of the Nugget Sandstone at **Mile 3.0**; red beds of the Carmel Formation and cliffs of the Entrada Sandstone are above. The Blue Mountain monocline is visible to the left in the distance. **Mile 3.5**: The Nugget-Carmel contact can be seen on the left. The high cliffs are the Entrada Sandstone. There is a good view across the eastern Uinta Basin to the right. An overlook on the right at **Mile 3.7** offers another view of the Uinta Basin.

Mile 3.9: The contact between the Carmel Formation and the Entrada Sandstone is to the left and right. **Mile 4.0**: The road crosses exposures of the Entrada Sandstone. A nice view of Blue Mountain is to the left. Clark Hat Butte trail is to the right at **Mile 4.8**; a picnic ground is to the left. The road here is on top of the Entrada Sandstone, which dips strongly toward the south. The Entrada-Carmel contact is seen again at **Mile 5.1**.

Mile 5.6: The area is heavily covered with sagebrush and grass. Blocks of scattered boulders across the countryside are conglomeratic. The road crosses a covered fault or series of faults, more or less on line with the Blue Mountain monocline where it crosses Round Top Mountain. There is a bed of conglomerate in a gully at **Mile 6.5** that is the Gartra Grit (Shinarump) Member of the Chinle Formation. **Mile 7.0**: Exposures of conglomerate of the Gartra Grit. Blue Mountain is still beautiful to the left. The road at **Mile 7.1** is crossing up through the Gartra Grit. **Mile 7.6**: The road is in the upper Chinle Formation.

Mile 8.0: Escalante Overlook, on the left, provides a view of the Uinta Basin. The mountainside is thoroughly covered with sagebrush and grass. Poor exposures in roadcuts at **Mile 9.4** look like Mancos Shale. Poor exposures of Gartra Grit are to the right and left at **Mile 9.7**, dipping southward. The road is running along a ridge at **Mile 10.5**, formed on the Gartra Grit.

Mile 11.0: A Moffat County road is to the right. Another Moffat County dirt road goes to the left at **Mile 11.7**. There are poor exposures of sandstone to the left at **Mile 13.6**.

Limestone exposures in roadcuts at **Mile 14.5** are in the "Park City Formation." The countryside is densely covered with sagebrush and grass. **Mile 14.7**: There is red sandstone in roadcuts on the right. A good view of the Uinta Mountains to the north is at **Mile 15.1** as the road crosses a crest in the hill. The roadcuts at **Mile 15.6** are in limestone of the "Park City Formation." **Mile 15.8**: More "Park City" limestone in the roadcuts and a few poor exposures.

A glimpse of a canyon to the right at **Mile 17.8** is rimmed with the white Weber Sandstone. There are poor exposures of sandstone along the road and on the left at **Mile 18.7.** Deep in the canyon to the right are exposures down to about Steamboat Rock. **Mile 19.3**: Canyon Overlook and Picnic Area on the right (take the loop). **Mile 20.0**: Picnic loop with a view into the canyons of the Green River. The massive white cliffs are in the Weber Sandstone (Pennsylvanian and Permian?) with exposures down into the Morgan (Middle Pennsylvanian); the Madison Limestone (Mississippian) is down in the canyon bottoms. There is a very large west-east fault down in the canyon, faulting the Weber against the Moenkopi Formation (Triassic). And there is a whopper of a fault between here and the canyon.

Mile 21.5: Back at the main road, definitely at the top of the Weber Sandstone. The Utah state line is at **Mile 23.4.** The poor exposures are in the top of the Weber Sandstone. The road to the left at **Mile 24.0** goes to the top of a hang glider launch point. **Mile 24.8**: The top of the Weber Sandstone is exposed on the left. Roadcuts at **Mile 25.0** are in the poorly exposed "Park City Formation." Roadcuts at **Mile 26.1** are in the "Park City Formation," and a corral is to the left.

Mile 27.1: Viewpoint on the right, overlooking the canyon of the Green River. The view is straight down on the Mitten Park fault; the white cliffs to the left are vertical Weber Sandstone, and the red buttes to the right are in the Moenkopi Formation, capped by the Shinarump (Gartra) Member. To the right of the Moenkopi-Shinarump buttes is the Yampa River fault, joining the Mitten Park fault. Low in the distance are the canyons of the Yampa and Green Rivers.

Mile 27.5: A dirt road down to Echo Park on the Green River (13 miles) is to the right. **Mile 27.7**: Entering Dinosaur National Monument (actually, the road for the entire route from the Visitor Center is under Dinosaur National Monument jurisdiction). Island Park Overlook is to the left at **Mile 28.2. Mile 28.4**: Back to the main road from the overlook.

The road to the right at **Mile 29.6** goes to the Iron Springs Overlook. **Mile 29.8**: The overlook gives a view down into the canyon of the Green River. The Weber Sandstone comprises the massive cliffs, with a covered top. Roadcuts at **Mile 30.2** are in thick gravel deposits of the Bishop Conglomerate of Late Tertiary age.

Mile 33.1: Echo Park Overlook. The dirt road going to Echo Park is obvious below, but Echo Park is hidden from view. The scene to the east is of the Yampa River Canyon country, and the Green River is visible just below Steamboat Prow. More roadcuts in the Bishop Conglomerate are at **Mile 33.6.**

Mile 34.3: Harpers Corner loop and parking area. Trailhead to Harpers Corner, 1 mile (seems farther), where there is a great view across the Mitten Park fault and rocks down to the Precambrian Uinta Mountain Group.

Figure 110. View toward the east into the Yampa River country from near Harpers Corner, Dinosaur National Monument. The light-colored rock that forms the myriad cliffs is in the Weber Sandstone.

Harpers Corner is on the Morgan Formation, capped by the Bishop Conglomerate. The view to the east is of the Yampa River Canyon country, exposing mostly the Weber Sandstone (Figure 110); to the west is Whirlpool Canyon on the Green River, where rocks from the Precambrian up through the Weber Sandstone are beautifully exposed. Return to U.S. 40 at Dinosaur, Colorado.

57. Vernal, Utah, to Grand Junction, Colorado, via U.S. 40, Colorado 64, and Colorado 139

U.S. 40, heading east from Vernal, Utah, follows the northern margin of the Uinta Basin, largely in the Mancos Shale of Late Cretaceous age. The Split Mountain and Blue Mountain anticlines, along the route to the north, bound the Uinta Basin, which formed in Early Tertiary time as a huge freshwater lake. Near Dinosaur, Colorado, the highway crosses ridges of Mesaverde sandstone units along the southern monocline of the Blue Mountain structure. Colorado 64 heads southeast from Dinosaur across the beautifully exposed Rangely anticline and giant oilfield. The road then turns south at Rangely to cross the eastern margin of the Uinta Basin along the Douglas Creek arch in canyons to Douglas Pass, largely through exposures of Mesaverde sandstone formations. A steep descent from Douglas Pass in rocks of the Mesaverde Group into the Grand Valley in the Mancos Shale brings us to Loma, Colorado, and Interstate 70, leading to Grand Junction, Colorado.

Mile 0.0: Main Street in downtown Vernal, Utah, on U.S. 40 and 500 East, stoplight. Head east on U.S. 40. **Mile 0.9: Milepost 147.**

Mile 1.1: Entering Naples, Utah. **Mile 3.8: Milepost 150:** Leaving the Vernal area across Ashley Valley. **Mile 4.7:** The view to the left is of Split Mountain at about 10:00 and Blue Mountain–Yampa Plateau at about 11:00. The gray lowlands to the left are in the Mancos Shale (Late Cretaceous). The low hills to the left at **Mile 7.5** are in the dark gray Mancos Shale.

Mile 8.8: Milepost 155. Crossing Ashley Creek at **Mile 8.9.** The Mancos Shale forms the valley walls. At **Mile 9.5** the highway ascends the Ashley Creek valley onto gray hills of the Mancos Shale.

Mile 11.3: Entering Jensen, Utah, in a broad valley of the Green River in farmland. Split Mountain is obvious and beautiful to the left at about 10:00, with the Jensen syncline separating Split Mountain from the Blue Mountain structure (the Section Ridge anticline) at about 11:00. The road to the left at **Mile 12.3** is Utah 149, leading north to Dinosaur National Monument and the Dinosaur Quarry (Roadlog 55). Continue on U.S. 40.

Mile 12.8: Bridge across the Green River. The Dominguez-Escalante Expedition of 1776 camped a short distance to the north. The road to the right at **Mile 13.0** goes to the Redwash Oilfield. Roadcuts are in the Mancos Shale, capped by loose sand. **Mile 13.8: Milepost 160:** On top of a sand flat.

Mile 15.0: The hills to the right, forming a hogback, are Mancos Shale, capped by the Frontier and Dakota Sandstones, dipping south along the south flank of the Section Ridge anticline. **Mile 18.7: Milepost 165:** The hogbacks on the right are the varicolored Morrison Formation, capped by the Cedar Mountain and Dakota Formations. The steeply dipping south flank of the Section Ridge anticline, here forming a monoclinal fold, is prominent to the left; the light-colored rock is the Weber Sandstone. Blue Mountain is the southern margin of the Uinta Mountains. The highway at **Mile 19.6** passes through the hogbacks formed by the Cedar Mountain and Dakota Formations and then goes back into the Mancos Shale. The hogbacks on the right at **Mile 20.7** are capped by the Mesaverde Group, and the low hills on the left are in the Mancos Shale.

Mile 23.7: Milepost 170: The road to the right goes to Redwash. The highway at **Mile 24.5** is passing between exposures of the hogback, formed by steeply dipping beds of the Mesaverde Group: the Castlegate and Sego Sandstones. **Mile 28.7: Milepost 175:** The highway crosses the Mesaverde hogbacks and goes back down into the Mancos Shale. The ridge to the left is capped by the Frontier Sandstone.

Mile 29.7: The Utah-Colorado state line, Moffat County, Colorado. **Mile 30.7: Milepost 1.** The flatiron on the left at **Mile 32.0** is the Frontier Sandstone.

Mile 32.2: Entering Dinosaur, Colorado, elevation 5,900 feet. Formerly named Artesia, for artesian water wells in the area, it was renamed Dinosaur in 1965 to capitalize on the tourist traffic headed for Dinosaur National Monument. The town boomed after World War II as the Rangely Oilfield was developed. **Mile 32.6**: Intersection, in the heart of "beautiful downtown Dinosaur," of U.S. 40 and Colorado 64; turn right (south) on 64 toward Rangely, Colorado, 18 miles. (To follow Roadlog 56, Dinosaur to Harpers Corner, Dinosaur National Monument, continue east on U.S. 40 a short distance.)

Mile 33.1: Leaving Dinosaur, heading southeast toward Rangely. The ridge to the left is the Sego Sandstone, the Buck Tongue of the Mancos Shale, and the Castlegate Sandstone (Mesaverde Group). **Mile 33.5: Milepost 1**.

At **Mile 33.9** the road crosses a hogback ridge of Mesaverde sandstone and the axis of the Redwash syncline, visible to the left at **Mile 34.0**. The syncline plunges to the west, and rocks of the Mesaverde Group are exposed along the crest. **Mile 34.3**: Entering Rio Blanco County. The top of the Castlegate Sandstone is at road level at **Mile 36.0** and forms the caprock on the canyon walls ahead. The road crosses the northwest nose of the Rangely anticline and then crosses down-section into the Mancos Shale. At **Mile 36.8** the road is on the axis of the westward plunging nose of the Rangely anticline. The Castlegate Sandstone caps the canyon walls. **Mile 37.4: Milepost 5**.

The road to the right at **Mile 38.2** goes to Bonanza, Utah. The Castlegate Sandstone forms the dipslope to the left. **Mile 39.2**: The Castlegate Sandstone is well exposed in the gully to the left, on the south flank of the Rangely anticline. The road at **Mile 40.4** is along the south flank of the Rangely anticline on the Castlegate Sandstone. Roadcuts are in the Mancos Shale at **Mile 40.9**, and good Mancos Shale exposures are on the left.

Mile 41.1: The road descends into a valley on the Mancos Shale along the crest of the anticline. Oilfield pump jacks and other oil facilities are everywhere. The huge Rangely Oilfield, the largest oilfield in Colorado, was discovered in 1902 when small amounts of oil were produced from shallow wells in the fractured Mancos Shale. In 1933 Standard Oil Company of California, now Chevron Oil Company, drilled a deep hole to the Weber Sandstone (Pennsylvanian) that produced 300 barrels of oil a day from below 6,300 feet. The well was shut in because of a lack of transportation facilities until 1943, when World War II increased the demand for oil and development of the field began. Today there are more than 400 producing wells in the field. Most production is from the Weber Sandstone reservoir, where the recoverable reserves are estimated at more than 1 billion barrels. Although shallow wells are still producing from fracture zones in the Mancos Shale, the bulk of the oil from the Weber Sandstone comes from a sedimentary transition zone between the porous Weber Sandstone and nonporous red beds of the Maroon Formation that coincides approximately with the anticline. Thus, the field is a combination structural-stratigraphic trap.

Mile 43.2: Oilfield facilities are abundant. **Mile 43.3: Milepost 11**: Oil storage tanks and a gathering facility are on the right and ahead on the left. **Mile 46.2: Milepost 14**: Chevron Oil Company offices are on the left. Other oil facilities are to the right and ahead.

The road to the left at **Mile 47.6** is a county road going north to Blue Mountain, Colorado. The surface rocks through here are all Mancos Shale. The anticline is ringed by exposures of Mesaverde sandstone beds.

Mile 49.2: Milepost 17: Welcome to Rangely, Colorado. **Mile 49.8**: Crossing the White River. **Mile 50.0**: Rangely city limits. **Mile 52.0**: Intersection of Colorado 64 and Colorado 139, which goes south toward Interstate 70 and Grand Junction, Colorado. Turn right (south) on 139 and reset trip odometer to **Mile 0.0**.

Mile 0.0: Milepost 72: The road enters the canyon of Douglas Creek; Mancos Shale forms the lower slopes, and Mesaverde sandstone beds (Castlegate) are above, dipping south along the south flank of the Rangely anticline.

Mile 1.3: Roadcut across the hogback of the Castlegate Sandstone into the Buck Tongue of the Mesaverde Group. The road crosses the Sego Sandstone of the Mesaverde Group at **Mile 1.5. Mile 2.0: Milepost 70**: The road is in the upper sandstone of the Sego. The red clinker beds to the left at **Mile 2.5** are naturally burned coal beds in the Mesaverde Group. The rocks are here pretty much flat-lying along the axis of the Uinta Basin. The road follows Douglas Creek. The cliffs at the skyline to the left at **Mile 3.6** are in the Green River Formation (Tertiary) beneath the Roan Plateau. Prominent Mesaverde sandstone ledges are on the left at **Mile 5.7.**

Mile 8.8: Milepost 63: The road is near the top of the Mesaverde Group sandstone beds. The road follows Douglas Creek for the next 10 miles in Pinyon Valley. From here southward to Douglas Pass, the valley follows the top of the Douglas Creek arch, a largely subsurface high structure that separates the Uinta Basin to the west from the Piceance Basin to the east.

The prominent sandstone cliff to the left at **Mile 9.6** is near the top of the Mesaverde Group. **Mile 10.0**: Massive sandstone cliffs of the Mesaverde Group line the canyon walls to the right and left. **Mile 14.0**: Oil and gas wells of the Douglas Creek–Dragon Trail gas field are to the right and left. **Mile 14.5**: The slope-forming rocks above the lower canyon walls of Mesaverde sandstones are in the Price River(?) Formation.

Mile 15.7: Milepost 56: A point-of-interest sign is on the left for Canyon Pintado. Father Escalante of the Dominguez-Escalante Expedition of 1776 named the canyon "Painted Canyon" for Fremont petroglyphs on the cliffs to the right of the road. East Douglas Creek Road is to the left at **Mile 17.0.** The road is high in the Mesaverde Group. Interbedded sandstone and mudstone beds above are in the Upper Cretaceous Price River(?) Formation as the valley widens.

Mile 18.9: Oil and gas wells on the right. The Buck Tongue of the Mancos Shale is exposed below the Sego Sandstone. These rocks occur along a highly faulted anticline on the Douglas Creek arch for the next 2 miles. Examples of fault disturbances are seen to the left along the way. **Mile 20.3**: The West Douglas hydrocarbon plant of the Western Gas Supply Company is to the right. **Mile 20.5**: COSO Field Services Company. Red clinker beds at **Mile 23.2** are in the Mesaverde Group.

Mile 26.7: Milepost 45: The road is back into the red beds, suggesting fault disturbances. The road just crossed a fault at **Mile 27.4**; Mesaverde sandstones are exposed in cliffs to the left, with prominent side canyons along the fault. The road is back into poorly exposed Mesaverde near the axis of the South Douglas Creek anticline and gas field. The valley has a peculiar broad, flat floor here at **Mile 28.1**, as if it had been glaciated. **Mile 29.5**: The valley is becoming heavily covered with brush, masking rock exposures. The countryside is still in the Mesaverde Group. **Mile 31.6: Milepost 40.** Entering Garfield County at **Mile 32.3.**

Mile 34.6: A massive sandstone of the Tertiary Wasatch Formation is in the roadcut on the left as the road begins its long, steep ascent to Douglas Pass. High, soft roadcuts to the left at **Mile 36.0** are in the Green River Formation.

Mile 36.7: Douglas Pass, named for Chief Douglas, a Ute Indian. Elevation 8,240 feet. The magnificent view to the south across the Roan and Book Cliffs includes the La Sal Mountains, the northern slope of the Uncompahgre Plateau, and Grand Valley. The Douglas Creek Member of the Green River Formation is exposed in a roadcut on the right. As the road begins its steep descent from the pass, massive roadcuts in the Green River Formation are on the right.

Look back and up toward Douglas Pass at **Mile 38.5** for a good view of the Green River Formation, which often slumps and slides here during spring thaws. The highly contorted soft sandstone on the left at **Mile 39.1** is formed by slumping. The rocks at **Mile 39.8** are stained white from spring seepage above. **Mile 40.0**: A rather massive sandstone is on the left in the Wasatch Formation above gray shale beds of the upper Mesaverde Group. **Mile 41.0**: Massive sandstone beds of the Mesaverde Group are exposed in cliffs to the left for the next 12 miles.

Mile 49.3: Milepost 22: Red rocks on the right are clinker beds in the

Hunter Canyon Formation of the Mesaverde Group. **Mile 53.1**: The top of the Mancos Shale is at about road level. There are good exposures of Mesaverde cliffs ahead and to the left. **Mile 53.3: Milepost 18**: The road is in the Mancos Shale at the foot of the Book Cliffs.

The La Sal Mountains are barely visible ahead at **Mile 54.8**, and the Uncompahgre Plateau is just to the left. The road is heading straight for the Uncompahgre Plateau. **Mile 64.2: Milepost 7.**

Mile 65.8: Entering the village of Loma, Colorado. **Mile 66.1**: Crossroad at Q Road. **Mile 69.6**: Downtown Loma. **Mile 69.9**: Stop sign, intersection with U.S. 6, and Union Pacific Railroad crossing.

Mile 71.1: Milepost 0: Interstate 70 interchange. Turn left on I-70 to Grand Junction, Colorado (13 miles).

Glossary

ANGULAR UNCONFORMITY	An unconformity, or break, between two series of rock layers such that rocks of the lower series underlie rocks of the upper series at an angle; the two series are not parallel. The lower series was deposited, then tilted and eroded prior to deposition of the upper layers.
ANTICLINE	An elongate fold in the rocks in which sides slope downward and away from the crest; an upfold.
ARKOSE	A sandstone containing a significant proportion of feldspar grains, usually signifying a source area composed of granite or gneiss.
BASE LEVEL	The level, actual or potential, toward which erosion constantly works to lower the land. Sea level is the general base level, but there may be local, temporary base levels such as lakes.
BASEMENT	The crust of the earth beneath sedimentary deposits, usually, but not necessarily, consisting of metamorphic and/or igneous rocks of Precambrian age.
BASEMENT FAULT	A fault that displaces basement rocks and originated prior to deposition of overlying sedimentary rocks. Such faults may or may not extend upward into overlying strata, depending on their history of rejuvenation.
BENTONITE	A rock composed of clay minerals and derived from the alteration of volcanic tuff or ash.
BRACHIOPOD	A type of shelled marine invertebrate now relatively rare but abundant in earlier periods of earth history. Brachiopods, common fossils in rocks of Paleozoic age, have a bivalve shell that is symmetrical right and left of center.
BRYOZOA	Tiny aquatic animals that build large colonial structures that are common fossils in rocks of Paleozoic age.

CARBON, OR RADIOCARBON, DATING	A method of determining an age in years by measuring the concentration of carbon 14 remaining in formerly living matter, based on the assumption that assimilation of carbon 14 ceased abruptly at the time of death and that the matter thereafter remained a closed system. A half-life of 5,570 ± 30 years for carbon 14 makes the method useful in determining ages in the range of 500 to 40,000 years.
CEPHALOPOD	A marine mollusk that secretes a chambered shell, usually coiled in a planospiral but occasionally straight. The main body of the animal is housed in the last open chamber; it maintains buoyancy by filling the enclosed chambers with gas, and it swims by jetting fluid. Modern examples are the nautilus and squid. Ammonitic cephalopods have complexly crinkled chamber walls that may be used to distinguish species. They are especially useful in dating Mesozoic rocks.
CHERT	A very dense siliceous rock usually found as nodular or concretionary masses, or as distinct beds, associated with limestones. Jasper is red chert containing iron-oxide impurities.
CLASTIC ROCKS	Deposits consisting of fragments of preexisting rocks; conglomerate, sandstone, and shale are examples.
CONGLOMERATE	The consolidated equivalent of gravel. The constituent rock and mineral fragments may be of varied composition and range widely in size. The rock fragments are rounded and smoothed from transportation by water.
CONTACT	The surface, often irregular, constituting the junction of two bodies of rock.
CONTINENTAL DEPOSITS	Deposits laid down on land or in bodies of water not connected with the ocean.
CORRELATION	The process of determining the position or time of occurrence of one geologic phenomenon in relation to others. Usually it means determining the equivalence of geologic formations in separated areas through a comparison and study of fossils or rock peculiarities.
CRINOID	A marine invertebrate animal, abundant as fossils in rocks of Paleozoic age. Most lived attached to the sea bottom by a jointed stalk, the "head" resembling a lily-like plant, hence the common name *sea lily*.
DIATREME	A breccia-filled volcanic pipe that was formed by a gaseous explosion.
DIKE	A sheet-like body of igneous rock that filled a fissure in older rock while in a molten state. Dikes that intrude layered rocks cut the beds at an angle.

DISCONFORMITY	A break in the orderly sequence of stratified rocks above and below which the beds are parallel. The break is usually marked by erosional channels, indicating a lapse of time or absence of part of the rock sequence.
DOLOMITE	A mineral composed of calcium and magnesium carbonate, or a rock composed chiefly of the mineral dolomite, formed by alteration of limestone.
DOME	An upfold in which strata dip downward in all directions from a central area; the opposite of a basin.
EOLIAN	Pertaining to wind. Designates rocks or soils whose constituents have been transported and deposited by wind, such as windblown sand and dust (loess) deposits.
EROSIONAL UNCONFORMITY	A break in the continuity of deposition of a series of rocks caused by an episode of erosion.
EXTRUSIVE ROCK	A rock that has solidified from molten material poured or thrown out onto the earth's surface by volcanic activity.
FACIES	A physical aspect or characteristic of a sedimentary rock, as related to adjacent strata. The term is usually applied to distinguish different aspects of the sediments in time-equivalent or laterally continuous beds. For example, the white sandstone facies of the Cedar Mesa Sandstone changes laterally to the age-equivalent red arkosic sandstone facies of the Cutler Group in Canyonlands Country. Such a change from one aspect to another is called a facies change.
FAULT	A break or fracture in rocks along which there has been movement, one side relative to the other. Displacement along a fault may be vertical (normal or reverse fault) or lateral (strike-slip or "wrench" fault).
FORAMINIFERA	Generally, microscopic one-celled animals (protozoans), almost entirely of marine origin, whose shells are sufficiently durable to be preserved as fossils. They are usually abundant in marine sediments and are small enough to be retrievable in drill cuttings and cores.
FORMATION	The fundamental unit in the local classification of layered rocks, consisting of a bed or beds of similar or closely related rock types and differing from strata above and below. A formation must be readily distinguishable, thick enough to be mappable, and of broad regional extent. A formation may be subdivided into two or more MEMBERS and/or combined with other closely related formations to form a GROUP.

FUSULINIDS	Small, spindle-shaped foraminifera occurring as elongate chambers enrolled into complex internal forms that resemble a jellyroll. They are found only in marine rocks of Pennsylvanian and Permian age, and they are excellent fossils for dating and correlating sedimentary rocks because of their rapid evolutionary history.
GNEISS	A banded metamorphic rock with alternating layers of usually elongated, tubular, unlike minerals.
GRABEN	A down-faulted block.
GRANITE	An intrusive igneous rock with visibly granular, interlocking, crystalline quartz, feldspar, and perhaps other minerals.
GROUP	A formal unit of stratigraphic nomenclature consisting of two or more FORMATIONS.
HALITE	Common rock salt (NaCl) precipitated from seawater under conditions of intense evaporation. Halite occurs as bedded, usually crystalline, white or red rock only in the subsurface and flows plastically under high confining pressure.
HORST	An up-faulted block.
IGNEOUS ROCK	Rocks formed by solidification of molten material (magma), including rocks crystallized from cooling magma at depth (INTRUSIVE) and those poured out onto the surface as lavas (EXTRUSIVE).
INTRUSIVE ROCK	Rock that has solidified from molten material within the earth's crust and has not reached the surface; it usually has a visibly crystalline texture.
LACCOLITH	An intrusive igneous body consisting of a circular feeder stock that branches outward along bedding planes in sedimentary rocks to form flat-bottomed, bulged-upward sills, shaped much like a mushroom.
LAW OF SUPERPOSITION	A concept stating that if structurally undisturbed, any sequence of sedimentary rocks will have the oldest beds at the base and the youngest at the top.
LIMESTONE	A bedded sedimentary deposit consisting chiefly of calcium carbonate, usually formed from the calcified hard parts of organisms.
METAMORPHIC ROCK	Rocks formed by the alteration of preexisting igneous or sedimentary rocks, usually by intense heat and/or pressure or mineralizing fluids.
MEMBER	A formal or informal unit of stratigraphic nomenclature that constitutes a subdivision of a FORMATION into two or more parts.

MONOCLINE	A sharp fold formed in sedimentary rocks where they are draped across a deep-seated fault, much like a carpet draped over a stair step.
OROGENY	Literally, the process of formation of mountains, but practically, the processes by which structures in mountainous regions were formed, including folding, thrusting, and faulting in the outer layers of the crust, and plastic folding, metamorphism, and plutonism (emplacement of magma) in the inner layers. An episode of structural deformation may be called an orogeny, e.g., the Laramide Orogeny.
"REEF"	Some topographic features on the Colorado Plateau are called "reefs," such as Capitol Reef and San Rafael Reef, named by pioneers in the 1800s. They thought of their covered wagons as "prairie schooners" and severe obstacles to travel as "reefs," using the nautical terms. These "reefs" are jagged ridges of sharply upturned Navajo and Wingate Sandstone along monoclinal folds.
SALT DOME	Usually, a circular uplift of sedimentary rocks caused by the pushing upward of a mass of salt and/or gypsum at depth. As in eastern Canyonlands country, the circular cells may coalesce into elongate walls aligned along deep-seated faults to form salt anticlines.
SANDSTONE	A consolidated rock composed of sand grains cemented together; usually composed predominantly of quartz, it may contain other sand-size fragments of rocks and/or minerals.
SCHIST	A crystalline metamorphic rock with closely spaced foliation (platy texture) that splits into thin flakes or slabs.
SEDIMENTARY ROCK	Rocks composed of sediments, usually aggregated through processes of water, wind, glacial ice, or organisms, derived from preexisting rocks. In the case of limestones, constituent particles are usually derived from organic processes.
SHALE	Solidified mud, clays, and silts that are fissile (split like paper) and break along original bedding planes.
SILL	A tabular body of igneous rock that was injected in the molten state along bedding planes between layers of preexisting rocks.
STRATIGRAPHY	The definition and interpretation of layered rocks, the conditions of their formation, their character, arrangements, sequence, age, distribution, and correlation, using fossils and other means.
STRATUM	A single layer of sedimentary rock, separated from adjacent layers, or strata, by surfaces of erosion, nondeposition, or abrupt changes in character.

SYNCLINE	An elongate, trough-like downfold in which the sides dip downward and inward toward the axis.
TECTONIC	Pertaining to rock structures formed by earth movements, especially those that are widespread.
TRILOBITE	A general term for a group of extinct arthropods that occur as fossils in rocks of Paleozoic age. The fossils consist of flattened, segmented shells with a distinct thoraxial lobe and paired appendages, usually found as fragments.
TYPE LOCALITY	The place from which the name of a geologic formation is taken and where the unique characteristics of the formation may be examined.
UNCONFORMITY	A general term for a surface of erosion or nondeposition separating sequences of layered rocks. Variations include ANGULAR UNCONFORMITY, DISCONFORMITY, nonconformity, and diastem.
UPWARP	A broad area where the layered rocks have been uplifted by internal forces; the classic example is the Monument Upwarp of southeastern Utah.

References

Abbot, W. 0.; Liscomb, R. L. 1956. "Stratigraphy of the Book Cliffs in East Central Utah." Inter-mountain Association of Petroleum Geologists, *7th Field Conference Guidebook,* pp. 120–23.

Akers, J. P.; Cooley, M. E.; Reppening, C. A. 1958. "Moenkopi and Chinle Formations of Black Mesa and Adjacent Areas." New Mexico Geological Society, *9th Field Conference Guidebook,* pp. 88–94.

Allen, J. E.; Balk, R. 1954. "Mineral Resources of Fort Defiance and Tohatchi Quadrangles, Arizona and New Mexico." *New Mexico Bureau of Mines and Mineral Resources, Bulletin* 36.

Baars, D. L. 1962a. "Permian Strata of Central New Mexico." New Mexico Geological Society, *12th Field Conference Guidebook,* pp. 113–20.

———. 1962b. "Permian System of Colorado Plateau." *American Association of Petroleum Geologists Bulletin* 46:149–218.

———. 1966. "Pre-Pennsylvanian Paleotectonics—Key to Basin Evolution and Petroleum Occurrences in Paradox Basin, Utah and Colorado." *American Association of Petroleum Geologists Bulletin* 50:2082–2111.

———. 1973. "Permianland: The Rocks of Monument Valley." In H. L. James, ed., *Guidebook of Monument Valley and Vicinity, Arizona and Utah: New Mexico Geological Society 24th Field Conference Guidebook,* pp. 68–71.

———. 1976. "The Colorado Plateau Aulacogen—Key to Continental Scale Basement Rifting." In M. H. Podwysocki and J. L. Earle, eds., *Proceedings of the Second International Conference on Basement Tectonics,* pp. 157–64.

———. 1988. "Triassic and Older Stratigraphy: Southern Rocky Mountains and Colorado Plateau." In L. L. Sloss, ed., *Sedimentary Cover: North American Craton,* U.S. Geological Society of America, *The Geology of North America* D-2:53–64.

———. 1991. "Redefinition of the Pennsylvanian and Permian Boundary in Kansas, Midcontinent, USA." Program and Abstracts, *International Congress on the Permian System of the World, Perm', USSR,* p. A3.

———. 1992. *The American Alps: The San Juan Mountains of Southwest Colorado.* University of New Mexico Press, Albuquerque.

———. 1993. *Canyonlands Country: Geology of Canyonlands and Arches National Parks.* University of Utah Press, Salt Lake City.

———. 1994. "Proposed Repositioning of the Pennsylvanian-Permian Boundary in Kansas." *Kansas Geological Survey, Bulletin* 231:5–11.

———. 1995. *Navajo Country: Geology and Natural History of the Four Corners Region*. University of New Mexico Press, Albuquerque.

———. 1998a. *Beyond the Spectacular in Monument Valley Navajo Tribal Park*. Cañon Publishers, Grand Junction, Colo.

———. 1998b. *The Incredible Story Behind Arches National Park*. Cañon Publishers, Grand Junction, Colo.

———. 1998c. *The Mind-Boggling Scenario of the Colorado National Monument*. Cañon Publishers, Grand Junction, Colo.

———. 1998d. *The Rocks and Times of Canyon de Chelly National Park*. Cañon Publishers, Grand Junction, Colo.

———. 2000. *The Colorado Plateau: A Geologic History*. Revised edition. University of New Mexico Press, Albuquerque.

Baars, D. L.; Doelling, H. H. 1987. "Moab Salt-Intruded Anticline, East-Central Utah." *Geological Society of America Centennial Field Guide—Rocky Mountain Ssection*, pp. 275–80.

Baars, D. L.; Seager, W. R. 1967. "Depositional Environment of White Rim Sandstone (Permian), Canyonlands National Park, Utah" (abstract). *American Association of Petroleum Geologists Bulletin* 51:453.

Baars, D. L.; Stevenson, G. M. 1977. "Permian Rocks of the San Juan Basin." In J. E. Fassett, ed., *New Mexico Geological Society 28th Field Conference Guidebook*, pp. 133–38.

———. 1981. "Tectonic Evolution of the Paradox Basin." In D. L. Weigand, ed., *Rocky Mountain Association of Geologists Guidebook*, pp. 23–31.

———. 1982. "Subtle Stratigraphic Traps in Paleozoic Rocks of the Paradox Basin." In M. Halbouty, ed., *American Association of Petroleum Geologists Memoir* 32:131–58.

Baars, D. L., and 15 others. 1988. "Basins of the Rocky Mountain Region." In L. L. Sloss, ed., *Sedimentary Cover: North American Craton*, U.S. Geological Society of America, *The Geology of North America* D-2:109–220.

Baars, D. L.; Thomas, W. A.; Drahovzal, J. A.; Gerhard, L. C. 1995. "Preliminary Ivestigations of Basement Tectonic Fabric of the Conterminous USA." In R. W. Ojakangas and others, eds., *Basement Tectonics* 10:149–58.

Baker, A. A. 1933. "Geology and Oil Possibilities of the Moab District, Grand and San Juan Counties, Utah." *U.S. Geological Survey Bulletin* 841.

———. 1935. "Geologic Structure of Southeastern Utah." *American Association of Petroleum Geologists Bulletin* 19:1472–1507.

———. 1936. "Geology of the Monument Valley—Navajo Mountain Region, San Juan County, Utah." *U.S. Geological Survey Bulletin* 865:106.

———. 1946. "Geology of the Green River Desert–Cataract Canyon Region, Emery, Wayne, and Garfield Counties, Utah." *U.S. Geological Survey Bulletin* 951:122.

Baker, A. A.; Dane, C. H.; McKnight, E. T. 1936. "Geology of the Monument Valley–Navajo Mountain Region, San Juan County, Utah." *U.S. Geological Survey Bulletin* 865.

Baker, A. A.; Dane, C. H.; Reeside, J. B., Jr. 1933. "Paradox Formation of Eastern Utah and Western Colorado." *American Association of Petroleum Geologists Bulletin* 17:963–80.

Baker, A. A.; Dobbin, C. E.; McKnight, E. T.; Reeside, J. B. 1927. "Notes on the Stratigraphy of the Moab Region, Utah." *American Association of Petroleum Geologists Bulletin* 11:785–808.

Benson, M. 1994. *1001 Colorado Place Names*. University Press of Kansas, Lawrence.

Beus, S. S.; Morales, M. 1990. *Grand Canyon Geology*. Oxford University Press, New York; Museum of Northern Arizona Press, Flagstaff.

Blagbrough, J. W. 1967. "Cenozoic Geology of the Chuska Mountains." In F. D. Trauger, ed., *New Mexico Geological Society 18th Field Conference Guidebook*, pp. 70–77.

Blakey, R. C. 1990. "Stratigraphy and Geologic History of Pennsylvanian and Permian Rocks, Mogollon Rim, Central Arizona and Vicinity." *Geological Society of America Bulletin* 102:1189–1217.

———. 1993. "Supai Group and Hermit Formation." In S. S. Beus and M. Morales, eds., *Grand Canyon Geology.* Oxford University Press, New York; Museum of Northern Arizona Press, Flagstaff, pp. 147–82.

Blakey, R. C.; Baars, D. L. 1987. "Monument Valley, Arizona and Utah." *Geological Society of America Centennial Field Guide—Rocky Mountain Section,* pp. 361–64.

Byerly, P. E.; Joesting, H. R. 1959. "Regional Geophysical Investigations of the Lisbon Valley Area, Utah and Colorado." *U.S. Geological Survey Professional Paper* 316-C:39–50.

Case, J. E.; Joesting, H. R.; Byerly, P. E. 1963. "Regional Geophysical Investigations in the La Sal Mountains Area, Utah and Colorado." *U.S. Geological Survey Professional Paper* 316-F:91–116.

Chuvashov, B. I. 1989. "The Carboniferous-Permian Boundary in the USSR." In B. R. Wardlaw, ed., *Working Group on the Carboniferous-Permian Boundary, 28th International Geological Congress Proceedings,* pp. 42–56.

Cole, R. D. 1987. "Cretaceous Rocks of the Dinosaur Triangle." In W. R. Averett, ed., *Paleontology and Geology of the Dinosaur Triangle,* The Museum of Western Colorado, Grand Juction, Colo., pp. 21–36.

Condon, S. M. 1989. "Modifications to Middle and Upper Jurassic Nomenclature in the Southeastern San Juan Basin, New Mexico." In O. J. Anderson and others, eds., *New Mexico Geological Society 40th Field Conference Guidebook,* pp. 231–38.

Cowie, J. W.; Bassett, M. G. 1989. "Global Stratigraphic Chart." International Union of Geological Sciences, Bureau of International Commission on Stratigraphy (ICS:IUGS), *Supplement to Episodes* 12(2).

Craig, L. C.; Cadigan, R. A. 1958. "The Morrison and Adjacent Formations in the Four Corners Area." Intermountain Association of Petroleum Geologists, *9th Field Conference Guidebook,* pp. 182–92.

Craig, L. C.; Dickey, D. D. 1956. "Jurassic Strata of Southeastern Utah and Southwestern Colorado." Intermountain Association of Petroleum Geologists, *7th Field Conference Guidebook,* pp. 93–104.

Dane, C. H. 1935. "Geology of Salt Valley Anticline and Adjacent Areas, Grand County, Utah." *U.S. Geological Survey Bulletin* 863:184.

Darton, N. H. 1922. "Geologic Structures of Parts of New Mexico." *U.S. Geological Survey Bulletin* 726-E:173–275.

———. 1926. "The Permian of Arizona and New Mexico." *American Association of Petroleum Geologists Bulletin* 10:189–252.

———. 1928. "Red Beds and Associated Formations in New Mexico." *U.S. Geological Survey Bulletin* 794:356.

Davis, G. H. 1978. "Monocline Fold Pattern of the Colorado Plateau." In V. Matthews, ed., *Geological Society of America Memoir* 151:215–33.

———. 1997. "Field Guide to Geologic Structures in the Zion–Bryce–Cedar Breaks Region, Utah." Geological Society of America, *Penrose Conference on Tectonics of Continental Interiors.*

Davydov, V. I. 1991. "The Carboniferous-Permian Boundary in the USSR and Its Correlation." Program and Abstracts, *International Congress on the Permian System of the World, Perm', USSR,* p. A3.

Dubiel, R. F. 1989. "Sedimentology and Revised Nomenclature for the Upper Part of the Upper Triassic Chinle Formation and the Lower Jurassic Wingate Sandstone, Northwestern

New Mexico and Northeastern Arizona." In O. J. Anderson and others, eds., *New Mexico Geological Society 40th Field Conference Guidebook*, pp. 213–23.

Dutton, C. E. 1881. "The Physical Geology of the Grand Canyon District." *Second Annual Report of the U.S. Geological Survey*, pp. 47–166.

———. 1882. "Tertiary History of the Grand Canyon District." *U.S. Geological Survey Monograph* 2:264.

Ellingson, J. A. 1973. "The Mule Ear Diatreme." In D. L. Baars, ed., *Four Corners Geological Society Guidebook*.

Elston, D. P.; Landis, E. R. 1960. "Pre-Cutler Unconformities and Early Growth of the Paradox Valley and Gypsum Valley Salt Anticlines, Colorado." *U.S. Geological Survey Professional Paper* 400-B, Art. 118:B261–65.

Elston, D. P.; Shoemaker, E. M.; Landis, E. R. 1962. "Uncompahgre Front and Salt Anticline Region of Paradox Basin, Colorado and Utah." *American Association of Petroleum Geologists Bulletin* 46(10):1857–78.

Elston, W. E. 1960. "Structural Development and Paleozoic Stratigraphy of Black Mesa Basin, Northeastern Arizona, and Surrounding Areas." *American Association of Petroleum Geologists Bulletin* 44:21–36.

Fassett, J. E.; Hinds, J. S. 1971. "Geology and Fuel Resources of the Fruitland Formation and Kirtland Shale of the San Juan Basin." *U.S. Geological Survey Professional Paper* 676.

Gilbert, G. K., 1877, "Report on the Geology of the Henry Mountains," *U.S. Geographical and Geological Society Survey, Rocky Mountain Region (Powell)*, p.160.

Gilluly, J.; Reeside, J. B., Jr. 1928. "Sedimentary Rocks of the San Rafael Swell and Some Adjacent Areas in Eastern Utah." *U.S. Geological Survey Professional Paper* 150.

Green, M. W.; Pierson, C. T. 1977. "A Summary of the Stratigraphy and Depositional Environments of Jurassic and Related Rocks in the San Juan Basin, Arizona, Colorado, and New Mexico." In J. E. Fassett and H. L. James, eds., *New Mexico Geological Society 28th Field Conference Guidebook*, pp. 147-52.

Gregory, H. E. 1917. "Geology of the Navajo Country." *U.S. Geological Survey Professional Paper* 93.

———. 1938. "The San Juan Country." *U.S. Geological Survey Professional Paper* 188.

———. 1952. "Geology and Geography of the Zion Park Region, Utah and Arizona." *U.S. Geological Survey Professional Paper* 220.

Gregory, H. E.; Moore, R. C. 1931. "The Kaiparowits Region: A Geographic and Geologic Reconnaissance of Parts of Utah and Arizona." *U.S. Geological Survey Professional Paper* 164.

Grundy, W. D.; Oertell, E. W. 1958. "Uranium Deposits in the White Canyon and Monument Valley Mining Districts, San Juan County, Utah, and Navajo and Apache Counties, Arizona." Intermountain Association of Petroleum Geologists, *9th Field Conference Guidebook*, pp. 197–207.

Hack, J. T. 1942. "Sedimentation and Vulcanism in the Hopi Buttes." *Geological Society of America Bulletin* 53:335–72.

Hamblin, W. K. 1990. "Late Cenozoic Lava Dams in the Western Grand Canyon." In S. S. Beus and M. Morales, eds., *Grand Canyon Geology*, Oxford University Press, New York; Museum of Northern Arizona Press, Flagstaff, pp. 385–434.

Harshbarger, J. W.; Repenning, C. A.; Irwin, J. H. 1958. "Stratigraphy of the Uppermost Triassic and the Jurassic Rocks of the Navajo Country." New Mexico Geological Society, *9th Field Conference Guidebook*, pp. 98–114.

Heaton, R. L. 1933. "Ancestral Rockies and Mesozoic and Paleozoic Stratigraphy of Rocky Mountain Region." *American Association of Petroleum Geologists Bulletin* 17:109–68.

Herman, G.; Sharps, S. L. 1956. "Pennsylvanian and Permian Stratigraphy of the Paradox Salt Embayment." Intermountain Association of Petroleum Geologists, *7th Field Conference Guidebook,* pp. 77–84.

Hintze, Lehi F. 1988. "Geologic History of Utah." *Brigham Young University Geology Studies Special Publication 7.*

Hite, R. J. 1960. "Stratigraphy of the Saline Facies of the Paradox Member of the Hermosa Formation of Southeastern Utah and Southwestern Colorado." Four Corners Geological Society, *3rd Field Conference Guidebook,* pp. 86–89.

Hopkins, R. L. "Kaibab Limestone." 1990. In S. S. Beus and M. Morales, eds., *Grand Canyon Geology,* Oxford University Press, New York; Museum of Northern Arizona Press, Flagstaff, pp. 225–46.

Hunt, A. P.; Lucas, S. G. 1992. "Stratigraphy, Paleontology, and Age of the Fruitland and Kirtland Formations (Upper Cretaceous), San Juan Basin, New Mexico." In S. G. Lucas and others, eds., *New Mexico Geological Society 43rd Field Conference Guidebook,* pp. 217-40.

Hunt, C. B. 1969. "Geologic History of the Colorado River." *U.S. Geological Survey Professional Paper* 669:59–130.

Hunt, C. B.; Averitt, P.; Miller, R. L. 1953. "Geology and Geography of the Henry Mountains Region, Utah." *U.S. Geological Survey Professional Paper* 228.

Irwin, J. H.; Stevens, P. R.; Cooley, M. E. 1971. "Geology of the Paleozoic Rocks, Navajo and Hopi Indian Reservations, Arizona, New Mexico, and Utah." *U.S. Geological Survey Professional Paper* 521-C.

Jackson, M. P. A.; Schultz-Ela, D. D.; Hudec, M. R.; Watson, I. A.; Porter, M. L. 1998. "Structure and Evolution of Upheaval Dome: A Pinched-Off Salt Diapir." *Geological Society of America Bulletin* 110:1547–73.

Jentgen, R. W. 1977. "Pennsylvanian Rocks in the San Juan Basin, New Mexico and Colorado." In J. E. Fassett and H. L. James, eds., *New Mexico Geological Society 28th Field Conference Guidebook,* pp. 129–32.

Joesting, H. R.; Byerly, P. E. 1958. "Regional Geophysical Investigations of the Uravan Area, Colorado." *U.S. Geological Survey Professional Paper* 316-A:1–17.

Joesting, H. R.; Case, J. E. 1960. "Salt Anticlines and Deep-Seated Structures in the Paradox Basin, Colorado and Utah." *U.S. Geological Survey Professional Paper* 400-B, Art. 114:B252–56.

———. 1962. "Regional Geophysical Studies in Salt Valley–Cisco Area, Utah and Colorado." *American Association of Petroleum Geologists Bulletin* 46:1879–89.

Jones, R. W. 1959. "Origin of Salt Anticlines of Paradox Basin." *American Association of Petroleum Geologists Bulletin* 43:1869–95.

Kelley, V. C. 1951. "Tectonics of the San Juan Basin." New Mexico Geological Society, *2nd Field Conference Guidebook,* pp. 124–31.

———. 1955. "Regional Tectonics of the Colorado Plateau and Relationship to the Origin and Distribution of Uranium." *University of New Mexico (Publication in Geology),* no. 5.

———. 1958. "Tectonics of the Region of the Paradox Basin." Intermountain Association of Petroleum Geologists, *9th Field Conference Guidebook.*

Kriens, B. J.; Shoemaker, E. M.; Herkenhoff, K. E. 1999. "Geology of the Upheaval Dome Impact Structure, Southeast Utah." *Journal of Geophysical Research* 104(E8):18,867–87.

Kunkel, R. P. 1958. "Permian Stratigraphy of the Paradox Basin." Intermountain Association of Petroleum Geologists, *9th Field Conference Guidebook,* pp. 163–68.

———. 1960. "Permian Stratigraphy in the Salt Anticline Region of Western Colorado and Eastern Utah." Four Corners Geological Society, *3rd Field Conference Guidebook,* pp. 91–97.

Lessentine, R. H. 1965. "Kaiparowits and Black Mesa Basins: Stratigraphic Synthesis." *American Association of Petroleum Geologists Bulletin* 49:1997–2019.

Likharev, B. K. 1959. "The Boundaries and Principal Subdivisions of the Permian System." *Soviet Geology* 66:12-30.

Linford, L. D. 2000. *Navajo Places.* University of Utah Press, Salt Lake City.

Lucas, S. G.; Hayden, S. N. 1989. "Triassic Stratigraphy of West-Central New Mexico." In O. J. Anderson and others, eds., New Mexico Geological Society, *40th Field Conference Guidebook,* pp. 191–211.

Lucchitta, I. 1990. "History of the Grand Canyon and of the Colorado River in Arizona." In S. S. Beus and M. Morales, eds., *Grand Canyon,* Oxford University Press, New York; Museum of Northern Arizona Press, Flagstaff, pp. 311–32.

Mallory, W. W. 1972. "Pennsylvanian Arkose and the Ancestral Rocky Mountains." In W. W. Mallory, ed., *Geologic Atlas of the Rocky Mountain Region*, Rocky Mountain Association of Geologists, pp. 131–2.

McKee, E. D. 1933. "The Coconino Sandstone: Its History and Origin." *Carnegie Institution of Washington Pub.* 440:77–115.

———. 1934. "Investigations of Light-colored Cross-bedded Sandstone of Canyon de Chelly, Arizona." *American Journal of Science,* 5th series, 28.

———. 1938. "The Environment and History of the Toroweap and Kaibab Formations of Northern Arizona and Southern Utah." *Carnegie Institution of Washington Pub.* 492.

———. 1945. "Cambrian History of the Grand Canyon Region." *Carnegie Institution of Washington Pub.* 563.

———. 1954. "Stratigraphy and History of the Moenkopi Formation of Triassic Age." *Geological Society of America Memoir* 61.

———. 1963. "Nomenclature for Lithologic Subdivisions of the Mississippian Redwall Limestone, Arizona." *U.S. Geological Survey Professional Paper* 475-C, Art. 65:C21–22.

———. 1982. "The Supai Group of Grand Canyon." *U.S. Geological Survey Professional Paper* 1173.

McKee, E. D.; Hamblin, W. K.; Damon, E. 1968. "K-Ar Age of Lava Dam in Grand Canyon." *Geological Society of America Bulletin* 79:133–36.

McKee, E. D.; Wilson, R. F.; Breed, W. J. 1964. *Evolution of the Colorado River in Arizona.* Museum of Northern Arizona, Flagstaff.

McKnight, E. T. 1940. "Geology of Area Between Green and Colorado Rivers, Grand and San Juan Counties, Utah." *U.S. Geological Survey Bulletin* 908:1–147.

McLemore, V. T.; Chenoweth, W. L. 1992. "Uranium Deposits in the Eastern San Juan Basin, Cibola, Sandoval, and Rio Arriba Counties, New Mexico." In S. G. Lucas and others, eds., *New Mexico Geological Society 43rd Field Conference Guidebook,* pp. 341-50.

Miser, H. D. 1924. "The San Juan Canyon, Southeastern Utah." *U.S. Geological Survey Water Supply Paper* 538.

Molenaar, C. M. 1977. "Stratigraphy and Depositional History of Upper Cretaceous Rocks of the San Juan Basin Area, New Mexico and Colorado, with a Note on Economic Resources." In J. E. Fassett and H. L. James, eds., *New Mexico Geological Society 28th Field Conference Guidebook,* pp. 159–66.

———. 1983. "Major Depositional Cycles and Regional Correlations of Upper Cretaceous Rocks, Southern Colorado Plateau and Adjacent Areas." In M. W. Reynolds and E. D. Dolly, eds., *Mesozoic Paleogeography of West-Central United States*, Rocky Mountain Section, Society of Economic Paleontologists and Mineralogists, pp. 201–24.

Molenaar, C. M.; Rice, E. D. 1988. "Cretaceous Rocks of the Western Interior Basin." In L. L. Sloss, ed., *Sedimentary Cover: North American Craton, U.S.* Geological Society of America, *The Geology of North America* D-2:77–82.

Murray, J. A. 1998. *The Colorado Plateau.* Northland Publishing, Flagstaff.

Noble, L. F. 1914. "The Shinumo Quadrangle, Grand Canyon District, Arizona." *U.S. Geological Survey Bulletin* 549.

Northrop, S. A.; Wood, G. H. 1946. "Geology of Nacimiento Mountains, San Pedro Mountains, and Adjacent Plateaus in Part of Sandoval and Rio Arriba Counties, New Mexico." *U.S. Geological Survey Preliminary Map* 57, Oil and Gas Investigations Series.

O'Sullivan, R. B. 1977. "Triassic Rocks in the San Juan Basin of New Mexico and Adjacent Areas." In J. E. Fassett and H. L. James, eds., *New Mexico Geological Society 28th Field Conference Guidebook,* pp. 139–46.

O'Sullivan, R. B.; Craig, L. C. 1973. "Jurassic Rocks of Northeast Arizona and Adjacent Areas." In H. L. James, ed., *New Mexico Geological Society 24th Field Conference Guidebook,* pp. 79–85.

O'Sullivan, R. B.; Green, M. W. 1973. "Triassic Rocks of Northeast Arizona and Adjacent Areas." In H. L. James, ed., *New Mexico Geological Society 24th Field Conference Guidebook,* pp. 72–78.

Page, H. G.; Repenning, C. A. 1958. "Late Cretaceous Stratigraphy of Black Mesa, Navajo and Hopi Indian Reservations, Arizona." New Mexico Geological Society, *9th Field Conference Guidebook,* pp. 115–22.

Peterson, F. 1988a. "Stratigraphy and Nomenclature of Middle and Upper Jurassic Rocks, Western Colorado Plateau, Utah and Arizona." *U.S. Geological Survey Bulletin* 1633-B:B13-56.

———. 1988b. "A Synthesis of the Jurassic System in the Southern Rocky Mountain Region." In L. L. Sloss, ed., *Sedimentary Cover: North American Craton, U.S.* Geological Society of America, *The Geology of North America* D-2:65–76.

Peterson, F.; Kirk, A. R. 1977. "Correlation of Cretaceous Rocks in the San Juan, Kaiparowits, and Henry Basins, Southern Colorado Plateau." In J. E. Fassett and H. L. James, eds., *New Mexico Geological Society 28th Field Conference Guidebook,* pp. 167–78.

Peterson, F.; Pipiringos, G. N. 1979. "Stratigraphic Relationships of the Navajo Sandstone to Middle Jurassic Formations, Southern Utah and Northern Arizona." *U.S. Geological Survey Professional Paper* 1035–B.

Pierce, H. W. 1958. "Permian Sedimentary Rocks of the Black Mesa Basin Area." New Mexico Geological Society, *9th Field Conference Guidebook,* pp. 82–87.

———. 1967. "Permian Stratigraphy of the Defiance Plateau." In E. D. Trauger, ed., *New Mexico Geological Society 18th Field Conference Guidebook,* pp. 57–62.

Potochnik, A. R.; Reynolds, S. J. 1990. "Side Canyons of the Colorado River, Grand Canyon." In S. S. Beus and M. Morales, eds., *Grand Canyon Geology,* Oxford University Press, New York; Museum of Northern Arizona Press, Flagstaff.

Reeside J. B., Jr.; Baker, A. A. 1929. "The Cretaceous Section in Black Mesa, Northern Arizona." *Washington Academy of Science Journal* 19:30 37.

Reeside, J. B., Jr.; Hunt, C. B.; Hendricks, T. A. 1941. "Transgressive and Regressive Cretaceous Deposits in Southern San Juan Basin, New Mexico." *U.S. Geological Survey Professional Paper* 193:101–21.

Reeside, J. B., Jr.; Knowlton, F. H. 1924. "Upper Cretaceous and Tertiary Formations of the Western Part of the San Juan Basin, Colorado and New Mexico." *U.S. Geological Survey Professional Paper* 134.

Repenning, C. A.; Irwin, J. H. 1954. "Bidahochi Formation of Arizona and New Mexico." *American Association of Petroleum Geologists Bulletin* 38:1821–26.

Roebuck, R. C. 1958. "Chinle and Moenkopi Formations, Southeastern Utah." Intermountain Association of Petroleum Geologists, *9th Field Conference Guidebook,* pp. 169–71.

Ross, C. A.; Ross, J. R. P. 1994. "The Need for a Bursumian Stage, Uppermost Carboniferous, North America." *Permophiles* 24:3-6.

———. 1998. "Bursumian Stage, Uppermost Carboniferous of Midcontinent and Southwestern North America." *Carboniferous Newsletter* 6:40–42.

Sears, J. D. 1956. "Geology of Comb Ridge and Vicinity North of San Juan River, San Juan County, Utah." *U.S. Geological Survey Bulletin* 1021-E.

Sears, J. D.; Hunt, C. B.; Hendricks, T. A. 1941. "Transgressive and Regressive Cretaceous Deposits in Southern San Juan Basin, New Mexico." *U.S. Geological Survey Professional Paper* 193-F:101–21.

Shelton, J. S. 1966. *Geology Illustrated.* W. H. Freeman, San Francisco.

Shoemaker, E. M. 1954. "Structural Features of Southeastern Utah and Adjacent Parts of Colorado, New Mexico, and Arizona." Utah Geological Society, *Guidebook to the Geology of Utah* 9:48–69.

Shoemaker, E. M.; Case, J. E.; Elston D. P. 1958. "Salt Anticlines of the Paradox Basin." Intermountain Association of Petroleum Geologists, *9th Field Conference Guidebook,* pp. 39–59.

Sloss, L. L., ed. 1988. *Sedimentary Cover: North American Craton, U.S.* Geological Society of America, *The Geology of North America* D-2.

Spieker, E. M. 1946. "Late Mesozoic and Early Cenozoic History of Central Utah." *U.S. Geological Survey Professional Paper* 205-D:117–61.

———. 1949. "The Transition Between the Colorado Plateaus and the Great Basin in Central Utah." Utah Geological Society, *Guidebook to the Geology of Utah* 4.

Stevenson, G. M.; Baars, D. L. 1986. "The Paradox: A Pull-Apart Basin of Pennsylvanian Age." In J. A. Peterson, ed., *Paleotectonics and Sedimentation in the Rocky Mountain Region, United States. American Association of Petroleum Geologists Memoir* 41:513-40.

Stewart, J. H. 1956. "Triassic Strata of Southeastern Utah and Southwestern Colorado." Intermountain Association of Petroleum Geologists, *7th Field Conference Guidebook,* pp. 85–92.

———. 1959. "Stratigraphic Relations of Hoskinnini Member (Triassic?) of Moenkopi Formation on Colorado Plateau." *American Association of Petroleum Geologists Bulletin* 43:1852–68.

Stewart, J. H.; Williams, G. A.; Albee, H. F.; Ralip, 0. B.; Cadigan, R. A. 1959. "Stratigraphy of Triassic and Associated Formations in Part of the Colorado Plateau." *U.S. Geological Survey Bulletin* 1046.

Stewart, J. H.; Wilson, R. F. 1960. "Triassic Strata of the Salt Anticline Region, Utah and Colorado." Four Corners Geological Society, *3rd Field Conference Guidebook*, pp. 98–106.

Stokes, W. L. 1944. "Morrison Formation and Related Deposits in and Adjacent to the Colorado Plateau." *Geological Society of America Bulletin* 55:951–92.

———. 1948. "Geology of the Utah-Colorado Salt Dome Region with Emphasis on Gypsum Valley, Colorado." Utah Geological Society, *Guidebook to the Geology of Utah,* no. 3.

———. 1949. "Triassic and Jurassic Rocks of Utah." *The Oil and Gas Possibilities of Utah,* Utah Geological and Mineralogical Survey, pp. 78–89.

———. 1952. "Lower Cretaceous in Colorado Plateau." *American Association of Petroleum Geologists Bulletin* 36:1766–76.

———. 1958. "Continental Sediments of the Colorado Plateau." Intermountain Association of Petroleum Geologists, *9th Field Conference Guidebook,* pp. 26–30.

———. 1986. *Geology of Utah.* Utah Museum of Natural History, University of Utah; Utah Geological and Mineral Survey, Department of Natural Resources.

Stokes, W. L.; Phoenix, D. A. 1948. "Geology of the Egnar–Gypsum Valley Area, San Miguel and Montrose Counties, Colorado." *U.S. Geological Survey Preliminary Map 93, Oil and Gas Investigations Series.*

Thomas, W. A.; Baars, D. L. 1995. "The Paradox Transcontinental Fault Zone." *Oklahoma Geological Survery Circular* 97:3-12.

Turner, C. H.; Fishman, N. S. 1991. "Jurassic Lake T'oo'dichi': A Large Alkaline, Saline Lake, Morrison Formation, Eastern Colorado Plateau." *Geological Society of America Bulletin* 103:538–58.

Van Cott, J. W. 1990. *Utah Place Names.* University of Utah Press.

Ver Wiebe, W. A. 1930. "Ancestral Rocky Mountains." *American Association of Petroleum Geologists Bulletin* 14:765–88.

Wanek, A. A. 1954. "Geologic Map of the Mesa Verde Area, Montezuma County, Colorado." *U.S. Geological Survey Map* OM-152.

Warner, L. A. 1978. "The Colorado Lineament: A Middle Precambrian Wrench Fault System." *Bulletin of the Geological Society of America* 89:161–71.

Wengerd, S. A. 1958. "Pennsylvanian Stratigraphy, Southwest Shelf, Paradox Basin." Intermountain Association of Petroleum Geologists, *9th Field Conference Guidebook,* pp. 109–34.

Wengerd, S. A.; Matheny, M. L. 1958. "Pennsylvanian System of Four Corners Region." *American Association of Petroleum Geologists Bulletin* 42:2048–2106.

Wengerd, S. A.; Strickland, J. W. 1954. "Pennsylvanian Stratigraphy of Paradox Salt Basin, Four Corners Region, Colorado and Utah." *American Association of Petroleum Geologists Bulletin* 38:2157–99.

White, D. 1929. "Flora of the Hermit Shale, Grand Canyon, Arizona." *Carnegie Institution of Washington Pub.* 405.

Wood, G. H.; Northrop, S. A. 1946. "Geology of Nacimiento Mountain, San Pedro Mountain, and Adjacent Plateaus in Parts of Sandoval and Rio Arriba Counties, New Mexico." *U.S. Geological Survey Preliminary Map 57, Oil and Gas Investigations Series.*

Wood, H. B.; Lekas, M. A. 1958. "Uranium Deposits of the Uravan Mineral Belt." Intermountain Association of Petroleum Geologists, *9th Field Conference Guidebook,* pp. 208–15.

Woodruff, E. G. 1910. "Geology of the San Juan Oil Field, Utah." *U.S. Geological Survey Bulletin* 471:76–104.

Woodward, L. A. 1973. "Structural Framework and Tectonic Evolution of the Four Corners Region of the Colorado Plateau." In H. L. James, ed., *New Mexico Geological Society 24th Field Conference Guidebook,* pp. 94-98.

Woodward, L. A.; Callender, J. F. 1977. "Tectonic Framework of the San Juan Basin." In J. E. Fassett and H. L. James, eds., *New Mexico Geological Society 28th Field Conference Guidebook*, pp. 209 12.

Young, R. G. 1955. "Sedimentary Facies and Intertonguing in the Upper Cretaceous of the Book Cliffs, Utah." *Geological Society of American Bulletin* 66:177–202.

———. 1973. "Cretaceous Stratigraphy of the Four Corners Area." In H. L. James, ed., *New Mexico Geological Society 24th Field Conference Guidebook,* pp. 86-93.

Geological Society Guidebooks

Written mostly for professional geologists, these guidebooks contain numerous significant technical papers and several roadlogs each. They are listed by geological society and date of publication.

Four Corners Geological Society

Geology of Parts of Paradox, Black Mesa, and San Juan Basins. Cooper, Jack C., editor, 217 pp., 1955.

Geology of Southwestern San Juan Basin. Little, C. J., and Gill, J. J., editors, 198 pp., 1957.

Geology of the Paradox Basin Fold and Fault Belt. Smith, K. G., editor, 173 pp., 1960.

Geology and Natural History of the Grand Canyon Region. Baars, D. L., editor, 212 pp., 1969.

Geology of Canyonlands and Cataract Canyon. Baars, D. L., and Molenaar, C. M., editors, 99 pp., 1971.

Cretaceous and Tertiary Rocks of the Southern Colorado Plateau. Fasset, James A., editor, 217 pp., 1973.

Canyonlands Country. Fassett, James A., editor, 281 pp., 1975.

Permianland. Baars, D. L., editor, 186 pp., 1979.

Geology of Cataract Canyon and Vicinity. Campbell, John A., editor, 200 pp., 1987.

Geological Society of America

Rocky Mountain Section: Centennial Field Guide, Volume 2. Beus, S. S., editor, 475 pp., 1987.

Rocky Mountain Section: 37th Annual Meeting, San Juan Mountains. Brew, D. C., editor, 209 pp., 1984.

Grand Junction Geological Society

Northern Paradox Basin. Averett, W. R., editor, 146 pp., 1983.

Intermountain Association of Petroleum Geologists	*Geology and Economic Deposits of East Central Utah.* Peterson, J. A., editor, 225 pp., 1956. *Guidebook to the Geology of the Paradox Basin.* Sanborn, A. F., editor, 308 pp., 1958.
Museum of Western Colorado	*Paleontology and Geology of the Dinosaur Triangle.* Averett, W. R., editor, 162 pp., 1987.
New Mexico Geological Society	*Southwestern San Juan Mountains.* Kottlowski, F. E., and Baldwin, B., editors, 258 pp., 1957. *Black Mesa Basin, Northeastern Arizona.* Anderson, R. Y., and Harshbarger, J. W., editors, 205 pp., 1958. *West-Central New Mexico.* Weir, J. E., and Baltz, E. H., editors, 162 pp., 1959. *Defiance–Zuni–Mt. Taylor Region.* Trauger, F. D., editor, 228 pp., 1967. *San Juan–San Miguel–La Plata Region.* Shomaker, J., editor, 211 pp., 1968. *Monument Valley and Vicinity.* James, H. L., editor, 206 pp., 1973. *San Juan Basin III.* Fassett, J. E., editor, 301 pp., 1977. *Western Slope Colorado.* Epis, R. C., and Callender, J. F., editors, 337 pp., 1981. *Albuquerque Country II.* Grambling, J. A., and Wells, S. G., editors, 370 pp., 1982. *Rio Grande Rift: Northern New Mexico.* Baldridge, W. S.; Dickerson, P. W.; Rieker, R. E.; and Zidek, J., editors, 380 pp., 1984. *Southeastern Colorado Plateau.* Anderson, O. J.; Lucas, S. G.; Love, D. W.; and Cather, S. M., editors, 246 pp., 1989. *San Juan Basin IV.* Lucas, S. G.; Kues, B. S.; Williamson, T. E.; and Hunt, A. P., editors, 411 pp., 1992.
Rocky Mountain Association of Geologists	*Symposium on Geology of the Cordilleran Hingeline.* Hill, J. G., editor, 407 pp., 1976. *Geology of the Paradox Basin.* Wiegand, D. L., editor, 285 pp., 1981.
Utah Geological Association	*Geology and Resources of the Paradox Basin.* Huffman, C. H.; Lund, W. R.; and Godwin, L. H., editors, with Four Corners Geological Society, 460 pp., 1996. *Geology of Utah's Parks and Monuments.* Sprinckle, D. A.; Chidsey, T. C.; and Anderson, P. B., editors, 644 pp., full color, 2000.

Acknowledgments

I hope this volume has provided readers accurate descriptions of the geology as seen along the myriad roads of the Colorado Plateau Province. This vast, beautiful region is crossed by many travelers of diverse backgrounds, including geology students, wandering professional geologists from other regions, teachers, the sight-seeing public, and vacationers driving between our national parks and monuments. Accuracy is the key word here, and no single person can be thoroughly knowledgeable about all aspects of the diverse geology found across such a broad region as the Colorado Plateau.

I owe considerable credit to those who undertook the staggering job of reviewing the original rough manuscript. One anonymous reviewer gave up after only a few pages of my rambling and uninteresting text in its earliest form. However, several noted geologists gave unrelenting service toward making this book a technically accurate account. To these friends and colleagues I owe my sincerest thanks and respect. Lehi Hintze, Professor Emeritus of Brigham Young University, made valuable contributions to the first draft. Bill Chenoweth, a most knowledgeable retired uranium geologist living in Grand Junction, Colorado, kept me honest on all aspects of uranium mining and history across the entire province. Bill Thomas, of the Geology Department at the University of Kentucky, a long-time friend and antagonist on geologic processes both great and small, asked important questions and taught me some proper English along the way. And last, but certainly not least, the vast knowledge and long-time experience of Fred ("Pete") Peterson, retired after decades of work on Mesozoic rocks of the Colorado Plateau for the USGS, was extremely helpful in correcting my many misconceptions (and typos) of the Jurassic and Cretaceous stratigraphy, especially along the western margin of the province. To these untiring critics I offer my special thanks and condolences. In spite of all this assistance, I take final responsibility for any errors.

Many of my colleagues, mentors, and pioneers of Colorado Plateau geology are now deceased. Among those who were most influential in my early career are Sherman A. ("Sherm") Wengerd, Edwin D. ("Eddie") McKee, Wm. Lee Stokes, and Cornelius M. ("K") Molenaar. To these modern "giants of geology" I dedicate this book.

Jeff Grathwohl, director and senior editor of the University of Utah Press, instigated the start of this overwhelming project and supported and encouraged me throughout the preparation of the manuscript. His untiring enthusiasm and sympathy propelled me to complete the project at times of my greatest discouragement. It is because of his dedication that the final product is better than anyone could hope.

Another person who helped immensely, enduring thousands of miles of driving, endless nights in sometimes seamy motels, hundreds of "delicious" café meals, and unrelenting weather conditions, is my wife, Renate. Without her endless search for often missing or obscure highway mileposts, her continual warnings of impending death by traffic, her dedicated checking of batteries in the voice recorder, and her patience in replacing her worn-out car, the preparations for this project would never have been completed. Oh yes, and she drafted and improved, with her trusty computer, most of the book's many illustrations. To her I extend my greatest thanks and admiration.